DYNAMIC MODELS IN CHEMISTRY

A WORKBOOK OF COMPUTER SIMULATIONS USING ELECTRONIC SPREADSHEETS

Diskettes containing the models described in this book are available for both MS DOS and Macintosh computers. Contact the publisher for information.

DYNAMIC MODELS IN CHEMISTRY

A Workbook of Computer Simulations Using Electronic Spreadsheets

Daniel E. Atkinson
University of California, Los Angeles

Douglas C. Brower
Catawba College

Ronald W. McClard
Reed College

David S. Barkley, Editor

N. Simonson & Company
Marina del Rey, California

Sponsoring Editor: David S. Barkley
Copyeditor: Betty Duncan-Todd
Cover Designer: Tarabochia & Hunt

Library of Congress Cataloging-in-Publication Data

Atkinson, Daniel E.
    Dynamic models in chemistry: a workbook of computer
simulations using electronic spreadsheets / Daniel E. Atkinson,
Douglas C. Brower, Ronald W. McClard; David S. Barkley, editor.
        p. cm.
    ISBN 0-9622556-2-9
    1. Chemistry–Data processing. 2. Electronic spreadsheets.
 I. Brower, Douglas C., 1958 -   . II. McClard, Ronald W., 1951 -
III. Title
QD39.3.E46A87   1990
540'.1'13--dc20   90-37688
                                                   CIP

N. Simonson & Company
13450 Maxella Avenue
Suite G225
Marina del Rey, California 90292-5631
(213) 301-2847

We dedicate this book to our
parents, teachers, and colleagues

N. Simonson & Company Titles of Related Interest

D. E. Atkinson, S. G. Clarke, and D. C. Rees
Dynamic Models in Biochemistry (1987)

F. Potter and C. W. Peck
Dynamic Models in Physics, Volume I: Mechanics (1989)

PREFACE

Students of the sciences must integrate large amounts of information into simple, coherent frameworks of insight. Students must then use this insight to solve new problems in which the driving principles have been obscured by Nature or by a devious instructor. Such insight is acquired by actively and repetitively engaging the mind in numerous problem-solving explorations — explorations in the laboratory, in assigned problem sets, in independently constructed thought experiments. Students must dissect many different kinds of problems and explore many alternative solutions until the barriers to insight are overcome and the student says "Aha!"

The pocket calculator revolutionized science instruction by permitting the design of problem sets that were not contrived to fit within the limitations of pencil, paper, and slide rule.

The personal computer is again revolutionizing science instruction by permitting the design of problem sets that enable students to build and interrogate actual simulations of physical events and to explore these simulations with a full complement of graphical tools.

We believe that the emphasis should be as much on the building of the model as on its interrogation, but traditional software tools are not designed for rapid, one-time prototyping of individual problems. We have therefore chosen to design this book around a powerful, ubiquitous, nonprocedural programming language that permits rapid and intuitive construction of sophisticated simulations — the electronic spreadsheet. Other software tools may eventually make their way to the marketplace in forms useful to undergraduate students, but the intuitive man–machine interface that constitutes the electronic spreadsheet will have a place in personal computing for decades to come.

Electronic spreadsheet programs can serve as construction kits with which to build and explore alternative representations of ideas. Students who build spreadsheet models that correspond to their understanding of physical events can correct errors in their understanding as they refine their models to make them work correctly. They can ask questions of their simulations that may not have been anticipated by their textbook authors and instructors and can verify their understanding of a textbook table or graph by reproducing it from basic principles.

As we designed and wrote this book, we found that many unexpected questions arose — questions that could be immediately answered by interrogating the model itself. We believe that this ability to enter into extended dialogs with computer simulations will be as valuable to students as it has been for us.

We hope and expect that an early introduction to model building and interrogation will provide students with a tool than can be used throughout a career — a simple device with which to revise and clarify ideas.

Scope of the Book

This book has been designed for students taking an introductory course in chemistry. Much of the material drills students on elementary topics. However, personal computers and computer simulations naturally promote independent study and exploration. Rather than cut off an exciting exploration, we have chosen to offer deeply interested students the opportunity to explore some issues in depth. Such pathways are often found in the *Problems* section of each exercise. Talented students may find insight and entertainment in these pathways and in the treatment of even elementary topics.

Using the Book

The book is organized as a series of exercises to be performed on personal computers running electronic spreadsheet software. There are no special requirements with respect to the brand of personal computer or electronic spreadsheet program, provided that the software has the capability of generating simple xy-plots. The models in this book are described using a notation similar to that found in Lotus 1-2-3, Super-calc, Excel, etc. (i.e., columns are identified by letters and rows are identified by numbers). Chapter 1 is designed to help students with no prior computer experience acquire the facility needed to make effective use of this book.

Students build each computer simulation by following the book's detailed directions while sitting at a personal computer. The exercises follow the same progression found in most major chemistry textbooks but are sufficiently modular that they can be sampled out of order where appropriate.

Each exercise begins with a brief introduction to the quantitative aspects of the topic to be explored and then provides detailed directions for building the spreadsheet model on virtually any personal computer and spreadsheet program. Exercises are divided into a *Lessons* section and a *Problems* section.

Lessons are tightly controlled and offer considerable guidance as models are interrogated, tables completed, and results graphed. *Problems* require varying degrees of student initiative and often permit the interested student to enter domains of genuine, independent exploration.

This book is a beginning and will grow and prosper only with your help, advice, and comments. Please write us with your experiences as you explore the book.

Acknowledgments

We have benefited from the experience and advice of many people as we have designed and written this book.

The idea for such books grew out of discussions with Neil Patterson and Peter Renz at W. H. Freeman & Company. An earlier book, *Dynamic Models in Biochemistry*, pioneered our understanding of the use of spreadsheets to learn and to think, and we are deeply indebted to two of its authors — Steven Clarke and Douglas Rees.

Sandra Lamb directed the early undergraduate testing of the book from her laboratory at UCLA and did much to evangelize this approach to teaching when such evangelism was most necessary.

Dr. Daniel P. Gerrity, Department of Chemistry, Reed College was of considerable help to one of us as this book was written.

David S. Barkley, Editor
Daniel E. Atkinson
Douglas C. Brower
Ronald W. McClard

CONTENTS

DYNAMIC MODELS IN CHEMISTRY

A WORKBOOK OF COMPUTER SIMULATIONS USING ELECTRONIC SPREADSHEETS

CHAPTER 1

GETTING STARTED

This special property of digital computers, that they can mimic any discrete machine, is described by saying that they are universal *machines. The existence of machines with this property has the important consequence that ... it is unnecessary to design various new machines.* — A. M. Turing

This book uses electronic spreadsheets to create computer simulations that can help you better understand the behavior of molecules and their chemical reactions. Your goal is to apply these capabilities to the subjects found in your introductory chemistry book.

You will learn to take the principles that underlie general chemistry and to represent them inside your personal computer where they can be experienced and explored. Your computer can serve as a potter's wheel; your ideas can be lumps of clay.

Digital computers are *universal machines* that can simulate any machine or any process that can be precisely described. Engineers use advanced computer systems to build and study working simulations of prototype spacecraft. Meteorologists build dynamic computer models of complete weather systems to verify their current knowledge and to search for new knowledge. With a computer, you can visit the eye of a hurricane or follow the jet stream around the world between breakfast and lunch.

Until recently, this ability to simulate physical processes with digital computers was limited to those people with special computer facilities. The introduction of affordable computing hardware and especially the introduction of powerful yet easy-to-learn nonprocedural programming languages, such as the electronic spreadsheet, now make remarkable computing powers accessible to all students.

A new model world is accessible, one much closer to the real world.

The Electronic Spreadsheet

The first electronic spreadsheet program Visicalc appeared in 1978 and quickly became the principal calculating tool of personal computer users in virtually every walk of life. With spreadsheets it is possible to

- Build models of dynamic processes in a tiny fraction of the time that it would take with a procedural language such as C, Pascal, BASIC, or FORTRAN.

- Change and refine models quickly and intuitively in order to explore alternative scenarios that might otherwise go unnoticed.

- Graph the results of your explorations on-screen and automatically change the graph by changing model assumptions.

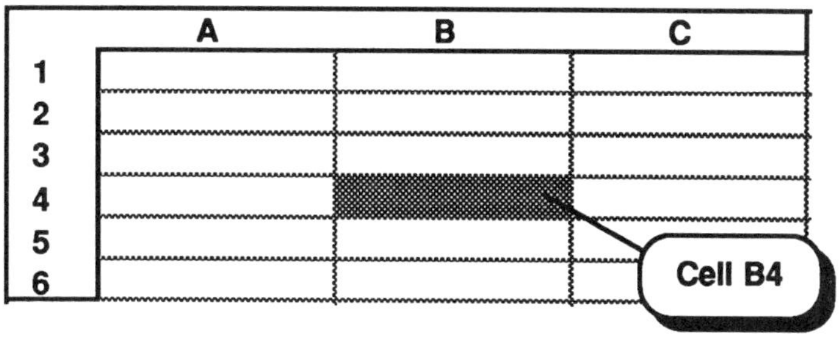

	A	B	C
1			
2			
3			
4			
5			
6			

Cell B4

Figure 1.1

Electronic spreadsheet programs appear on your computer screen as a rectangular array of cells. Each cell is uniquely named according to the column and row in which it is located. Cell B4, for example, is the only cell that resides in both column B and row 4. Only a few spreadsheet programs exist that do not label columns with letters and rows with numbers. If your program has its own convention, you can adapt the models of this book for your program.

You may already know how to use an electronic spreadsheet program. If you've never worked with such a program, prepare yourself for an exciting experience. You will be building sophisticated models in a few hours.

The concept is simple. A large, two-dimensional array of rectangular cells is displayed on a computer terminal with column and row headings that provide each cell with a unique name. Although a few exceptions remain, the vast majority of spreadsheet programs have adopted the convention of identifying columns with letters and rows with numbers. Each cell in the spreadsheet is named according to its column and row location (Figure 1.1).

Spreadsheet cells may be blank or may contain either *labels*, *values*, or *formulas*. Collections of labels, values, and formulas work together to generate models. Figure 1.2 is a model that calculates the molecular weight of H_2O from the frequency and atomic weight of each atom.

- *Labels* are text that describe a spreadsheet; identify the entries in a column, row, or cell; or name regions within a spreadsheet. They are not used in calculations. Figure 1.2 contains a label in cell B1 that identifies all numbers beneath it as atomic weights.

- *Values* are data to be used by your model. Cell B2 in Figure 1.2 contains the atomic weight of hydrogen. Most spreadsheets can perform calculations on numbers and strings of letters. In this book, calculations are numeric.

- *Formulas* perform calculations. Figure 1.2 contains a formula in cell D2 that reads =C2*B2. This formula indicates that cell D2 contains the product of the contents of cells C2 and B2. (* is often used by computers to signify multiplication. The other basic operations – addition, subtraction, division and exponentiation – are represented by +, –, /, and ^, respectively.) Although your spreadsheet program may represent this formula slightly differently, all formulas are easy to read and understand.

Figure 1.2

Each cell in a spreadsheet array may be either a *label*, *value*, or *formula*. Labels are passive signposts that tell you where you are (e.g., cell B1). Values are usually numbers required by the model for its calculations (e.g., cell B2). Formulas are algebraic expressions that use the addresses of other cells as arguments (e.g., cell D2).

Labels, values, and formulas are entered into cells by first pointing to the cell with your computer's pointing device (typically either an arrow key on your keyboard or a mouse) and then typing in the required string of characters.

	A	B	C	D
1	Atom	At. Wt.	No.	
2	H	1.008	2	=C2*B2
3	O	16	1	=C3*B3
4				
5			Mol. Wt. =	=D2+D3

Cells that contain formulas usually display the result of the calculation rather than the formula itself. Of course, if you wish to look behind the calculated result to the formula, your spreadsheet will always have a method for doing so.

The principal elegance of spreadsheet models resides in their response to change. Suppose that acid (H^+) is added to a beaker of H_2O and a fraction of the water molecules become protonated to H_3O^+. The model in Figure 1.2 serves to represent the molecular weight of any compound that contains only hydrogen and oxygen. If the value in cell C2 is changed from 2 to 3, the molecular weight changes from that of H_2O to that of H_3O^+ (Figure 1.3).

Figure 1.2 represents a system that you can calculate in your head. Once you get the feel for the notation, however, you can build models of complete chemical systems and encounter quite unexpected behaviors.

Most modern spreadsheet programs also include the capability to represent model behavior graphically. Although the model described in Figure 1.2 has no obviously interesting graphical counterpart, most of your studies will make extensive use of this capability, as you will soon see in the exercises at the end of this chapter.

Using Electronic Spreadsheets

Now is the time to get out the instruction manual that came with your electronic spreadsheet program, turn on your computer, and start to work.

Many spreadsheets come with tutorials that will get you up to speed quickly. You may find that the power of your program is so tantalizing that you will spend hours seeking out exotic features and hidden opportunities. For all exercises in this book, however, you need only learn how to

- *Point to a cell and write into it:* Point to a cell by using either the arrow keys on your keyboard or an alternative pointing device

	A	B	C	D
1	Atom	At. Wt.	No.	
2	H	1.008	2	2.016
3	O	16	1	16
4				
5			Mol. Wt. =	18.016

Change the value of
cell C2 from 2 to 3

	A	B	C	D
1	Atom	At. Wt.	No.	
2	H	1.008	3	3.024
3	O	16	1	16
4				
5			Mol. Wt. =	19.024

Figure 1.3

Electronic spreadsheet models are dynamic. When the value of a cell is changed, all formulas that use that value are recalculated. When the content of cell C2 is changed from 2 to 3, the calculated results displayed in cells D2 and D5 also change. A model that calculates the molecular weight of H_2O becomes a model that calculates the molecular weight of H_3O^+.

such as a mouse. Write to a cell by typing the appropriate string of characters from your keyboard.

Special "built-in" functions on spreadsheets permit you to calculate logarithms, sums of lists of numbers, etc. Take a quick look at these functions in your program's instruction manual and get a sense of how to enter them. Formulas that contain such "built-in" functions are easy to spot — e.g., =A3+SUM(B1:B10). The precise form used by your program may be slightly different.

When preparing formulas, it is not necessary to manually type in addresses such as B3. Instead, you can simply point to cell B3 with your cursor (using the arrow keys, mouse, or other pointing device) when you come to the appropriate place in the construction of your formula. Take the time to explore and practice thoroughly this technique using the exercises in your spreadsheet manual or tutorial. This elegant feature makes flawless formula building both easy and intuitive.

• *Edit a cell:* Spreadsheets provide simple editing capabilities that allow you to change a portion of a formula without rewriting the entire formula.

• *Format a cell:* You will discover that spreadsheets permit you to define the manner in which cell contents are displayed. Do you want your label left- or right-justified? Do you want your values displayed in floating point or scientific notation and to how many decimal places?

- *Copy a cell or range of cells to another location using absolute and relative variables:* Once you've written a complex formula, you need not rewrite it into the next 100 cells. *Copy it instead.* You will learn about absolute and relative variables shortly. (Briefly, relative variables change when copied to new locations — can you predict when this might be important? — but absolute variables remain unchanged.)

- *Edit a spreadsheet:* A spreadsheet is a highly malleable medium, and you will sometimes want to rearrange a spreadsheet without completely rewriting it. Spreadsheets provide simple yet powerful editing capabilities that allow you to insert or delete rows and columns, move groups of cells to new locations, etc.

- *Save and recall a model from diskette or hard disk:* When you turn off your computer, it forgets. Hard disks and diskettes are used to save work between sessions.

- *Graph a model:* You should learn how to make simple xy-plots using your spreadsheet program.

You do not need to know about such optional features as keyboard macros or database management, although you may find that macros are an easy way of "automating" your spreadsheets.

If you have never used an electronic spreadsheet program, turn now to the training and reference manuals that were distributed with your program and spend an hour or so learning the basics. Once you are comfortable with the basic functions of your program, proceed to the exercises that follow.

These exercises show you how to build models using the descriptions in this book and give you an opportunity to become comfortable with your electronic spreadsheet program. Just for fun, a simple relationship important in acid–base chemistry (the Henderson–Hasselbalch equation) is used to explore the behavior of weak acids and bases. If you haven't studied acid–base chemistry yet, don't worry. Just follow along, plug in the numbers, and watch the model run. You'll come back to this subject later in the book.

ENTERING LABELS, VALUES, AND FORMULAS

Electronic spreadsheet programs are designed and written by companies that want to sell their products to the largest possible audience. They are designed to be easy to use. However, even if you are familiar with electronic spreadsheet programs and their operation, you may want to skim these exercises quickly to become familiar with the format of this book and the way it presents electronic spreadsheet models.

All electronic spreadsheets work essentially the same way. To build a model you must

1. Point to a cell.
2. Write into a cell.
3. Repeat on a new cell.

You can build any model using these steps, but there are shortcuts. For example, often you will find it convenient to generate tables of data by placing the same (or very similar) formula in an entire column of cells. Electronic spreadsheets have special commands that permit you to copy the contents of one or more cells to new locations. Therefore, you can clone formulas as many times as you wish. You can clone identical copies of formulas, or you can instruct your program to make subtle changes in the formula to reflect its new location in the spreadsheet. In

TO GET THE MOST OUT OF THIS
EXERCISE, YOU SHOULD
ALREADY KNOW

The definition of the terms:
cell
row
column
label
value
formula
model

How to build simple models using
your electronic spreadsheet program.

How to use your software reference
manual and on-line help to learn how
to use your program.

(You can get this information from the
Introduction to this chapter and from
the tutorial and reference manuals
accompanying your electronic
spreadsheet program.)

this exercise, you learn how to enter labels, values, and formulas into your program to build a simple model. In the following exercise, you learn how to COPY formulas to multiple locations.

Because the rules governing syntax for the contents of a cell are not consistent from program to program, identify some of the specific syntax requirements of your program and compare them with the syntax that is used throughout this book.[1] For example,

1. Are formulas identified by a preceding =, or are formulas identified by some other mechanism? (All Microsoft Excel formulas contain a preceding =. Lotus 1-2-3 formulas begin with a number or with an operator such as + or –.)

   In this book, we precede formulas with =, as in =B3+B5. The equivalent Lotus 1-2-3 formula would be +B3+B5. If you're using Lotus 1-2-3, be sure you understand the difference.

2. Do built-in functions such as LOG, SIN, EXP require a preceding @ (e.g., Lotus 1-2-3)?

3. Is the logarithm to base 10 of the contents of cell B1 divided by the contents of cell B2 expressed by the formula @LOG(B1/B2) (e.g., Lotus 1-2-3) or by the formula LOG10(B1/B2) (e.g., Microsoft Excel). Notice that @ isn't the only difference between the two formulas.

   Other built-in functions tend to retain their form across different electronic spreadsheet programs (except, of course, for the "@" notation). This book uses LOG10 for log to base 10 and LN for log to base e.

You might find that your first few sessions at the keyboard are frustrating. There are only a few things to learn, but they all seem to pile up at the beginning.

LESSONS

Later in this book, you will explore the dissociation of a weak acid as an example of a simple but very important equilibrium relationship (Figure 1.4). One equation that can predict the concentrations of the ionic species in this equilibrium is the Henderson–Hasselbalch equation (Equation 1.1):

Figure 1.4

$$HB \rightleftharpoons H^+ + B^-$$

Suppose the compound HB can dissociate into a proton H^+ and a negatively charged ion B^-. The relative concentrations of HB, H^+, and B^- are determined by physical properties summarized in the compound's dissociation constant K_{eq} and can be calculated using the Henderson-Hasselbalch Equation (Equation 1.1).

$$pH = pK + \log \frac{[B^-]}{[HB]} \qquad (1.1)$$

where pH is $-\log[H^+]$, pK is $-\log[K_{eq}]$, and $[H^+]$, $[B^-]$, $[HB]$ are the molar concentrations of H^+, B^-, HB, respectively. Terms such as pH, pK, and molar concentration may or may not mean anything to you right now. Suffice it to say that "molar concentration" is a number that characterizes the number of molecules present in a specified volume; pK is a number that characterizes how tightly H^+ and B^- "stick" together; and pH is a number that is used as an alternative to $[H^+]$ to measure the concentration of H^+. As noted earlier, it is the negative logarithm of $[H^+]$. An aqueous solution with a pH of 2 has $[H^+]$ equal to $10^{-2}M$. An aqueous solution with a pH of 3 has $[H^+]$ equal to $10^{-3}M$. Therefore, as the pH of a solution increases, the concentration of H^+, called its acidity, decreases.

Let's use Equation 1.1 to build a simple model capable of exploring the behavior of weak acids and bases.

1. Review the instruction manual of your electronic spreadsheet program and its tutorials.

2. Turn to the end of this exercise and briefly examine Model 1.1. Model 1.1 is a very simple model that calculates pH from the pK of a weak acid and the concentrations of its dissociated and undissociated forms. Although you can easily perform these calculations on a pocket calculator, this model is a good place to learn about elementary spreadsheet manipulations.

Most models consist of two parts:

• A figure that defines the contents of the labels, values, and formulas contained in the model

• A legend containing verbal instructions for building and using the model

The figure requires a bit of explanation. All figures that describe the structure of a spreadsheet model have two parts: a formula block and a value block. (An optional, third component, the formula set block, is described in the next exercise.)

• *Formula blocks* show the actual contents of each cell (i.e., the keystrokes entered into each cell). Cells B1–B3 contain the values of [B⁻], [HB], and pK. Cell B5 contains a formula representing Equation 1.1.

(We built our models using Microsoft Excel; if your computer is running that program, the formula should be entered precisely as shown. If you are using a different program, small changes may be required. The differences are very small, and you will soon find yourself making the translations automatically and without conscious thought.)[2]

- *Value blocks* show how your screen actually appears when the model is complete and correct. Cells B1–B3 remain unchanged. Cell B5 contains the result of calculating Equation 1.1 using the arguments in cells B1–B3. Use value blocks to *verify* your model after it has been built. Compare the calculated results on your screen with the calculated results in the value block. Complex formulas can contain errors when first prepared or copied.

2. Build Model 1.1 on your computer by pointing the on-screen cursor to the appropriate cell and typing in a label, value, or formula that conforms to the syntax requirements of your program. If you run into trouble, watch the screen as you type in the offending characters — helpful error messages may appear. Also, ask yourself what the model is trying to do (e.g., take a logarithm to base 10) and then look in the index of your program's reference manual for your program's rules in performing this task. Better yet, learn how to use the on-line help function of your program. The syntax of all built-in functions is often found there.

3. Save the model to a data diskette, if you wish.

4. Examine the model on your computer screen. Does it correspond to Model 1.1 (VALUES)? If it does, you have probably entered the model successfully. If it doesn't, either you did not enter the correct keystrokes or you are seeing differences in display format (i.e., 5.1979 vs 5.198). Ignore display-format differences now and throughout this book.

   a. How does your program identify formulas? A preceding =? Letters of the alphabet? Numbers or operators (e.g., $0 \ldots 9, +, -$)?

   b. Does your program assume that all character sequences that begin with numbers are values? What about special characters such as ., (, @?

   c. What built-in functions does your program support (e.g., logarithms, trigonometric functions, etc.)?

Whenever instructions call for you to build and verify a model, build the model by using the formula block and formula sets to enter labels, values, and formulas into the appropriate cells. Verify the model by comparing your on-screen values with the values in the value block.

5. When strong acid is added to a solution of the salt of a weak acid, the added protons H^+ can combine with B^- to form HB (therefore increasing [HB] and decreasing [B$^-$]). Verify that the change in pH with changes in [HB] (or [B$^-$]) is not linear by completing the table below and plotting it on a piece of graph paper.

pK = 4.8

[B⁻]	[HB]	pH
0.20	0.01	_____
0.19	0.02	_____
0.18	0.03	_____
0.17	0.04	_____
0.16	0.05	_____
0.15	0.06	_____
0.14	0.07	_____
0.13	0.08	_____
0.12	0.09	_____
0.11	0.10	_____
0.10	0.11	_____
0.09	0.12	_____
0.08	0.13	_____
0.07	0.14	_____
0.06	0.15	_____
0.05	0.16	_____
0.04	0.17	_____
0.03	0.18	_____
0.02	0.19	3.82
0.01	0.20	3.50

Can you identify a range of pH values where the addition of strong acid will change pH the least?

6. In the next exercise, you see how this model can be changed and your spreadsheet software used to plot [HB] vs pH automatically.

7. Before moving on, however, take one last look at Model 1.1. There is a powerful technique for building formulas that can help eliminate

typing errors. Lesson 2 asked you to enter the formula by entering the appropriate keystrokes. Now try this instead:

- If you use Lotus 1-2-3 or similar packages, position your cursor at cell B5 and type the single character +. Now use the arrow keys on your keyboard to move the cursor to cell B3 where the pK value is located. Do you see what's happening in cell B5? Once you start the formula by entering +, 1-2-3 knows your intentions and you can enter cell addresses by "pointing" instead of typing.

  Continue with the keystrokes +@LOG(and then use your arrow keys to "point" to cell B1, type /, "point" to cell B2, type), and finally press <ENTER>.

- If you use Microsoft Excel or similar packages, position your cursor at cell B5 and type the single character =. Now use your arrow keys (or mouse) to move the cursor to cell B3 where the pK value is located. Do you see what's happening in cell B5? Once you start the formula by entering =, Excel knows your intentions and you can enter cell addresses by "pointing" instead of typing.

  Continue with the keystrokes +LOG10(and then "point" to cell B1, type /, "point" to cell B2, type), and finally press <ENTER>.

Building formulas in this way can reduce your errors because you are fetching cell addresses in an intuitive fashion.

PROBLEMS

1. When strong acid is added to a solution of the salt of a weak acid, [B⁻] decreases by the same amount that [HB] increases. How might you modify your model so that [HB] changes automatically with changes in [B⁻]?

2. Most spreadsheet programs support *data tables*. Data tables are built by issuing commands to your spreadsheet that automatically insert a series of inputs into a value-containing cell of your model and automatically tabulate the results. Find this feature in your software reference manual and use it to build the table in Lesson 5.

MODELS

FORMULAS

	A	B
1	[B] =	0.05
2	[HB] =	0.02
3	pK =	4.8
4		
5	pH =	=B3+LOG10(B1/B2)

VALUES

	A	B
1	[B] =	0.05
2	[HB] =	0.02
3	pK =	4.8
4		
5	pH =	5.19794001

Model 1.1

ENTER data into cells B1–B3. Cells B1 and B2 contain the concentrations of B- and HB, respectively, while cell B3 contains the pK value of the weak acid.

READ results from cell B5 (pH).

NOTICE that the formula in cell B5 was written to support the syntax of Microsoft Excel. The equivalent formula for Lotus 1-2-3 would be +B3+@LOG(B1/B2).

COPYING FORMULAS:
RELATIVE AND ABSOLUTE VARIABLES

Although the cycle of pointing to a cell, writing into a cell, and repeating on a new cell ultimately builds a model of any complexity, an electronic spreadsheet program has features that ease the task of building large models. You may want to learn about all of them by reading your program's instruction manual. In this experiment, study one feature that you will use frequently: the COPY command.

Look back at Exercise 1.1 and its implementation of the Henderson–Hasselbalch equation (Equation 1.1). Model 1.1 calculates the pH of an aqueous solution containing a weak acid from the arguments pK, [B⁻], and [HB]. Near the end of Exercise 1.1, you observed how pH changes with added strong acid by repeatedly decreasing the value of [B⁻] (cell B1) while increasing the value of [HB] (cell B2) by the same amount. You can automate this exercise by building a model of different design (Figure 1.5). There are several features of this model that you should notice:

- The model is organized as a table of values, beginning in row 7, that is *driven* by a parameter table located in cells A1–B5.

- The table itself uses multiple copies of the Henderson–Hasselbalch equation (column C) to compute the

FORMULAS

	A	B	C
1	pK =	4.8	
2	[B] + [HB] =	0.21	
3			
4	[B]o =	0.2	
5	Δ[B] =	-0.005	
6			
7	[B]	[HB]	pH
8	=B4	=B2-A8	=B1+LOG10(A8/B8)
9	=A8+B5	=B2-A9	=B1+LOG10(A9/B9)
10	=A9+B5	=B2-A10	=B1+LOG10(A10/B10)
11	=A10+B5	=B2-A11	=B1+LOG10(A11/B11)

	A	B	C
44	=A43+B5	=B2-A44	=B1+LOG10(A44/B44)
45	=A44+B5	=B2-A45	=B1+LOG10(A45/B45)
46	=A45+B5	=B2-A46	=B1+LOG10(A46/B46)

Figure 1.5

Formulas in electronic spreadsheets can be copied to new locations. If the cell address in a formula remains unchanged during a COPY, the variable contains an *absolute* cell address. If the cell address in a formula changes changes during a COPY, the variable contains a *relative* cell address. Specifically, if the formula is copied to a location three rows down and one column to the right, relative cell addresses change the same way.

pH of the solution from a range of values of [B-] (column A), [HB] (column B), and the pK of the weak acid (cell B1).

• The range of values of [B-] and [HB] in columns A and B are *calculated*, and they always sum to the concentration of total acid (cell B2): The first table value for [B-] (cell A8) takes its starting value from cell B4 in the parameter table; each cell below B8 (B9, B10, etc.) increments the value in the previous cell by the contents of cell B5. The different values for [HB] calculated for cells B8–B46 are determined by subtracting [B-] from the concentration of total weak acid (cell B2), thus ensuring that [B-] + [HB] remains constant.

Take a moment to explore the model and its structures and features. You could build this model now using your electronic spreadsheet program by simply typing in all appropriate labels, values, and formulas, but a moment's thought tells you that this is a lengthy task. Your spreadsheet's COPY command offers an easier way.

LESSONS

1. Examine the formulas in column C of Figure 1.5 and notice that, although the reference to the pK value in cell B1 remains unchanged in all formulas, references to [B-] and [HB] in columns A and B change. The change, however, is predictable in that references to [HB] and [B-] are always one and two cells, respectively, to the left of the formula-containing cell.

- The reference to cell B1 is called an *absolute reference* because it never changes.

- The reference to the cells in columns A and B are *relative references* because they change relative to the location of the formula that uses them as arguments (e.g., one cell to the left).

In many spreadsheet programs, formulas can be written such that their variables are defined with either relative or absolute cell addresses. When these formulas are copied using the COPY command, variables containing absolute addresses never change, whereas variables containing relative addresses change relative to their new location. In Lotus 1-2-3, for example, the formula +$B$1+@LOG(A10/B10) uses dollar signs to indicate that B1 is an absolute address. In Microsoft Excel, this formula would be written =$B$1+LOG10(A10/B10).[1] Your spreadsheet software may or may not use a similar technique. Some programs identify absolute and relative addressing by selecting options from within the COPY command itself.

2. Examine Model 1.2 at the end of this exercise and notice that column C of the model is defined by Formula Set I. Beneath the formula and value blocks, you can find the Formula Set I block. It contains two pieces of information — a formula, =B1+LOG10(A8/B8), and the cell in which the formula should be placed (the prototype cell, C8). *The boldface type used for B1 in the formula indicates an absolute address.* Formula Set I is an instruction that says "place the formula =B1+LOG10(A8/B8) into cell C8 and copy it into every cell that belongs to Formula Set I" (Figure 1.6).

   Compare the description of the contents of column C found in Model 1.2 with the description found in Figure 1.5 and assure yourself that both descriptions represent the same model.

   You may also wish to return to your software reference manual and study the subject of relative and absolute addressing further.

3. Different spreadsheet users have different model-building styles. Until you develop your own, use the following sequence:

   a. Enter all labels from Model 1.2 (FORMULAS) into your computer. When complete your screen should look like Figure 1.7.

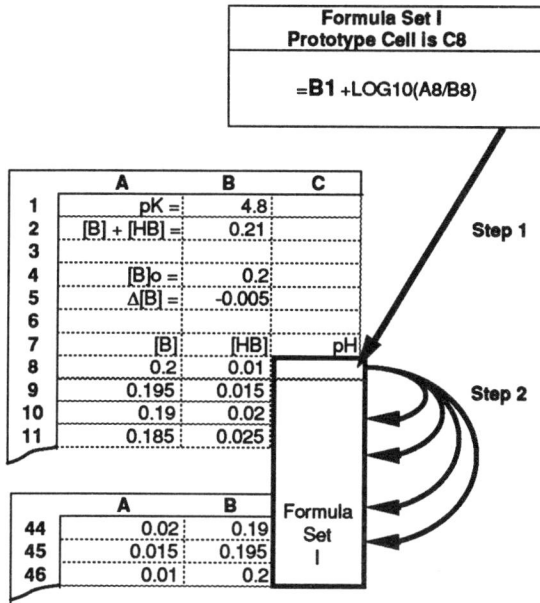

Formula Set I
Prototype Cell is C8

=**B1** +LOG10(A8/B8)

	A	B	C	
1	pK =	4.8		
2	[B] + [HB] =	0.21		Step 1
3				
4	[B]o =	0.2		
5	Δ[B] =	-0.005		
6				
7	[B]	[HB]	pH	
8	0.2	0.01		
9	0.195	0.015		Step 2
10	0.19	0.02		
11	0.185	0.025		

	A	B	
44	0.02	0.19	Formula Set I
45	0.015	0.195	
46	0.01	0.2	

Figure 1.6

Formula sets are used to describe formulas that will not conveniently fit into the formula block of a model description.

Step 1: The formula in a formula set is constructed in the prototype cell. (Absolute variables are displayed in boldface, relative variables in plain face.)

Step 2: The formula is copied from the prototype cell to all member cells of the formula set. The distinction between *relative* and *absolute* is relevant only during a COPY (and not a MOVE/CUT and PASTE) operation.

Labels alone can give you a general sense of your screen layouts. You then have an opportunity to edit this layout before proceeding.

b. Enter all values from Model 2.1 (FORMULAS) into your computer. When complete your screen should look like Figure 1.8.

c. Enter the formulas located in cells A8, A9, B8, and C8 into your computer (remember that the formula in cell C8 is defined by Formula Set I and that one of the addresses points to an absolute address).[2]

When complete your screen should look like Figure 1.9.

	A	B	C
1	pK =		
2	[B] + [HB] =		
3			
4	[B]o =		
5	Δ[B] =		
6			
7	[B]	[HB]	pH
8			
9			
10			
11			

Figure 1.7

Many people design spreadsheet models by first entering just the labels. With the labels in place, you can get a sense of overall design and, if necessary, make changes in the design to suit your needs.

Figure 1.8

It's usually a good idea to enter values before formulas. Then, as you enter formulas, you can examine the calculated result of the formula and apply rule-of-thumb tests for reasonableness.

	A	B	C
1	pK =	4.8	
2	[B] + [HB] =	0.21	
3			
4	[B]o =	0.2	
5	Δ[B] =	-0.005	
6			
7	[B]	[HB]	pH
8			
9			
10			
11			

d. Use your spreadsheet's COPY command to copy the contents of A9 into cells A9–A46, the contents of B8 into cells B8–B46, and the contents of C8 into cells C8–C46. If you don't recall how to use the COPY command, either look in your software reference manual or access your spreadsheet's on-line help facility.

e. Verify the correctness of your model by comparing the on-screen values with the values of Model 2.1 (VALUES).

f. On graph paper, plot pH vs [HB] for two weak acids with pK values of 4.8 and 9.6, respectively.

PROBLEM

The Henderson–Hasselbalch Equation (Equation 1.1) can be rearranged so that [B⁻]/[HB] can be calculated from pH and pK (Equation 1.2):

$$\frac{[B^-]}{[HB]} = \frac{10^{pH}}{10^{pK}} \tag{1.2}$$

Design, build, and verify a model that generates a table of values in which [B⁻]/[HB] is calculated as a function of pH for a weak acid of known pK.

Figure 1.9

Many models apply different arguments to multiple copies of the same formula. You need only enter the formulas once (as shown) and then COPY them to all appropriate locations.

	A	B	C
1	pK =	4.8	
2	[B] + [HB] =	0.21	
3			
4	[B]o =	0.2	
5	Δ[B] =	-0.005	
6			
7	[B]	[HB]	pH
8	0.2	0.01	6.10103
9	0.195		
10			
11			

MODELS

FORMULAS

	A	B	C
1	pK =	4.8	
2	[B] + [HB] =	0.21	
3			
4	[B]o =	0.2	
5	Δ[B] =	-0.005	
6			
7	[B]	[HB]	pH
8	=B4	=B2-A8	
9	=A8+B5	=B2-A9	
10	=A9+B5	=B2-A10	
11	=A10+B5	=B2-A11	

Formula Set I

	A	B
44	=A43+B5	=B2-A44
45	=A44+B5	=B2-A45
46	=A45+B5	=B2-A46

VALUES

	A	B	C
1	pK =	4.8	
2	[B] + [HB] =	0.21	
3			
4	[B]o =	0.2	
5	Δ[B] =	-0.005	
6			
7	[B]	[HB]	pH
8	0.2	0.01	6.10103
9	0.195	0.015	5.91394
10	0.19	0.02	5.77772
11	0.185	0.025	5.66923

	A	B	C
44	0.02	0.19	3.82228
45	0.015	0.195	3.68606
46	0.01	0.2	3.49897

Formula Set I
Prototype Cell is C8

=**B1**+LOG10(A8/B8)

Model 1.2

ENTER data into cells B1 and B2, B4 and B5. Cells B1 and B2 contain the pK and total concentration of weak acid, respectively. Cells B4 and B5 determine the initial value and incremental value of [B-] placed in the table in the lower half of the model.

READ results from columns A, B, and C beginning at row 8.

NOTICE that the table is organized so that [B-] + [HB] always totals to the value entered into cell B2 and that the range of values calculated in the table is determined by the contents of cells B4 and B5. Furthermore, the boldface type used for address B1 in Formula Set I indicates an absolute address.

PLOTTING MODEL DATA

TO GET THE MOST OUT OF THIS
EXERCISE, YOU SHOULD
ALREADY KNOW

How to design spreadsheet models
that encompass tables of data.

How to use the graphing capabilities
of your electronic spreadsheet
program.

(These matters are covered in
Exercise 1.2 and in the tutorial and
reference manuals accompanying
your electronic spreadsheet program.)

Most modern spreadsheet programs also include the capability to
graphically represent model behavior. Suppose, for example, that you
wish to plot pH vs [B⁻] for weak acids with different pK values (Figure
1.10). Your spreadsheet program probably calls such a plot either an
xy–plot or a scatter plot, and it can represent such plots in a variety of
formats. In this exercise, you learn how to build such plots.

LESSONS

1. Suppose you wish to plot the function $y = x/2$ at the points

x	y
0	0
2	1
4	2
8	4
16	8

Two different different approaches are available to plot this data,
only one of which is valid for this book. Notice that y is a linear
function of x and that this fact is clearly represented in Figure 1.11a.
What happened in Figure 1.11b? *Figure 1.11b plots the x-data as if
they were categories (apples, oranges, pears) and spaces the values*

0, 2, 4, 8, and 16 evenly across the x-*axis.* Plotting commands clearly labeled "scatter plot" or "*xy*-plot" never do this, but a command named "line plot" should be examined carefully to determine which convention it follows.[1] In this book, you will almost always be interested in exploring *xy*-plots of your data. Don't mislead yourself by using the wrong plotting commands!

Open your tutorial or reference manual that comes with your specific spreadsheet program and learn how to draw simple graphs. Lesson 1a below allows you to explore the difference between line and *xy*-plots using Lotus 1-2-3. Lesson 1b accomplishes the same goal using Microsoft Excel. If you are using neither of these programs, it should still be easy to identify the command sequence that gives the appropriate (for these data, linear) plot.

a. If you are using Lotus 1-2-3 (or any of its closest clones such as Quattro), launch the program and enter the values shown in Figure 1.12 into the appropriate cells. To create an *xy*-plot of these data (*x*-data in column A, *y*-data in column B), enter the following command sequence from the READY prompt:

Keystrokes	Command (optional explanation)
/ G T X	COMMAND GRAPH TYPE XY
X A2.A6 <ENTER>	X A2.A6 (Specifies *x*-range)
A B2.B6 <ENTER>	Y B2.B6 (Specifies a single *y*-range)
V	VIEW

(You could also, of course, select the *X*- and *Y*-ranges by "pointing" with your arrow keys.)

Your screen should now contain a linear plot of *x* vs *y* similar to Figure 1.11a.

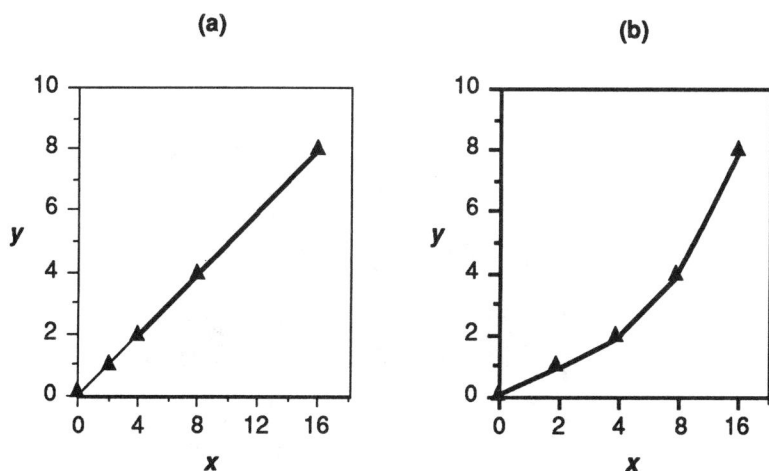

(a)

(b)

Figure 1.10

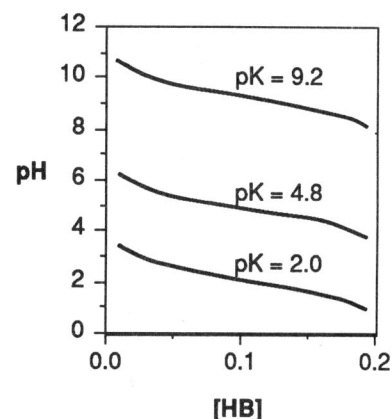

Most modern electronic spreadsheet programs can represent calculated data in graphic form. These data were plotted using Model 1.3.

Figure 1.11

The term *line chart* can be ambiguous. In (a) both the *x*- and *y*-axes represent lines in a real number system, and the linear function $y = x/2$ graphs as a straight line. In (b), however, the *x*-axis simply lists the categories 0, 2, 4, 8, and 16. They are distributed evenly along the axis, and the graph is no longer a straight line. Make sure that you invoke the graphics commands in your spreadsheet program that generate plots like those in (a).

Figure 1.12

	A	B
1	x	y
2	0	0
3	2	1
4	4	2
5	8	4
6	16	8

You can explore the difference between line plots and *xy*-plots by graphing the values shown in this simple spreadsheet model.

Remove the graph from your screen by pressing any key once. Your spreadsheet should return, and the command line should be in the graph submenu with the cursor over VIEW. You can see the difference between an *xy*-plot and a line plot by entering the following command sequence:

Keystrokes	**Command (optional explanation)**
T L V	TYPE LINE VIEW

Notice that the plot is no longer linear (as in Figure 1.11b).

The line-plot option is of limited value for most exercises in this book because it always spaces *x*-axis data evenly along the *x*-axis. You will, however, use the *xy*-option frequently.

b. If you are using Microsoft Excel, launch the program and enter the values shown in Figure 1.12 into the appropriate cells. To create an *xy*-plot of these data (*x*-data in column A, *y*-data in column B), select the block of cells you wish to plot such that the *x*-values constitute the left-hand column of data (Figure 1.13) and place the contents of this block on your clipboard by using the COPY command. Open a new graphics window by selecting the NEW option in the FILE menu and requesting a "Chart" when asked. Use the EDIT menu to PASTE SPECIAL your data into the chart window, making sure that "Categories in First Column" has been checked (Figure 1.14). *Don't select "Series Names in First Row" unless the first row of your selection includes identifying labels.* Use the GALLERY menu to select an appropriate scatter chart.

Your chart window should now be occupied by a linear plot of *x* vs *y* similar to Figure 1.11a.

Figure 1.13

In Excel, graphs are generated by selecting a block of data in your "worksheet" window and then opening a "chart" window. To generate *xy*-plots, you must additionally select your data block such that the left-hand column contains data destined for the *x*-axis and place it on your clipboard with a COPY command. (Excel also supports a convention in which *x*-axis data can be in the top row, but no models in this book are organized in this fashion.)

	A	B	
1	x	y	
2	0	0	
3	2	1	
4	4	2	
5	8	4	
6	16	8	
7			
8			

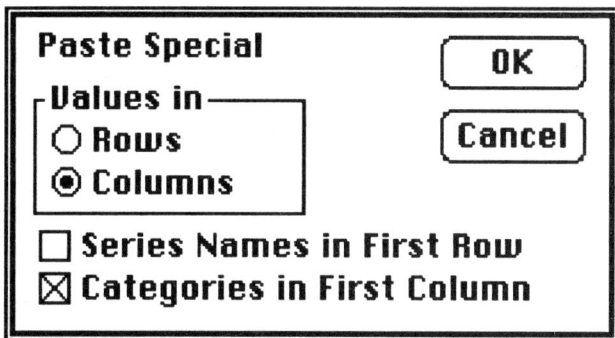

Figure 1.14

The PASTE SPECIAL command serves to specify how your data should be organized in an *xy*-plot. For a data block such as that shown in Figure 1.13, tell Excel that your data are organized in columns and that your categories (values destined for the *x*-axis) are in the first column. If you fail to check "Categories in First Column," your first column of data will be treated as one more set of *y*-axis data.

You can see the difference between an *xy*-plot and a line plot by making certain that your chart window is selected and then using the GALLERY menu to select a line chart. Notice that the plot is no longer linear (as in Figure 1.11b).

The line-plot option is of limited value for most exercises in this book because it always spaces *x*-axis data evenly along the *x*-axis. You will, however, use the *xy*-option frequently.

First-time Excel users often neglect to COPY and PASTE SPECIAL. They get a chart, but the x-axis fails to get specified!

If your spreadsheet is organized such that the data destined for your x-axis lies to the right (or very far to the left) of your y-axis data, use the above instructions to plot your x-axis against any arbitrary data (that happens to be to the right) and then edit your formula bar. These instructions may seem confusing to you if you haven't recently explored your users manual or worked through Excel's tutorial. We recommend these to you with some emphasis. Excel is a powerful but subtle program.

2. Build and verify Model 1.3. Build an on-screen graphic that plots [HB] vs pH for all three weak acids. Figure 1.15 shows the format in which such plots will be defined throughout this book.

   Notice that multiple data sets are plotted against the *y*-axis. The process is quite similar to what has already been described (if you're using Lotus 1-2-3 or Microsoft Excel), and its extension to multiple data sets is fully described in your software reference manual and on-line help facility.

3. Your spreadsheet program may have the capability of representing *xy*-plots in a number of different formats (Figure 1.16). In Lotus 1-2-3 the starting point for such explorations is the keystroke sequence

Figure 1.15

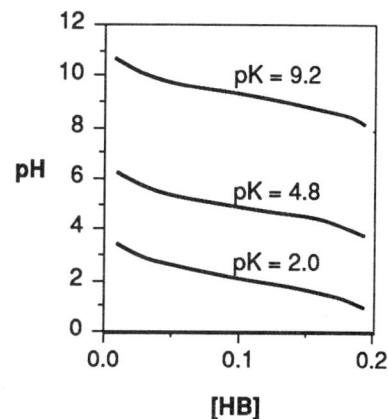

XY–PLOT

<u>Data</u>
Horizontal (*x*) axis: cells B8–B46 ([HB])
Vertical (*y*) axis: cells C8–C46 (pK = 2), D8–D46 (pK = 4.8), and E8–E46 (pK = 9.2)

Figure 1.16

Your spreadsheet program may be able to format your *xy*-plots in several ways:

(a) Data points with lines

(b) Data points alone

(c) Lines alone

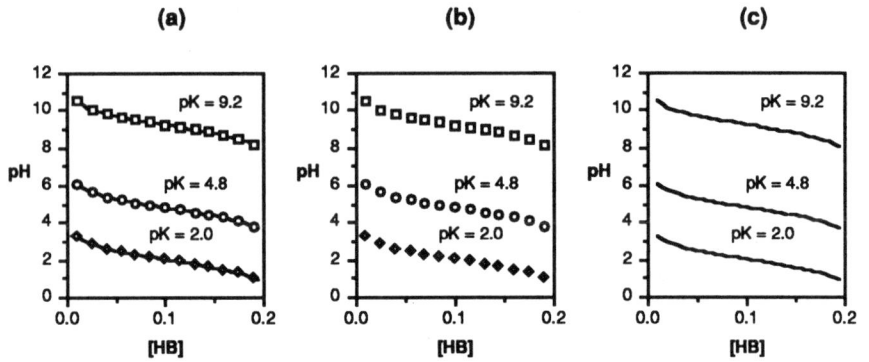

(a)   (b)   (c)

/ G O F (COMMAND GRAPH OPTIONS FORMAT). In Microsoft Excel, you can click on the graph element you wish to modify and then select the PATTERNS option from the FORMAT menu.

Explore the formatting options your spreadsheet program supports.

MODELS

Model 1.3

BUILD and verify Model 1.3.

ENTER data into cells B1, B3, and B4. Cell B1 contains the total concentration of weak acid, whereas cells B3 and B45 determine the initial value and incremental value of [B-] placed in the table in the lower portion of the model. Also ENTER the pK values of three different weak acids into cells C6, D6, and E6.

READ results from columns A, B, C, D, and E beginning at row 8.

NOTICE that the table is to permit pH vs [HB] to be plotted simultaneously for three different weak acids.

FORMULAS

	A	B	C	D	E
1	[B] + [HB] =	0.21			
2					
3	[B]o =	0.01			
4	Δ[B] =	0.005			
5					
6		pK =	2	4.8	9.2
7	[B]	[HB]	pH	pH	pH
8	=B3	=B1-A8	Formula Set I	Formula Set II	Formula Set III
9	=A8+B4	=B1-A9			
10	=A9+B4	=B1-A10			
11	=A10+B4	=B1-A11			
12	=A11+B4	=B1-A12			

VALUES

	A	B	C	D	E
1	[B] + [HB] =	0.21			
2					
3	[B]o =	0.01			
4	Δ[B] =	0.005			
5					
6		pK =	2	4.8	9.2
7	[B]	[HB]	pH	pH	pH
8	0.01	0.2	0.69897	3.49897	7.89897
9	0.015	0.195	0.88606	3.68606	8.08606
10	0.02	0.19	1.02228	3.82228	8.22228
11	0.025	0.185	1.13077	3.93077	8.33077
12	0.03	0.18	1.22185	4.02185	8.42185

Formula Set I Prototype Cell is C8	Formula Set II Prototype Cell is D8
=C6 +LOG10(A8/B8)	=D6 +LOG10(A8/B8)

Formula Set III Prototype Cell is E8
=E6 +LOG10(A8/B8)

CHAPTER 2

STOICHIOMETRY

Chemists work in two worlds. They often must think about chemical reactions in terms of numbers of molecules, but usually perform these reactions by measuring out grams of solid and milliliters of liquid.

Stoichiometry bridges these two worlds and provides the framework necessary to make quantitative chemical measurements. In the exercises that follow, you have an opportunity to build automatic computer machines that can perform stoichiometric calculations.

Stoichiometric calculations are usually performed with a pocket calculator or with pencil and paper. One way to test your knowledge of a subject, however, is to teach it to someone (or something) else. In this chapter, you use your spreadsheet as a very literal-minded student that will only act as you instruct it to act. If you can teach a spreadsheet the fundamentals of stoichiometry, you can be sure that you understand them yourself.

In subsequent chapters, you will discover that spreadsheet models can behave much like a physical experiment and the emphasis of each exercise will be on a study of the model's behavior. In this chapter, the emphasis is on building the model.

MOLECULAR WEIGHTS AND MOLES

The molecular weight of a molecule is the ratio of the mass of the molecule to one-twelfth of the mass of an atom of ^{12}C. Atomic weights are calculated similarly. Experimental study has shown that there are 6.023×10^{23} atoms of ^{12}C in 12 grams (g) of ^{12}C. Because ^1H has a mass that is one-twelfth of ^{12}C, it follows that there are 6.023×10^{23} atoms of ^1H in 1 g of ^1H. Similarly, there are 6.023×10^{23} atoms of ^{16}O present in 16 g of ^{16}O.

6.023×10^{23} is called *Avogadro's number*. A collection of 6.023×10^{23} objects (atoms, molecules, eggs) is called a *mole* of objects, and, in the laboratory, the mole provides a more convenient working scale. Indeed, a working definition of the molecular (atomic) weight of a molecule (atom) is the mass in grams of 1 mole (mol) of molecules (atoms). You can calculate the molecular weight of a substance by adding together the atomic weights of each atom in the molecule (Figure 2.1).

Pause here and work through all of this again. *You can do nothing else in this book until these relationships are completely clear!*

"Molecular weight" is a misnomer when describing a substance such as a salt because, properly speaking, strongly ionized substances are not organized as molecules. Older textbooks often use the term *formula*

TO GET THE MOST OUT OF THIS EXERCISE, YOU SHOULD ALREADY KNOW

The definition of the terms:
molecular formula
molecular weight
mole

How to calculate the molecular weight of a compound from its molecular formula.

How to convert from moles of a substance to mass and vice versa.

Figure 2.1

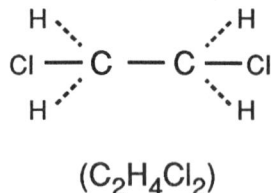

$(C_2H_4Cl_2)$

Atom	#	Atom. Wt.	Weight
H	4	1.008	4.032
C	2	12.01	24.02
Cl	2	35.45	70.90
		Mol. Wt. =	98.95

One molecule of dichloroethane contains 2 carbon atoms, 2 chlorine atoms, and 4 hydrogen atoms. One mol of dichloroethane contains 2 mol carbon atoms, 2 mol chlorine atoms, and 4 mol hydrogen atoms.

Since 1 mol of carbon atoms has a mass of 12.01 g, 2 mol of carbon atoms has a mass of 24.02 g (2 x 12.01). Therefore, 1 mol of dichloroethane contains 24.02 g of carbon. Similarly, 1 mol of dichloroethane contains 4.032 g of hydrogen and 70.90 g of chlorine.

One mol of dichloroethane weighs 98.95 g (24.02 + 4.032 + 70.90 rounded off to the appropriate number of significant figures).

weight in such instances. Formula weights are calculated in the same fashion as molecular weights (Figure 2.2).

Once you know the molecular weight of a substance, you can use Equation 2.1 to calculate

- The mass in grams of any arbitrary number of moles of substance.

- The number of moles of substance in any arbitrary weight of substance.

$$\text{molecular weight} = \frac{\text{mass in grams}}{\text{number of moles}} \qquad (2.1)$$

LESSONS

1. Figure 2.3 shows a simple spreadsheet model to calculate molecular weight of a compound. (A slightly modified version of this model is described fully as Model 2.1, but you're cheating if you look at it.) Cells B2–B19 contain the atomic weights of an arbitrary selection of common elements. Cells C2–C19 represent the number of atoms of each element present in the compound. Cells D2–D19 calculate the weight in grams of each element in 1 mol of the compound, and cell D21 calculates the molecular weight of the whole compound.

   a. What compound is represented by the model in Figure 2.3?

   b. What formulas are required in cells D2–D19?

   c. What formula is required in cell D21?

   d. Build the model described in Figure 2.3 in your computer (Model 2.1 is our answer).

2. Use the model that you built in Lesson 1 to calculate the molecular weights of the following compounds:

   a. $K_2S_2O_3$

   b. KrF_4

   c. $Al(CH_3)_3$

   d. $C_9H_{15}O_8P$ (bomyl, an insecticide)

3. Modify your model (Lesson 1) by adding new elements to the list of atomic weights and find the molecular weights of the following compounds:

   a. UCl_3

   b. $YBa_2Cu_3O_7$ (a high-temperature superconductor)

   c. $Co_3(AsO_4)_2 \cdot 8H_2O$

   d. $W(CO)_4I_3$

4. Once the molecular weight of a compound is known, it is a simple matter to translate between the weight of a sample and the number of moles in a sample (Equation 2.1). Modify the model you built for Lesson 1 so that it can perform this conversion (Figure 2.4). Our answer resides in Model 2.1.

   a. How many moles are present in 0.2 g of $K_2S_2O_3$?

   b. How many moles are present in 0.175 g of $Al(CH_3)_3$?

   c. Which sample contains the larger number of molecules: 1 g of $Al(CH_3)_3$ or 0.5 g of UCl_3?

Figure 2.2

$CuSO_4 \cdot H_2O$

Atom	#	Atom. Wt.	Weight
H	2	1.008	2.016
Cu	1	63.55	63.55
S	1	32.06	32.06
O	5	16.00	80.00
		Form. Wt. =	177.63

Formula weights are calculated the same way as molecular weights. Indeed, the term *formula weight* is gradually being replaced by *molecular weight* as the all-encompassing term. The formula for copper sulfate in its usual state has been written as $CuSO_4 \cdot H_2O$ rather than $CuSO_5H_2$ in order to convey more structural information about the compound.

The water molecule in $CuSO_4 \cdot H_2O$ may be removed by gentle heating – a process that transforms the compound from a gemlike blue crystal to a white powder. This experiment (or another like it) is often performed in introductory chemistry laboratories. If a 177.63-g sample of $CuSO_4 \cdot H_2O$ is dehydrated by heat to a white powder, what will be the final weight of the sample?

Figure 2.3

Spreadsheets should be designed to clearly display their structure and internal logic. This spreadsheet to calculate molecular weights uses the same format seen in Figures 2.1 and 2.2.

	A	B	C	D
	Atom/Group	At. Wt.	#	Weight
1				
2	Al	26.98		0
3	B	10.81		0
4	C	12.01		0
5	Ca	40.08		0
6	Cl	35.45		0
7	Cu	63.54		0
8	F	19.00		0
9	H	1.008	2	2.016
10	K	39.10		0
11	Kr	83.80		0
12	Mg	24.31		0
13	Mn	54.94		0
14	Na	22.99		0
15	Ni	58.71		0
16	O	16.00	1	16
17	P	30.97		0
18	S	32.06		0
19	H2O	18.02		0
20				
21		Mol. Wt. (g/mol) =		18.016

Figure 2.4

Once you calculate the molecular weight of a compound, you can convert between mass and number of atoms with ease (Equation 2.1). Cell B24 accepts a value (sample mass in grams) that is used to calculate the number of moles. Cell B28 accepts a value (number of moles in sample) to calculate sample mass in grams. Notice that two separate formulas must be used for these two separate calculations.

	A	B	C	D
1	Atom/Group	At. Wt.	#	Weight
2	Al	26.98		0
3	B	10.81		0
4	C	12.01		0
5	Ca	40.08		0
6	Cl	35.45		0
7	Cu	63.54		0
8	F	19.00		0
9	H	1.008	2	2.016
10	K	39.10		0
11	Kr	83.80		0
12	Mg	24.31		0
13	Mn	54.94		0
14	Na	22.99		0
15	Ni	58.71		0
16	O	16.00	1	16
17	P	30.97		0
18	S	32.06		0
19	H2O	18.02		0
20				
21		Mol. Wt. (g/mol) =		18.016
22				
23	MASS -> # ATOMS			
24	Sample mass in grams =	0.5		
25	# moles (mol) =	0.02775		
26				
27	# ATOMS -> MASS			
28	# moles (mol) =	0.02775		
29	Sample mass in grams =	0.5		

d. What is the mass in grams of of 0.04 mol of H_2O?

e. What is the mass in grams of 0.875 mol of $YBa_2Cu_3O_7$?

f. Which sample contains the larger mass: 0.01 mol of $YBa_2Cu_3O_7$ or 0.01 mol UCl_3?

5. Many crystalline salts can exist in alternate hydration states. Suppose you are given a 0.3061-g sample of a hydrate of $CrCl_2$. In this state, the solid is dark blue. You first heat the solid carefully to 50°C and notice a change in color to light blue. After permitting the sample to cool in a desiccator, you find that it now weighs 0.2778 g. Reheating the sample to 100°C, you observe a color change to light green. The sample now weighs 0.2495 g. Finally, heating to 120°C produces a colorless solid weighing only 0.1929 g.

Determine the composition of the dark blue, light blue, light green, and colorless solids, assuming that only H_2O is driven off by heating.

PROBLEMS

1. A 1.150-g sample of magnesium sulfate ($MgSO_4$) is gently heated in a desiccator after which the sample weighs 0.562 g. Of this sample, 0.100-g aliquots are placed in five separate containers at different relative humidities. After several days, the samples are removed, and their weights are found to be 0.103 g, 0.113 g, 0.115 g, 0.116 g, and 0.206 g. What do these data tell you about the allowable hydration states of $MgSO_4$?

2. A 4.00-g sample of anhydrous $MnSO_4$ is brought to equilibrium at a range of relative humidities and weighed. The results are displayed in the table below. What are the allowable hydration stages of $MnSO_4$?

Sample No.	Weight	Sample No.	Weight
1	4.02	11	5.90
2	3.99	12	5.89
3	4.48	13	6.37
4	4.51	14	6.37
5	4.50	15	6.38
6	4.93	16	6.85
7	4.97	17	6.87
8	5.40	18	7.30
9	5.45	19	7.33
10	5.44	20	7.35

3. A chemist prepares a sample of crystalline nickel(II) formate, $Ni(HCO_2)_2$, from an aqueous solution. She wishes to determine the degree of solvation in the solid: $Ni(HCO_2)_2 \cdot XH_2O$, where X is an unknown number. A sample of the solid weighing 0.3950 g is carefully ground up and heated to 130–140°C. After cooling, the dehydrated solid weighs 0.3177 g. Use algebra to determine the value of X in the formula $Ni(HCO_2)_2 \cdot XH_2O$ and then confirm your answer by building a computer simulation of the situation.

4. A chemist wishes to treat a sample of $CuSO_4$ weighing 2.614 g with barium chloride, $BaCl_2$, according to the following reaction:

$$CuSO_4 + BaCl_2 \rightarrow BaSO_4 + CuCl_2$$

The reaction goes to completion because of the insolubility of $BaSO_4$.

a. How many grams of $BaCl_2$ are required for a complete reaction?

b. Suppose the copper sulfate is hydrated (i.e., $CuSO_4 \cdot 5H_2O$ instead of $CuSO_4$). If the chemist does not catch this error, will too much or too little $BaCl_2$ be used? Explain your answer.

MODELS

Model 2.1

BUILD and verify Model 2.1.

ENTER data using cells A2–B19, B24, and B28. Cells A2–B19 constitute a table of data for the names and atomic weights of elements and groups of elements. Enter sample mass into cell B24 in order to read the number of moles from cell B25. Enter the number of moles into cell B28 in order to read sample mass from cell B29.

READ output from the model at cell D21 (molecular weight), cell B25 (the number of moles, given sample mass), and cell B29 (sample mass, given number of moles).

NOTICE that the data table can contain groups of elements as well as elements (e.g., H_2O in cells A19–B19). This is often useful when exploring closely related compounds such as salts in different hydration states.

FORMULAS

	A	B	C	D
1	Atom/Group	At. Wt.	#	Weight
2	Al	26.98		=B2*C2
3	B	10.81		=B3*C3
4	C	12.01		=B4*C4
5	Ca	40.08		=B5*C5
6	Cl	35.45		=B6*C6
7	Cu	63.54		=B7*C7
8	F	19.00		=B8*C8
9	H	1.008	2	=B9*C9
10	K	39.10		=B10*C10
11	Kr	83.80		=B11*C11
12	Mg	24.31		=B12*C12
13	Mn	54.94		=B13*C13
14	Na	22.99		=B14*C14
15	Ni	58.71		=B15*C15
16	O	16.00	1	=B16*C16
17	P	30.97		=B17*C17
18	S	32.06		=B18*C18
19	H2O	18.02		=B19*C19
20				
21			Mol. Wt. (g/mol) =	=SUM(D2:D19)
22				
23	MASS -> # ATOMS			
24	Sample mass in grams =	0.5		
25	# moles (mol) =	=B24/D21		
26				
27	# ATOMS -> MASS			
28	# moles (mol) =	0.02775		
29	Sample mass in grams =	=B28*D21		

VALUES

	A	B	C	D
1	Atom/Group	At. Wt.	#	Weight
2	Al	26.98		0
3	B	10.81		0
4	C	12.01		0
5	Ca	40.08		0
6	Cl	35.45		0
7	Cu	63.54		0
8	F	19.00		0
9	H	1.008	2	2.016
10	K	39.10		0
11	Kr	83.80		0
12	Mg	24.31		0
13	Mn	54.94		0
14	Na	22.99		0
15	Ni	58.71		0
16	O	16.00	1	16
17	P	30.97		0
18	S	32.06		0
19	H2O	18.02		0
20				
21			Mol. Wt. (g/mol) =	18.016
22				
23	MASS -> # ATOMS			
24	Sample mass in grams =	0.5		
25	# moles (mol) =	0.02775		
26				
27	# ATOMS -> MASS			
28	# moles (mol) =	0.02775		
29	Sample mass in grams =	0.5		

MOLE FRACTIONS

The mole fraction of a substance in a mixture of substances is defined by Equation 2.2:

$$x_i = \frac{n_i}{\Sigma n} \tag{2.2}$$

where x_i = mole fraction of substance i, n_i = number of moles of substance i, Σn = sum of the number of moles of all substances in the mixture. The concept of a mole fraction is used extensively in the study of both solutions and gases.

Suppose, for example, that a solution is prepared by mixing 1.50 mol of benzene, 0.50 mol of toluene, and 0.1 mol of xylene. If x_b is the mole fraction of benzene, x_t is the mole fraction of toluene, and x_x is the mole fraction of xylene, then

$$x_b = \frac{n_b}{\Sigma n} = \frac{1.50}{1.50 + 0.50 + 0.1} = 0.71 \tag{2.3}$$

$$x_t = \frac{n_t}{\Sigma n} = \frac{0.50}{1.50 + 0.50 + 0.1} = 0.24 \tag{2.4}$$

$$x_x = \frac{n_x}{\Sigma n} = \frac{0.10}{1.50 + 0.50 + 0.1} = 0.05 \tag{2.5}$$

TO GET THE MOST OUT OF THIS
EXERCISE, YOU SHOULD
ALREADY KNOW

The definition of the terms:
molecular formula
molecular weight
mole

How to calculate the molecular weight
of a compound from its molecular
formula.

How to convert from moles of a
substance to mass and vice versa.

Several properties of mole fractions are worth noticing. Mole fractions are always positive numbers between 0 and 1. The sum of mole fractions of all constituents of a mixture sums to 1. Mole fractions are dimensionless (the calculations cancel out the units).

LESSONS

1. Figure 2.5 is an example of a spreadsheet model that might be used to calculate mole fractions of four constituents in a mixture. The model uses the same design principles applied to the model in Exercise 2.1. After you have "taught" your spreadsheet to calculate mole fractions, verify your results with Model 2.2.

   The following step-by-step directions may help you navigate through the design of Figure 2.5, but perhaps you can build the model without help. [NOTE: If you get error messages in any of the cells of rows 21–23, see (g) below.]

   a. Enter all labels of Figure 2.5 into your spreadsheet program. Notice cell C2 contains a label that stretches across most of the spreadsheet.

   b. Cells C1–F1 contain the sample weight in grams for each of the four constituents of the mixture (the constituents are identified as [1], [2], [3], and [4] in row 3).

   c. In column B, enter the atomic weights associated with the elements listed in column A.

Figure 2.5

	A	B	C	D	E	F	G	H	I	J
1	Weight in grams =		1	2	0.5	0				
2				-----Number of Atoms/Molecule----	-----			------Intermediate Calculations------		
3	Atom	At. Wt.	[1]	[2]	[3]	[4]	[1]	[2]	[3]	[4]
4	Al	26.98					0	0	0	0
5	B	10.81					0	0	0	0
6	C	12.01			1		0	0	12.01	0
7	Ca	40.08					0	0	0	0
8	Cl	35.45			4		0	0	141.8	0
9	Cu	63.54					0	0	0	0
10	H	1.008	3	2			3.024	2.016	0	0
11	K	39.10					0	0	0	0
12	Kr	83.80					0	0	0	0
13	N	14.01					0	0	0	0
14	Na	22.99					0	0	0	0
15	O	16.00	4	1			64	16	0	0
16	P	30.97	1				30.97	0	0	0
17	S	32.06					0	0	0	0
18	Ti	47.90					0	0	0	0
19										
20					Mol. Wt. (g/mol) =		97.994	18.016	153.81	0
21					# moles (mol) =		0.010205	0.111012	0.003251	0
22					# molecules =		6.15E+21	6.69E+22	1.96E+21	0
23					Mole fraction =		0.081987	0.891896	0.026117	0
24					Sum, Mole fractions =		1			

d. Columns C–F contain the number of atoms of each element present in each of four compounds. Enter the numbers used in Figure 2.5 (so that you can verify your model at the end).

e. Columns G–J contain formulas that calculate the mass that each element in each compound contributes to the total mass of the compound (e.g., cell G10 calculates the mass of the three hydrogen atoms in compound 1, and cell I8 calculates the mass of the four chlorine atoms in compound 3).

What formulas are required by these cells?

Enter these formulas into your program (don't forget your COPY command).

f. Cells G20–J20 calculate the molecular weight of each of the four compounds.

What formulas are required by these cells?

Enter these formulas into your program. Use your COPY command.

g. Cells G21–J21 calculate the number of moles of each constituent contained in the sample weights entered in cells C1–F1.

What is the formula that should be entered in cell G21?

If you answered =C1/G20, your chemistry is perfect. However, if you copy this formula into cells H21–J21, you get an error message in cell J21 if you use the data shown in Figure 2.5. Compound 4 has not been defined and has no molecular weight. Cell J21 is trying to divide by 0! If all compounds have nonzero molecular weights your model will work fine using the formula =C1/G20.

The correct formula for cell G21 (if you wish your model to work for mixtures with less than four constituents) should say: "**If** C1/G20 causes an error, **then** the answer is 0; **otherwise** the answer is C1/G20." (See Figure 2.6.)

In Excel, the language used in this book, the formula in cell J21 should read =IF(ISERROR(C1/G20),0,C1/G20). Your spreadsheet program may use a different syntax.[1]

Figure 2.6

Predicate

IF(...   ,   ...   ,   ...   )

Do this if
predicate is true

Do this if
predicate is false

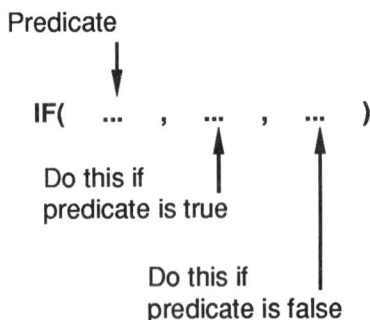

Spreadsheet programs have a powerful built-in function, the IF() function, that can be used to make decisions. IF() functions contain three parts:

1. A predicate (a statement that is either true or false [e.g., C1+C2=4])

2. A formula to be executed if the predicate is true

3. A formula to be executed if the predicate is false

A typical IF() function might read: IF(A1>=0,SQRT(A1),0). The function translates: "**If** the contents of cell A1 is greater than or equal to zero, **then** calculate the square root of A1; **otherwise** insert 0."

h. Cells G22–J22 contain formulas to calculate the number of molecules as an integer number rather than as a number of moles.

What formulas are required by these cells?

Enter these formulas into your program. Use your COPY command.

i. Cells G23–J23 contain formulas to calculate the mole fraction of each constituent.

What formulas are required by these cells?

Enter these formulas into your program. Use your COPY command.

j. Place the sum of mole fractions in cells G21–J21 in cell G24.

2. In a sealed container 1 g of O_2 and 4 g of N_2 are mixed together. What are the mole fractions of O_2 and N_2 in this gas mixture?

3. A beaker of ethyl alcohol, C_2H_5OH, contains 5% H_2O by weight. What are the mole fractions of alcohol and water in this mixture?

PROBLEM

A mixture is prepared containing 0.15 mol methylene chloride (CH_2Cl_2), 0.05 mol toluene (C_7H_8), and 0.30 mol ethyl acetate ($C_4H_8O_2$) (common components of commercial paint remover). Calculate the mole fractions of these components in the mixture. (NOTE: You can temporarily "overwrite" the formulas in cells G21–J21 with values.)

MODELS

FORMULAS

	A	B	C	D	E	F	G	H	I	J
1	Weight in grams =		1	2	0.5	0				
2				—Number of Atoms/Molecule—				—Intermediate Calculations—			
3	Atom	At. Wt.	[1]	[2]	[3]	[4]	[1]	[2]	[3]	[4]
4	Al	26.98					=B4*C4	=B4*D4	=B4*E4	=B4*F4
5	B	10.81					=B5*C5	=B5*D5	=B5*E5	=B5*F5
6	C	12.01			1		=B6*C6	=B6*D6	=B6*E6	=B6*F6
7	Ca	40.08					=B7*C7	=B7*D7	=B7*E7	=B7*F7
8	Cl	35.45			4		=B8*C8	=B8*D8	=B8*E8	=B8*F8
9	Cu	63.54					=B9*C9	=B9*D9	=B9*E9	=B9*F9
10	H	1.008	3	2			=B10*C10	=B10*D10	=B10*E10	=B10*F10
11	K	39.10					=B11*C11	=B11*D11	=B11*E11	=B11*F11
12	Kr	83.80					=B12*C12	=B12*D12	=B12*E12	=B12*F12
13	N	14.01					=B13*C13	=B13*D13	=B13*E13	=B13*F13
14	Na	22.99					=B14*C14	=B14*D14	=B14*E14	=B14*F14
15	O	16.00	4	1			=B15*C15	=B15*D15	=B15*E15	=B15*F15
16	P	30.97	1				=B16*C16	=B16*D16	=B16*E16	=B16*F16
17	S	32.06					=B17*C17	=B17*D17	=B17*E17	=B17*F17
18	Ti	47.90					=B18*C18	=B18*D18	=B18*E18	=B18*F18
19										
20					Mol. Wt. (g/mol) =		Formula Set I			
21					# moles (mol) =		Formula Set II			
22					# molecules =		Formula Set III			
23					Mole fraction =		Formula Set IV			
24					Sum, Mole fractions =		=SUM(G23:J23)			

VALUES

	A	B	C	D	E	F	G	H	I	J
1	Weight in grams =		1	2	0.5	0				
2				—Number of Atoms/Molecule—				—Intermediate Calculations—			
3	Atom	At. Wt.	[1]	[2]	[3]	[4]	[1]	[2]	[3]	[4]
4	Al	26.98					0	0	0	0
5	B	10.81					0	0	0	0
6	C	12.01			1		0	0	12.01	0
7	Ca	40.08					0	0	0	0
8	Cl	35.45			4		0	0	141.8	0
9	Cu	63.54					0	0	0	0
10	H	1.008	3	2			3.024	2.016	0	0
11	K	39.10					0	0	0	0
12	Kr	83.80					0	0	0	0
13	N	14.01					0	0	0	0
14	Na	22.99					0	0	0	0
15	O	16.00	4	1			64	16	0	0
16	P	30.97	1				30.97	0	0	0
17	S	32.06					0	0	0	0
18	Ti	47.90					0	0	0	0
19										
20					Mol. Wt. (g/mol) =		97.994	18.016	153.81	0
21					# moles (mol) =		0.010205	0.111012	0.003251	0
22					# molecules =		6.15E+21	6.69E+22	1.96E+21	0
23					Mole fraction =		0.081987	0.891896	0.026117	0
24					Sum, Mole fractions =		1			

Formula Set I Prototype Cell is G20	Formula Set II Prototype Cell is G21	Formula Set III Prototype Cell is G22
=SUM(G4:G18)	=IF(ISERROR(C1/G20),0,C1/G20)	=6.023E+23*G21

Formula Set IV Prototype Cell is G23
=G21/SUM(G21:J21)

Model 2.2

BUILD and verify Model 2.2.

ENTER data using cells C1–F1 and C4–F18. Cells C1–F1 contain the sample weight in grams of each of the constituents of the mixture. Cells C4–F18 contain the numbers of atoms of each element in each constituent.

READ output from the model at cells G20–J23. The contents are self-explanatory.

NOTICE that "Sum, Mole fractions" always equals 1 in a correctly functioning model.

LIMITING REAGENTS AND THEORETICAL YIELDS

TO GET THE MOST OUT OF THIS
EXERCISE, YOU SHOULD
ALREADY KNOW

The definition of the terms:
molecular formula
molecular weight
mole

How to calculate the molecular weight
of a compound from its molecular
formula.

How to convert from moles of a
substance to mass and vice versa.

How to balance chemical equations
and interpret their meaning.

(The model for this exercise is more
complex than many in this book. If
you have trouble you may wish to put
off this exercise to the end of the
chapter or until you have more
experience with electronic spreadsheets.)

Sometimes reactants are mixed together in exactly the correct propor-
tions so that they are entirely consumed during a reaction and only
product(s) remain. Consider, for example, the reaction

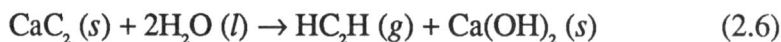

$$CaC_2\ (s) + 2H_2O\ (l) \rightarrow HC_2H\ (g) + Ca(OH)_2\ (s) \qquad (2.6)$$

All water (H_2O) will react with all calcium carbide (CaC_2) if and only
if 1 mol of CaC_2 is mixed with 2 mole of H_2O. Usually, however, one
reactant is in short supply. In a laboratory or an industrial setting, one
reactant is often deliberately in short supply because of high costs or
other reasons. Such a reagent is called the *limiting reagent*. When the
limiting reagent is depleted, the reaction must stop even if plentiful
amounts of other reactants are still present.

The amount of product produced if all the limiting reagent is used is
called the *theoretical yield* (expressed in either grams or moles). In the
laboratory, theoretical yields are rarely achieved. If the actual yield of
a laboratory reaction is known, the *percent yield* may be calculated
(Equation 2.7). Either grams or moles can be used in the calculation.

$$\% \text{ yield} = \frac{\text{actual yield (g or mol)}}{\text{theoretical yield (g or mol)}} \times 100 \qquad (2.7)$$

For example, if 5 mol of H_2O are gradually added to only 2 mol of CaC_2, 1 mol of H_2O will remain at the end of the reaction, and 2 mol of both HC_2H and $Ca(OH)_2$ will be produced.[1] The theoretical yield of HC_2H for this reaction is 2 mol. If laboratory measurements show an actual yield of 1.9 moles of HC_2H, the percent yield is 95%. You could, of course, also calculate a percent yield based on theoretical and actual yields expressed in grams.

In this exercise, you have an opportunity to build a computing machine that can identify limiting reagents and calculate theoretical yields. (The model is more complex than many that have come before and will come after. Don't let it intimidate you.)

First consider the methods that chemists use to identify limiting reagents. Suppose that a 0.122-g sample of silica (SiO_2) is treated with 0.342 g of bromine trifluoride (BrF_3) (Equation 2.8):

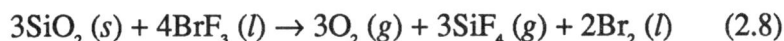

$$3SiO_2\ (s) + 4BrF_3\ (l) \rightarrow 3O_2\ (g) + 3SiF_4\ (g) + 2Br_2\ (l) \qquad (2.8)$$

Equation 2.8 declares that both reactants will be totally consumed in the reaction if and only if the mole ratio of SiO_2 to BrF_3 is 3:4 (Equations 2.9 and 2.10):

$$\frac{\text{moles of } SiO_2}{\text{moles of } BrF_3} = \frac{3}{4} \qquad (2.9)$$

or

$$\frac{\text{moles of } SiO_2}{3} = \frac{\text{moles of } BrF_3}{4} \qquad (2.10)$$

Equation 2.10 conveniently represents the situation in terms of *reactant ratios* (i.e., the ratio of the number of moles of a reactant to its coefficient in a balanced chemical equation). If the reactant ratios of all reactants are equal, there is no limiting reagent, and all reactants will be consumed. If all reactant ratios are not equal, then *the reagent with the smallest reactant ratio is the limiting reagent.* Study Equations 2.9 and 2.10 and confirm that this must be so.

Since 0.122 g of SiO_2 equals 0.00203 mol and 0.342 g of BrF_3 equals 0.00250 mol (Exercise 2.1), it is easy to see that BrF_3 is the limiting reagent (0.00203/3 > 0.00250/4). Some SiO_2 will remain after all BrF_3 has been depleted.[2]

Notice that you must consider *both* the number of moles of all reactants *and* the balanced equation for the reaction in order to identify the limiting reagent. Often you must also convert from weight or mass to moles.

- You cannot use the masses of the reactants alone to determine the limiting reagent. In the example, BrF_3 is present in greater mass than SiO_2 but is nevertheless the limiting reagent.

- You cannot use the number of moles of reactants alone to determine the limiting reagent. In the example, 3.2 mol of SiO_2 would not be the limiting reagent in the presence of 4 mol of BrF_3.

Once the limiting reagent has been identified, the theoretical yield is calculated from the ratio of moles of product produced to moles of limiting reagent used in the balanced chemical equation (the *coefficient ratio*). In the example, 2 mol of Br_2 is produced from 4 mol of BrF_3 (a coefficient ratio of 1:2). Therefore, 0.00125 mol of Br_2 would be obtained from the complete reaction of 0.00250 mol of BrF_3 with SiO_2. The theoretical yield of Br_2, given 0.00250 mol of BrF_3, is 0.00125.

LESSONS

1. Figure 2.7 depicts a model that can calculate the molecular weight and the number of moles of reactant and product, given sample weight in grams and the molecular formulas of all reagents. The model is very similar in design to models already described in Exercises 2.1 and 2.2.

   a. Cells I4–I18 store the contribution that each element in columns A and B make to the molecular weight of the reactant R1.

   What formulas belong in cells I4–I18?

   b. Cells J4–N18 store the contributions that each element makes to the molecular weights of the reactants and products R2, R3, P1, P2, and P3.

   What formulas belong in cells J4–N18?

   c. Cells I20–N20 calculate the molecular weights of each of the reactants and compounds identified in the model.

   What formulas belong in cells I20–N20?

d. Cells I21–N21 calculate the number of moles of each of the reactants and compounds identified in the model, given the sample weights entered into cells C1–H1.

What formulas belong in cells I21–N21?

e. Build the model described in Figure 2.7 and verify that it functions correctly by entering several different compounds of known molecular weight.

Model 2.3 contains a complete description of the correct formulas. [NOTE: The "# moles" calculations in cells I21–N21 are written so that molecular weights of zero do not force a "division by zero" error (see the legend to Model 2.3 for additional details). You may or may not wish to incorporate this feature at this time.]

2. Modify your model by adding the four additional rows of labels and formulas implied by Figure 2.8. Cells I24–N24 are used to enter the coefficients of a balanced chemical equation, and cells I25–N25 use these coefficients (and the results calculated in row 21) to compute the reactant ratios described by Equation 2.10.

Verify that your model works correctly and then answer the following questions. (The correct formulas for Figure 2.8 are completely described in Model 2.3). Add elements to your model as needed.

	A	B	C	D	E	F	G	H	I	J	K	L	M	N
1	Mass in grams =		0.12	0.34										
2			—Number of Atoms/Molecule—						—Intermediate Calculations—					
3	Atom	At. Wt.	[R1]	[R2]	[R3]	[P1]	[P2]	[P3]	[R1]	[R2]	[R3]	[P1]	[P2]	[P3]
4	Al	26.98							0	0	0	0	0	0
5	Br	79.91		1				2	0	79.91	0	0	0	159.82
6	C	12.01							0	0	0	0	0	0
7	Ca	40.08							0	0	0	0	0	0
8	Cl	35.45							0	0	0	0	0	0
9	F	19.00		3			4		0	57	0	0	76	0
10	H	1.008							0	0	0	0	0	0
11	K	39.10							0	0	0	0	0	0
12	Kr	83.80							0	0	0	0	0	0
13	Mg	24.31							0	0	0	0	0	0
14	Na	22.99							0	0	0	0	0	0
15	O	16.00	2			2			32	0	0	32	0	0
16	P	30.97							0	0	0	0	0	0
17	Si	28.09	1			1			28.09	0	0	0	28.09	0
18	Ti	47.90							0	0	0	0	0	0
19														
20						Mol. Wt. (g/mol) =			60.09	136.91	0	32	104.09	159.82
21						# moles (mol) =			0.0020303	0.002498	0	0	0	0

Figure 2.7

This model calculates the molecular weight (cells I20–N20) and number of moles (cells I21–N21) for a maximum of three reactants and three products of a chemical reaction. The amounts of initial reactants are determined by the mass in grams entered into cells C1–E1. The model is very similar to models already described in Exercises 2.1 and 2.2 and you should refer to these exercises if you have difficulty following its structure. Columns A and B constitute a table of atomic weights; cells C4–H18 are used to enter the number of atoms per molecule of each element in three reactants (R1, R2, R3) and three products (P1, P2, P3).

Figure 2.8

The model implied by Figure 2.7 can be supplemented with four additional rows of labels and formulas to calculate *reactant ratios.* Coefficients of the balanced chemical equation are entered into cells I24–N24 and the reactant ratios are read from cells I25–N25.

	A	B	C	D	E	F	G	H	I	J	K	L	M	N
22														
23	LIMITING REAGENT CALCULATOR:													
24				Coefficients in balanced equation =					3	4		3	3	2
25					Reactant ratio =		0.0006768	0.0006245	1E+09					

a. In a closed vessel, 7.00 g of calcium (Ca) is added to 5.40 g of vanadium pentoxide (V_2O_5) and the following reaction occurs:

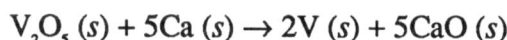

$$V_2O_5 \ (s) + 5Ca \ (s) \rightarrow 2V \ (s) + 5CaO \ (s)$$

Find the limiting reagent in this reaction.

b. Molybdenum sulfide $(MoS_2, 300 \ g)$ is roasted with oxygen gas $(O_2, 250 \ g)$ with the result

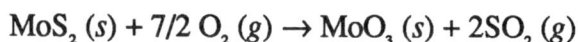

$$MoS_2 \ (s) + 7/2 \ O_2 \ (g) \rightarrow MoO_3 \ (s) + 2SO_2 \ (g)$$

What reactant limits the extent of this reaction?

c. FeS (3.11 g), CaC_2 (2.49 g), and CaO (2.70 g) are mixed together, and they react according to the following equation:

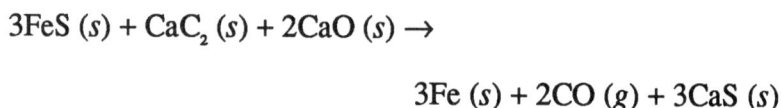

$$3FeS \ (s) + CaC_2 \ (s) + 2CaO \ (s) \rightarrow$$

$$3Fe \ (s) + 2CO \ (g) + 3CaS \ (s)$$

What is the limiting reagent?

3. Theoretical yield calculations are easily performed once the limiting reagent is known. The precise calculations performed, however, are different depending on which compound is the limiting reagent. You can automate this process with your spreadsheet program (see Problem 1), but often it is just as easy to "eyeball" an answer.

Use your model (Figures 2.7 and 2.8) to answer the following questions. According to the reaction, 20.0 g of $TiCl_4$ is reduced with 3.10 g of Mg:

$$TiCl_4 \ (l) + 2Mg \ (s) \rightarrow Ti \ (s) + 2MgCl_2 \ (s)$$

a. What is the limiting reagent in this reaction?

b. How many moles of magnesium are consumed in this reaction?

c. How many moles of titanium are produced in this reaction (i.e., what is the theoretical yield of titanium)?

PROBLEMS

1. Theoretical yield calculations require your spreadsheet program to
    "make decisions." *IF* reactant R1 is the limiting reagent, *THEN*
    Calculation I must be performed. *IF* reactant R2 is the limiting
    reagent, *THEN* Calculation II must be performed.

    It's easy to program your spreadsheet to make decisions, but
    programming isn't for everyone. Thus, this last component of your
    theoretical yield spreadsheet model is presented as a problem rather
    than an lesson. If your sole interest in this book is to understand
    chemistry, you can skip this problem. If your interest in this book
    extends to spreadsheets and how to use them, read on.

    Consider the addition to your model (Figures 2.7 and 2.8) contained
    in Figure 2.9 (and Model 2.3). Figure 2.9 follows the calculations
    associated with the product P2 when reactant R2 is the limiting
    reagent. Refer to Figure 2.7, row 3, to verify that column J is used
    for calculations associated with R2 and column M is used for
    calculations associated with P2.

    Step E of Figure 2.9 shows the formulas used to calculate the
    theoretical yield (in both moles and grams) for product P2. The
    theoretical yield in moles, for example, is found by multiplying
    the moles of limiting reagent (cell I32) by the ratio formed from
    the coefficients of P2 and the limiting reagent (cell M24 divided
    by cell I31). It's clear where the coefficient for P2 comes from
    (you wrote it into M24 yourself), but how did the the informa-
    tion about the limiting reagent (coefficient and number of moles
    used) come to reside in cells I31 and I32?

    Step D of Figure 2.9 shows that the coefficient of the limiting
    reagent (cell I31) is found by taking the largest value residing in
    cells I28–K28. The number of moles of limiting reagent (cell
    I32) is found by taking the largest value residing in cells
    I29–K29. The magic seems to be in the two ranges I28–K28 and
    I29–K29. These cells contain zero unless they belong to the
    limiting reagent.

    Steps B and C of Figure 2.9 are where the decisions are made.

    Step B is the easiest to understand. Cell J28 asks the question,
    Is the value of the reactant ratio cell in J25 the smallest value of
    any of the reactant ratios?[3] Cell J28 then makes a decision. If cell
    J25 contains the smallest reactant ratio, then cell J24 contains

Figure 2.9

Theoretical yield calculations require spreadsheet models to decide first which reagent is the limiting reagent. The model must perform different sets of calculations depending on which reagent is limiting. Decision making is a powerful feature of spreadsheet programs.

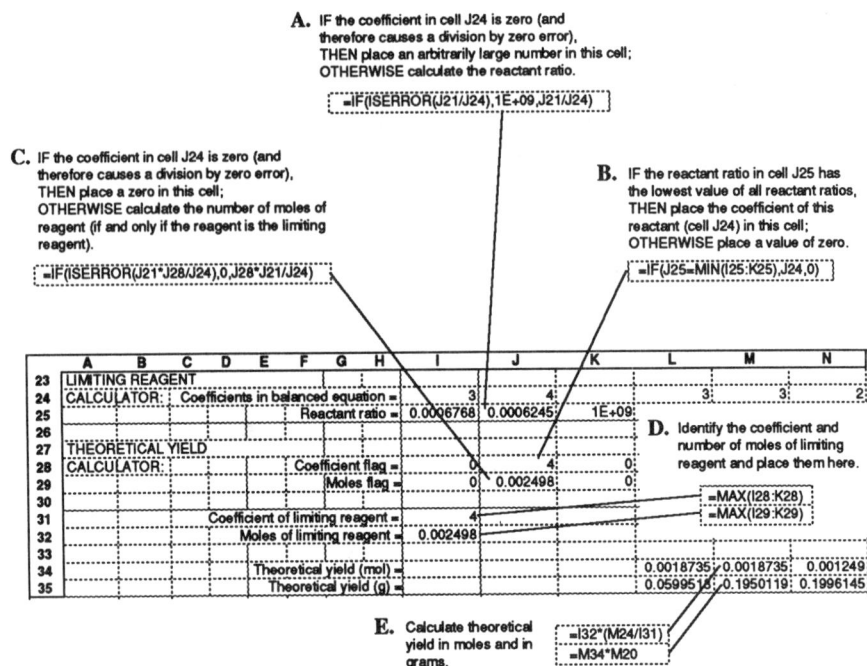

A. IF the coefficient in cell J24 is zero (and therefore causes a division by zero error), THEN place an arbitrarily large number in this cell; OTHERWISE calculate the reactant ratio.

=IF(ISERROR(J21/J24),1E+09,J21/J24)

C. IF the coefficient in cell J24 is zero (and therefore causes a division by zero error), THEN place a zero in this cell; OTHERWISE calculate the number of moles of reagent (if and only if the reagent is the limiting reagent).

=IF(ISERROR(J21*J28/J24),0,J28*J21/J24)

B. IF the reactant ratio in cell J25 has the lowest value of all reactant ratios, THEN place the coefficient of this reactant (cell J24) in this cell; OTHERWISE place a value of zero.

=IF(J25=MIN(I25:K25),J24,0)

	A	B	C	D	E	F	G	H	I	J	K	L	M	N
23	LIMITING REAGENT													
24	CALCULATOR:			Coefficients in balanced equation =					3	4		3	3	2
25							Reactant ratio =		0.0006768	0.0006245	1E+09			
26														
27	THEORETICAL YIELD													
28	CALCULATOR:						Coefficient flag =		0	4	0			
29							Moles flag =		0	0.002498	0			
30														
31						Coefficient of limiting reagent =			4					
32						Moles of limiting reagent =			0.002498					
33														
34							Theoretical yield (mol) =					0.0018735	0.0018735	0.001249
35							Theoretical yield (g) =					0.0599518	0.1950119	0.1996145

D. Identify the coefficient and number of moles of limiting reagent and place them here.

=MAX(I28:K28)
=MAX(I29:K29)

E. Calculate theoretical yield in moles and in grams.

=I32*(M24/I31)
=M34*M20

the coefficient of the limiting reagent, and the contents of cell J24 should be repeated in cell J28. Otherwise, cell J28 should contain a value of zero. Thus, the largest number in the range I28–K28 will always be the coefficient of the limiting reagent.

Step C is a bit more complex. Cell J29 asks the question, Is there a number equal to zero — or no number at all — in cell J24 (i.e., has a division by zero error occurred)? If an error has occurred, then a zero should be placed in cell J29 because R2 is undefined and doesn't enter into the reaction at all. If an error has not occurred, then the results of the calculation J21*J28/J24 should be placed in cell J29. The calculation J21*J28/J24 will always equal zero if R2 is not the limiting reagent (J28/J24 equals zero because J28 equals zero), the calculation J21*J28/J24 will always equal the contents of cell J21 if R2 is the limiting reagent (J28/J24 equals 1 because J28 equals J24). Thus, the largest number in the range I28–K28 will always be the number of moles of the limiting reagent.

Step A simply ensures that the reactant ratio (J21/J24) will be set to an arbitrarily large number if cell J24 equals zero or is left blank. Therefore, Step B will always find the correct, minimum reactant ratio.

Trace the model through and play with it until you understand how it works. Verify the model's correct behavior by working the following exercise:

a. Ask your new model to identify the limiting reagent and the theoretical yield of the reaction in Lesson 3.

2. Build the Model 2.3 from scratch, without reference to the materials in this book.

3. Consider the chemical reaction

$$2NF_3 + Cu \rightarrow N_2F_4 + CuF_2$$

Suppose that you have only 16 g of nitrogen trifluoride (NF_3) at your disposal but an unlimited amount of metallic copper (Cu). Use Model 2.3 to complete the table below and identify the theoretical yield of CuF_2 (in grams) corresponding to different amounts of copper. Graph your results in Figure 2.10.

16 g of NF_3

Cu (g)	CuF_2 (g)
0	_____
2	_____
4	_____
6	_____
8	_____
10	_____

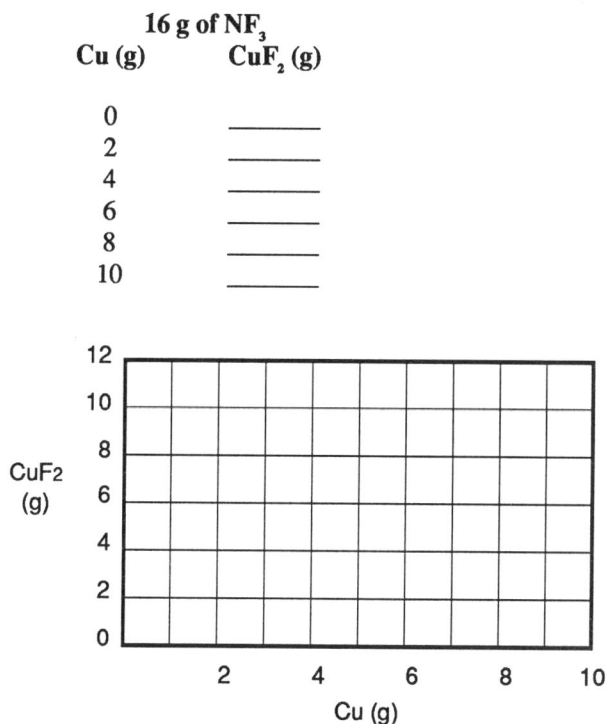

Figure 2.10

a. What is the maximum amount (in grams) of CuF_2 that can be produced from 16 g of NF_3?

b. How many grams of Cu are used when the maximum amount of CuF_2 is produced?

c. How many grams of Cu are used when 10 g of CuF_2 are produced?

4. In Problem 3, you should have found that approximately 6.1 g of Cu are required to produce 10 g of CuF_2 (if the reaction is 100% efficient). Complete the table below to determine the amount of NF_3 that can be efficiently used in the production of 10 g of CuF_2. Graph your results in Figure 2.11.

6.3 g of Cu

NF_3 (g)	CuF_2 (g)
0	_____
2	_____
4	_____
6	_____
8	_____
10	_____

Figure 2.11

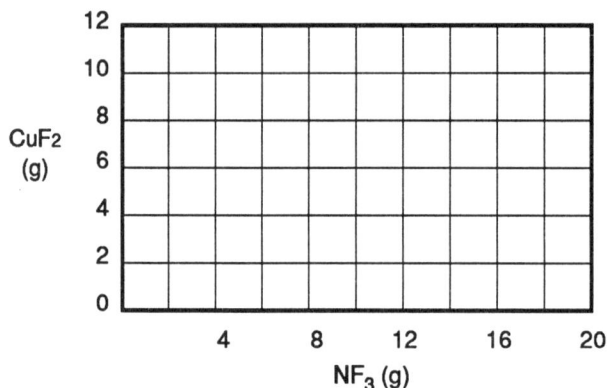

5. Consider the chemical reaction

$$Mo(CO)_6 + Br_2 \rightarrow Mo(CO)_4Br_2 + 2CO$$

Plot the yield of $Mo(CO)_4Br_2$ vs Br_2 in the presence of an arbitrary excess of $Mo(CO)_6$.

a. How many grams of bromine are required for a theoretical yield of 25.0 g of $Mo(CO)_4Br_2$?

b. Plot the yield of $Mo(CO)_4Br_2$ vs Br_2 in the presence of the amount of bromine that just yields 25.00 g of $Mo(CO)_4Br_2$. How many grams of $Mo(CO)_4Br_2$ can be used efficiently?

MODELS

FORMULAS

	A	B	C	D	E	F	G	H	I	J	K	L	M	N
1	Wgt in grams =		0.12	0.34										
2			---Number of Atoms/Molecule---						---Intermediate Calculations---					
3	Atom	At. Wt.	[R1]	[R2]	[R3]	[P1]	[P2]	[P3]	[R1]	[R2]	[R3]	[P1]	[P2]	[P3]
4	Al	26.98							=B4*C4	=B4*D4	=B4*E4	=B4*F4	=B4*G4	=B4*H4
5	Br	79.91		1				2	=B5*C5	=B5*D5	=B5*E5	=B5*F5	=B5*G5	=B5*H5
6	C	12.01							=B6*C6	=B6*D6	=B6*E6	=B6*F6	=B6*G6	=B6*H6
7	Ca	40.08							=B7*C7	=B7*D7	=B7*E7	=B7*F7	=B7*G7	=B7*H7
8	Cl	35.45							=B8*C8	=B8*D8	=B8*E8	=B8*F8	=B8*G8	=B8*H8
9	F	19.00		3			4		=B9*C9	=B9*D9	=B9*E9	=B9*F9	=B9*G9	=B9*H9
10	H	1.008							=B10*C10	=B10*D10	=B10*E10	=B10*F10	=B10*G10	=B10*H10
11	K	39.10							=B11*C11	=B11*D11	=B11*E11	=B11*F11	=B11*G11	=B11*H11
12	Kr	83.80							=B12*C12	=B12*D12	=B12*E12	=B12*F12	=B12*G12	=B12*H12
13	Mg	24.31							=B13*C13	=B13*D13	=B13*E13	=B13*F13	=B13*G13	=B13*H13
14	Na	22.99							=B14*C14	=B14*D14	=B14*E14	=B14*F14	=B14*G14	=B14*H14
15	O	16.00	2			2			=B15*C15	=B15*D15	=B15*E15	=B15*F15	=B15*G15	=B15*H15
16	P	30.97							=B16*C16	=B16*D16	=B16*E16	=B16*F16	=B16*G16	=B16*H16
17	Si	28.09	1			1			=B17*C17	=B17*D17	=B17*E17	=B17*F17	=B17*G17	=B17*H17
18	Ti	47.90							=B18*C18	=B18*D18	=B18*E18	=B18*F18	=B18*G18	=B18*H18
19														
20					Mol. Wt. (g/mol) =				Formula Set I					
21					# moles (mol) =				Formula Set II					
22														
23	LIMITING REAGENT CALCULATOR:													
24				Coefficients in balanced equation =					3	4		3	3	2
25					Reactant ratio =				Formula Set III					
26														
27	THEORETICAL YIELD CALCULATOR:													
28					Coefficient flag =				Formula Set IV					
29					Moles flag =				Formula Set V					
30														
31				Coefficient of limiting reagent =					=MAX(I28:K28)					
32				Moles of limiting reagent =					=MAX(I29:K29)					
33														
34					Theoretical yield (mol) =							Formula Set VI		
35					Theoretical yield (g) =							=L34*L20	=M34*M20	=N34*N20

VALUES

	A	B	C	D	E	F	G	H	I	J	K	L	M	N
1	Wgt in grams =		0.12	0.34										
2			---Number of Atoms/Molecule---						---Intermediate Calculations---					
3	Atom	At. Wt.	[R1]	[R2]	[R3]	[P1]	[P2]	[P3]	[R1]	[R2]	[R3]	[P1]	[P2]	[P3]
4	Al	26.98							0	0	0	0	0	0
5	Br	79.91		1				2	0	79.91	0	0	0	159.82
6	C	12.01							0	0	0	0	0	0
7	Ca	40.08							0	0	0	0	0	0
8	Cl	35.45							0	0	0	0	0	0
9	F	19.00		3			4		0	57	0	0	76	0
10	H	1.008							0	0	0	0	0	0
11	K	39.10							0	0	0	0	0	0
12	Kr	83.80							0	0	0	0	0	0
13	Mg	24.31							0	0	0	0	0	0
14	Na	22.99							0	0	0	0	0	0
15	O	16.00	2			2			32	0	0	32	0	0
16	P	30.97							0	0	0	0	0	0
17	Si	28.09	1			1			28.09	0	0	28.09	0	0
18	Ti	47.90							0	0	0	0	0	0
19														
20					Mol. Wt. (g/mol) =				60.09	136.91	0	32	104.09	159.82
21					# moles (mol) =				0.0020303	0.002498	0	0	0	0
22														
23	LIMITING REAGENT CALCULATOR:													
24				Coefficients in balanced equation =					3	4		3	3	2
25					Reactant ratio =				0.0006768	0.0006245	1E+09			
26														
27	THEORETICAL YIELD CALCULATOR:													
28					Coefficient flag =				0	4	0			
29					Moles flag =				0	0.002498	0			
30														
31				Coefficient of limiting reagent =					4					
32				Moles of limiting reagent =					0.002498					
33														
34					Theoretical yield (mol) =							0.0018735	0.0018735	0.001249
35					Theoretical yield (g) =							0.0599518	0.1950119	0.1996145

Formula Set I Prototype Cell is I20	Formula Set II Prototype Cell is I21	Formula Set III Prototype Cell is I25
=SUM(I4:I18)	=IF(ISERROR(C1/I20),0,C1/I20)	=IF(ISERROR(I21/I24),1E+09,I21/I24)

Formula Set IV Prototype Cell is I28	Formula Set V Prototype Cell is I29	Formula Set VI Prototype Cell is L34
=IF(I25=MIN(I25:K25),I24,0)	=IF(ISERROR(I21*I28/I24),0,I28*I21/I24)	=I32*(L24/I31)

Model 2.3

BUILD and verify Model 2.3.

ENTER data using cells C1–E1 (mass in grams of the starting reagents), cells C4–H18 (the number of atoms per molecule of each element in each compound), and cells I24–N24 (the coefficients of the balanced chemical equations.

READ output from the model at cells L35–N35 (the theoretical yield in moles and grams). Other portions of the model also offer useful information.

NOTICE that cells I21–N21 contain a formula that places a 0 in the cell if the "# moles" calculation leads to an error (e.g., Mol. Wt. in cells I20–N 21 is 0 or undefined). It is often necessary to "protect" a cell from erroneous calculations, especially if other cells make calculations based on the contents of the "protected" cell.

EXERCISE **2.4**

PERCENT COMPOSITION

TO GET THE MOST OUT OF THIS EXERCISE, YOU SHOULD ALREADY KNOW

The definition of the terms:
molecular formula
molecular weight
mole

How to calculate the molecular weight of a compound from its molecular formula.

How to convert from moles of a substance to mass and vice versa.

In this exercise, you examine how a chemist determines the composition by weight of a chemical compound from its empirical formula. You also learn how to judge the purity of a sample from an analysis of its composition. The arithmetic is simple.

Consider, for example, potassium chlorate ($KClO_3$). One mol of $KClO_3$ has a total mass equal to the sum of the masses of each of its constituent elements. From a chart of atomic weights, you can determine that

- 1 mol of K has a mass of 39.102 g
- 1 mol of Cl has a mass of 35.453 g
- 1 mole of O has a mass of 15.9994 g

From this information, you can determine the weight of 1 mol of $KClO_3$ and the percent composition by weight of each of its constituent elements in a straightforward fashion (Figure 2.12).

Chemists use analytic data regarding the percent composition by weight of the elements in a compound to determine the purity of an isolated substance. Suppose, for example, that you are given a sample that you believe to be glucose, $C_6H_{12}O_6$. Then, from the kinds of calculations displayed in Figure 2.12, you can determine the expected percent

	I	II	I x II	$\frac{\text{I x II}}{\text{Total}}$ x 100
Element	At. Mass	Moles/mole of compound	Contrib. to total mass	% of total
K	39.102	1	39.102	31.906
Cl	35.453	1	35.453	28.929
O	15.9994	3	47.998	39.165
			Total = 122.553	

Figure 2.12

The percent composition by weight of each of the elements in a pure compound can be calculated from a knowledge of the atomic mass of each element and the number of atoms of each element per molecule.

composition by weight of each of the elements in your sample. Impurities in the sample would cause your analysis to deviate from the values expected. Such impurities fall into two categories:

a. Impurities that do not contain any of the elements in the expected compound: Such impurities will change the percent composition of each element in the analyzed substance, but elements of the expected compound will change their percent compositions *proportionately*.

b. Impurities that do contain some or all of the elements in the expected compound: Such impurities will change the percent composition of each element in the analyzed substance and will change their proportionate compositions.

LESSONS

1. Figure 2.13 shows the output of a model designed around Figure 2.12 (blank cells have been deliberately left in the model for modification in Lesson 2). Build and verify this model. The solution is contained in Model 2.4.

    a. Determine the percent composition of C, H, and O for glucose, $C_6H_{12}O_6$.

    b. Determine the percent composition of Na, C, and H for sodium stearate, $NaC_{18}H_{35}O_2$, a common constituent of soaps.

2. Modify your model (Lesson 1) to include the ability to calculate the actual mass in grams of an element given the mass of the compound (Figure 2.14).

    a. A 2.3-g sample of glucose, $C_6H_{12}O_6$, is analyzed. Predict the mass in grams of C, H, and O that will be found in the sample.

Figure 2.13

The percent composition of a compound can be calculated using the instructions outlined in Figure 2.12. This electronic spreadsheet display was designed around these instructions. Besides a list of elements and atomic weights, columns A and B of the model include combinations of elements (H_2O and C_2O_2). You will find these combinations useful in later exercises.

	A	B	C	D	E	F
1						
2						
3	At/Mol	At/Mol Wt	#	Mass	% Comp.	
4	C	12.011		0	0	
5	H	1.00794		0	0	
6	O	15.9994	3	47.9982	39.1664	
7	Br	79.904		0	0	
8	Cl	35.453	1	35.453	28.9295	
9	Cu	63.546		0	0	
10	F	18.998		0	0	
11	I	126.9045		0	0	
12	N	14.0067		0	0	
13	P	30.97376		0	0	
14	K	39.0983	1	39.0983	31.9041	
15	Na	22.98977		0	0	
16	S	32.06		0	0	
17	H20	18.01528		0	0	
18	C202	56.0208		0	0	
19						
20		Total =		122.55	100	

   b. A 9.6-g sample of sodium stearate, $NaC_{18}H_{35}O_2$, is analyzed. Predict the mass in grams of Na, C, and O that will be found in the sample.

3. Build and verify Model 2.5. This model works identically to the model described in Figure 2.14 (Model 2.4), except that data for two separate compounds can be entered. Cells C6–C20 are used to enter

Figure 2.14

	A	B	C	D	E	F
1	Total grams =	1				
2						
3	At/Mol	At/Mol Wt	#	Mass	% Comp.	grams
4	C	12.011		0	0	0
5	H	1.00794		0	0	0
6	O	15.9994	3	47.9982	39.1664	0.39166
7	Br	79.904		0	0	0
8	Cl	35.453	1	35.453	28.9295	0.2893
9	Cu	63.546		0	0	0
10	F	18.998		0	0	0
11	I	126.9045		0	0	0
12	N	14.0067		0	0	0
13	P	30.97376		0	0	0
14	K	39.0983	1	39.0983	31.9041	0.31904
15	Na	22.98977		0	0	0
16	S	32.06		0	0	0
17	H20	18.01528		0	0	0
18	C202	56.0208		0	0	0
19						
20		Total =		122.55	100	1

the number of occurrences of a specific element (or group of elements) in a molecule of Compound 1, and cells D6–D20 are used to enter the number of occurrences of a specific element (or group of elements) in a molecule of Compound 2. The actual weights of the two compounds are entered into cells C1 and C2, respectively. Columns K and L have been added to this model in order to determine the percent composition of elements in various mixtures of these two compounds. A small region at the bottom of the model (cells E25–F27) is used for "scratch calculations" that display the ratios of the percent composition of certain elements.

Before you begin, confirm that the model performs the same calculations presented in Figure 2.14 and study the formulas in columns K and L to see how percent composition is calculated for the elements in a mixture of two compounds.

Suppose that a sample of supposedly "pure" $KClO_3$ has been unknowingly contaminated with NaI. Complete the table below and explore the the consequences of NaI contamination on the percent composition of the elements in $KClO_3$. Use the "scratch" area of your model to find the ratios of O to Cl, O to K, and Cl to K.

Grams		Apparent % Composition				Ratios		
$KClO_3$	NaI	K	Cl	O	O/Cl	O/K	Cl/K
90	10	____	____	____	____	____	____
80	20	____	____	____	____	____	____
70	30	____	____	____	____	____	____

What happens to the percent composition values for the elements of $KClO_3$ as the NaI contamination increases?

Examine the ratios of the the percent compositions of the elements of $KClO_3$ one to the other. Is the change in percent composition proportionate among K, Cl, and O?

4. Suppose that your preparation of the compound $C_{20}H_{19}IO_5P_2$ contains significant amounts of the contaminant (unreacted precursor) $C_8H_7I_2O_2P$. Use the Model 2.4 to complete the following table. You must modify the "scratch" area to calculate the appropriate ratios of the percent composition of specific elements.

	Grams		Apparent % Composition				
$C_{20}H_{19}IO_5P_2$	$C_8H_7I_2O_2P$	C	H	I	O	P	
90	10	___	___	___	___	___	
80	20	___	___	___	___	___	
70	30	___	___	___	___	___	

	Grams		Ratios				
$C_{20}H_{19}IO_5P_2$	$C_8H_7I_2O_2P$	C/H	C/I	H/I	I/O	O/P	
90	10	___	___	___	___	___	
80	20	___	___	___	___	___	
70	30	___	___	___	___	___	

When the contaminant in a sample contains elements in common with the primary substance, do the percent compositions of the elements of the primary substance change proportionately?

5. A presumably pure sample of $C_{20}H_{19}IO_5P_2$ was sent out for a C, H, I, P analysis. The following results came back:

$$\%C = 42.53 \quad \%H = 3.43 \quad \%I = 28.72 \quad \%P = 11.16$$

Assuming that the only contaminant is $C_8H_7I_2O_2P$, how pure is the sample? (HINT: By trial and error, change the weights for the two compounds entered in your model and "home in" on a pair of numbers that matches the percent composition values. Your job will be even easier if you only choose pairs of numbers that sum to 100. Do you see why?)

PROBLEMS

1. A simple method for synthesizing the compound potassium oxalatocuprate (II) dihydrate is:

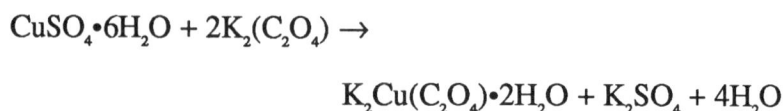

$$CuSO_4 \cdot 6H_2O + 2K_2(C_2O_4) \rightarrow$$

$$K_2Cu(C_2O_4) \cdot 2H_2O + K_2SO_4 + 4H_2O$$

The product's formula can be verified by analyzing for percent K, percent Cu, percent oxalate $[(C_2O_4)^{2-}]$, and percent H_2O. Invariably, however, students discover that one or more of their analyses do not match that predicted by theory. There can be two reasons for that result (other than botched laboratory technique):

a. The compound has some other formula than that assumed.

b. Contaminants in the sample have distorted the analyses.

Do not jump to conclusions about new structures until you have eliminated the possibility of contamination.

The product, $K_2Cu(C_2O_4) \cdot 2H_2O$, is made as insoluble crystals that are washed with cold water to remove any unreacted potassium oxalate, copper sulfate (the resulting solution is always bluish), and the by-product potassium sulfate — all of which are soluble in water. The most likely contaminant is potassium sulfate because it is made mole for mole with the product.

Use the model described in Figure 2.14 to complete the table below and predict the analyses of products that contain different amounts of K_2SO_4 contaminant.

Grams		Apparent % Composition			
$K_2Cu(C_2O_4) \cdot 2H_2O$	K_2SO_4	K	Cu	C_2O_4	H_2O
95	5	___	___	___	___
90	10	___	___	___	___
85	15	___	___	___	___

You are certain that K_2SO_4 is the only contaminant in a certain product. What is the percent contamination by K_2SO_4 in this product if the analysis shows 34.59% K, 24.46% Cu, 21.56% $[C_2O_4]^{2-}$ and 13.87% H_2O.

2. Build and verify a model that can be used to predict the percent composition of samples of $K_2Cu(C_2O_4) \cdot 2H_2O$ in presence of contamination from both K_2SO_4 and $CuSO_4 \cdot 6H_2O$.

MODELS

Model 2.4

BUILD and verify Model 2.4.

ENTER data using cells A4–A18 (symbol), B4–B18 (atomic or molecular weight), and C4–C18 [number of atoms (groups of atoms) of this type].

READ output from the model at cells D4–D18 (mass of atom or group of atoms), E4–D18 (percent composition of atom or group of atoms), and F4–F18 (mass in grams of atoms or group of atoms contained in a sample with the mass noted in cell B1).

NOTICE that other useful information and checksums are contained in cells D20–F20.

FORMULAS

	A	B	C	D	E	F
1	Total grams =	1				
2						
3	At/Mol	At/Mol Wt	#	Mass	% Comp.	grams
4	C	12.011		=C4*B4	=D4/D20*100	=(E4/100)*B1
5	H	1.00794		=C5*B5	=D5/D20*100	=(E5/100)*B1
6	O	15.9994	3	=C6*B6	=D6/D20*100	=(E6/100)*B1
7	Br	79.904		=C7*B7	=D7/D20*100	=(E7/100)*B1
8	Cl	35.453	1	=C8*B8	=D8/D20*100	=(E8/100)*B1
9	Cu	63.546		=C9*B9	=D9/D20*100	=(E9/100)*B1
10	F	18.998		=C10*B10	=D10/D20*100	=(E10/100)*B1
11	I	126.9045		=C11*B11	=D11/D20*100	=(E11/100)*B1
12	N	14.0067		=C12*B12	=D12/D20*100	=(E12/100)*B1
13	P	30.97376		=C13*B13	=D13/D20*100	=(E13/100)*B1
14	K	39.0983	1	=C14*B14	=D14/D20*100	=(E14/100)*B1
15	Na	22.98977		=C15*B15	=D15/D20*100	=(E15/100)*B1
16	S	32.06		=C16*B16	=D16/D20*100	=(E16/100)*B1
17	H2O	18.01528		=C17*B17	=D17/D20*100	=(E17/100)*B1
18	C2O2	56.0208		=C18*B18	=D18/D20*100	=(E18/100)*B1
19						
20			Total =	=SUM(D4:D18)	=SUM(E4:E18)	=SUM(F4:F18)

VALUES

	A	B	C	D	E	F
1	Total grams =	1				
2						
3	At/Mol	At/Mol Wt	#	Mass	% Comp.	grams
4	C	12.011		0	0	0
5	H	1.00794		0	0	0
6	O	15.9994	3	47.9982	39.1664	0.39166
7	Br	79.904		0	0	0
8	Cl	35.453	1	35.453	28.9295	0.2893
9	Cu	63.546		0	0	0
10	F	18.998		0	0	0
11	I	126.9045		0	0	0
12	N	14.0067		0	0	0
13	P	30.97376		0	0	0
14	K	39.0983	1	39.0983	31.9041	0.31904
15	Na	22.98977		0	0	0
16	S	32.06		0	0	0
17	H2O	18.01528		0	0	0
18	C2O2	56.0208		0	0	0
19						
20			Total =	122.55	100	1

FORMULAS

	A	B	C	D	E	F	G	H	I	J	K	L
1	Compound 1 grams =		90									
2	Compound 2 grams =		10									
3			# At/Mol			-----[1]-----			-----[2]-----			--[1] + [2]--	
4						%			%			%
5	At/Mol	At/Mol Wt	[1]	[2]	Mass	Comp.	grams	Mass	Comp.	grams	grams	Comp.
6	C	12.011										
7	H	1.00794										
8	O	15.9994	3		Form.	Form.	Form.	Form.	Form.	Form.	Form.	Form.
9	Br	79.904			Set	Set	Set	Set	Set	Set	Set	Set
10	Cl	35.453	1		I	II	III	IV	V	VI	VII	VIII
11	Cu	63.546										
12	F	18.998										
13	I	126.9045		1								
14	N	14.0067										
15	P	30.97376										
16	K	39.0983		1								
17	Na	22.98977		1								
18	S	32.06										
19	H2O	18.01528										
20	C2O2	56.0208										
21												
22		Total =					Formula Set IX					
23												
24					Scratch Area							
25					O/Cl =	=L8/L10						
26					O/K =	=L8/L16						
27					Cl/K =	=L10/L16						

VALUES

	A	B	C	D	E	F	G	H	I	J	K	L
1	Compound 1 grams =		90									
2	Compound 2 grams =		10									
3			# At/Mol			-----[1]-----			-----[2]-----			--[1] + [2]--	
4						%			%			%
5	At/Mol	At/Mol Wt	[1]	[2]	Mass	Comp.	grams	Mass	Comp.	grams	grams	Comp.
6	C	12.011			0	0	0	0	0	0	0	0
7	H	1.00794			0	0	0	0	0	0	0	0
8	O	15.9994	3		47.998	39.166	35.25	0	0	0	35.25	35.25
9	Br	79.904			0	0	0	0	0	0	0	0
10	Cl	35.453	1		35.453	28.93	26.037	0	0	0	26.037	26.037
11	Cu	63.546			0	0	0	0	0	0	0	0
12	F	18.998			0	0	0	0	0	0	0	0
13	I	126.9045		1	0	0	0	126.9	84.663	8.4663	8.4663	8.4663
14	N	14.0067			0	0	0	0	0	0	0	0
15	P	30.97376			0	0	0	0	0	0	0	0
16	K	39.0983		1	39.098	31.904	28.714	0	0	0	28.714	28.714
17	Na	22.98977		1	0	0	0	22.99	15.337	1.5337	1.5337	1.5337
18	S	32.06			0	0	0	0	0	0	0	0
19	H2O	18.01528			0	0	0	0	0	0	0	0
20	C2O2	56.0208			0	0	0	0	0	0	0	0
21												
22		Total =			122.55	100	90	149.89	100	10	100	100
23												
24					Scratch Area							
25					O/Cl =	1.3539						
26					O/K =	1.2276						
27					Cl/K =	0.9068						

Formula Set I Prototype Cell is E6	Formula Set IV Prototype Cell is H6	Formula Set VII Prototype Cell is K6
=C6*B6	=B6*D6	=G6+J6

Formula Set II Prototype Cell is F6	Formula Set V Prototype Cell is I6	Formula Set VIII Prototype Cell is L6
=E6/E22 *100	=(H6/H22)*100	=K6/K22 *100

Formula Set III Prototype Cell is G6	Formula Set VI Prototype Cell is J6	Formula Set IX Prototype Cell is E22
=(F6/100)*C1	=(I6/100)*C2	=SUM(E6:E20)

Model 2.5

BUILD and verify Model 2.5. This model is similar to Model 2.4, except that the composition of two different compounds (and mixtures of these two compounds) can be studied. This can be useful when one compound is presumed to be contaminated with another compound.

ENTER data using cells A6–A20 (symbol), B6–B20 (atomic or molecular weight), C6–C20 [number of atoms (groups of atoms) of this type in Compound 1], and D6–D20 [number of atoms (groups of atoms) of this type in Compound 2].

READ output from the model at cells E6–E20 (mass of atom or group of atoms in Compound 1), F6–F20 (percent composition of atom or group of atoms in Compound 1), and G6–G20 (mass in grams of atoms or group of atoms contained in a sample of Compound 1 with the mass noted in cell B1). Cells H6–H20, I6–I20, and J6–J20 calculate similar results for Compound 2. Cells K6- K20 and L6–L20 calculates the mass contributions and percent composition of a mixture of Compounds 1 and 2.

NOTICE that the scratch area below row 24 can be used to calculate the ratios of the percent composition of certain elements.

EMPIRICAL FORMULAS

An *empirical formula* is a concise statement of the molar ratio of elements in a chemical compound. The empirical formula H_2O, for example, expresses the fact that water contains two atoms of hydrogen (H) to every one atom of oxygen (O).[1]

Exercise 2.4 offered you the opportunity to build a computing machine capable of calculating the percent composition of a compound from its empirical formula. In this exercise, you explore the reverse process and discover how to obtain the empirical formula of a compound from its percent composition by weight.

Every new compound should receive an elemental analysis if only to verify that the new compound is actually the intended compound. Elemental analysis is usually expressed as percent composition by weight, and it becomes your task to first convert your data to percent composition in molar units and then to an empirical formula.

The process can be shown by example: Suppose a compound containing only carbon (C), hydrogen (H), and oxygen (O) is prepared and elemental analyses show the results seen in column A of Table 2.1.

- First, convert the data for each element to moles per 100 g of sample (Equation 2.11 and column B of Table 2.1).

$$\text{number of moles} = \frac{\text{mass in grams}}{\text{atomic weight}} \qquad (2.11)$$

- Second, find the simplest ratio of these values such that the lowest value in the range of values is 1 (Equation 2.12 and column C of Table 2.1). These numbers will be the first approximations of the subscripts of the empirical equation

$$\text{trial subscript} = \frac{\text{moles per 100 g}}{\text{MIN(moles per 100 g)}} \qquad (2.12)$$

where MIN(moles per 100 g) = the smallest value among the range of values of moles per 100 g of each element in the compound.

- Finally, multiply the trial subscript by some factor such that all subscripts approximate whole numbers (column D). In Table 2.1, each value in column C is multiplied by 2.

Because experimental analyses yield only approximate results, the empirical formula of the CHO compound analyzed in Table 2.1 is $C_3H_2O_2$ (propynoic acid, perhaps).

LESSONS

1. Each of the steps identified in Table 2.1 can be placed conveniently into a spreadsheet model (Figure 2.15).

   Column C of your model provides a region in which to enter the analytic data. If your total sample weight (cell B1) is 100 g, enter "percent composition" directly into this range of cells. Column D calculates percent composition (Figure 2.12), and column E converts the data to moles per 100 g of sample (Equation 2.11). Column F divides column E by the smallest value among the range of values in column E (Equation 2.12), to generate a trial subscript. Inspect column F for this value and manually enter it into cell B2. Inspection of the values in column F should give insight into the proper multiplier (cell B3) that will yield the actual subscripts (column G).

Table 2.1

Element	[A] % by Weight (g per 100 g)	[B] Moles per 100 g	[C] Trial Subscript	[D] Actual Subscript
C	51.3	4.27	1.50	3.00
H	2.90	2.88	1.01	2.02
O	45.7	2.86	1.00	2.00

Figure 2.15

	A	B	C	D	E	F	G
1	Total grams =	100					
2	Divisor =	2.855	<== Minimum value from column E				
3	Multiplier =	2	<==Start with 1 and increase till column G approximates integers				
4							
5					Moles per	Trial	Actual
6	Atom	At. Wt.	Grams	% Comp.	100 grams	Subscript	Subscript
7	H	1.008	2.9	2.9	2.8769841	1.00770022	2.01540044
8	B	10.81		0	0	0	0
9	C	12.01	51.3	51.3	4.2714405	1.496126258	2.99225252
10	N	14.01		0	0	0	0
11	O	16.00	45.7	45.7	2.85625	1.000437828	2.00087566
12	Na	22.99		0	0	0	0
13	Mg	24.31		0	0	0	0
14	Si	28.09		0	0	0	0
15	P	30.97		0	0	0	0
16	S	32.06		0	0	0	0
17	Cl	35.45		0	0	0	0
18	Ca	40.08		0	0	0	0
19	Ti	47.90		0	0	0	0
20	Cu	63.54		0	0	0	0
21	Au	197.97		0	0	0	0

Try to build this model without referring to its complete description (Model 2.6) and then test your model on the following exercises.

2. A 100-g sample of saccharin yields the following elemental analysis:

    C: 45.89 g
    H: 2.75 g
    O: 26.20 g
    N: 7.65 g
    S: 17.50 g

    What is the empirical formula of saccharine?

3. Aluminum silicate is used in dental cement. The compound contains only aluminum, silicon, and oxygen. A sample of aluminum silicate gave the following elemental analysis:

    Al: 33.30%
    Si: 17.33%

    Elemental analysis for oxygen is difficult and was not performed. What is the empirical formula of aluminum silicate?

PROBLEMS

The molecular formula is related to the empirical formula by the equation

$$\text{molecular formula} = (\text{empirical formula})_w \qquad (2.13)$$

where w is an integer that can be calculated

$$w = \frac{\text{molecular weight}}{\text{empirical formula weight}}$$
(2.14)

The formula weight is the sum of the atomic weights of the empirical formula. The molecular weight can be determined by means such as those described in Exercise 2.1.

Modify your spreadsheet to calculate the molecular formula of an analyzed compound.

1. Ethyl acetate is a common organic solvent with a molecular weight of 88.10 g/mol. The compound gives the following analysis:

    C: 54.53%
    H: 9.15%
    O: 36.32%

    What are the empirical and molecular formulas of ethyl acetate?

2. Valinomycin is an antibiotic produced by *Streptomyces fulvissimus* with a molecular weight of 1111.36 g/mol. Analysis gives the following results:

    C: 58.36%
    H: 8.16%
    O: 25.92%
    N: 7.56%

    What is the molecular formula of valinomycin?

Model 2.6

BUILD and verify Model 2.6.

ENTER data using cells B1–B3. Cell B1 contains sample mass in grams. Cell B2 should contain the smallest value found in column E. Cell B3 should contain a whole number such that all values in column G approximate integers.

READ output from the model at cells F6–F20 (to determine the appropriate multiplier for cell B3) and cells G6–G20 to determine the subscripts of the empirical formula (once the multiplier in cell B3 is correct).

NOTICE that portions of this model could be automated (e.g., automatic determination of the minimum value in the cell range E6–E20). You are welcome to incorporate such refinements if you wish.

MODELS

FORMULAS

	A	B	C	D	E	F	G
1	Total grams =	100					
2	Divisor =	2.855	<== Minimum value from column E				
3	Multiplier =	2	<==Start with 1 and increase till column G approximates integers				
4							
5					Moles per	Trial	Actual
6	Atom	At. Wt.	Grams	% Comp.	100 grams	Subscript	Subscript
7	H	1.008	2.9				
8	B	10.81					
9	C	12.01	51.3				
10	N	14.01					
11	O	16.00	45.7	Formula Set I	Formula Set II	Formula Set III	Formula Set IV
12	Na	22.99					
13	Mg	24.31					
14	Si	28.09					
15	P	30.97					
16	S	32.06					
17	Cl	35.45					
18	Ca	40.08					
19	Ti	47.90					
20	Cu	63.54					
21	Au	197.97					

VALUES

	A	B	C	D	E	F	G
1	Total grams =	100					
2	Divisor =	2.855	<== Minimum value from column E				
3	Multiplier =	2	<==Start with 1 and increase till column G approximates integers				
4							
5					Moles per	Trial	Actual
6	Atom	At. Wt.	Grams	% Comp.	100 grams	Subscript	Subscript
7	H	1.008	2.9	2.9	2.8769841	1.00770022	2.01540044
8	B	10.81		0	0	0	0
9	C	12.01	51.3	51.3	4.2714405	1.496126258	2.99225252
10	N	14.01		0	0	0	0
11	O	16.00	45.7	45.7	2.85625	1.000437828	2.00087566
12	Na	22.99		0	0	0	0
13	Mg	24.31		0	0	0	0
14	Si	28.09		0	0	0	0
15	P	30.97		0	0	0	0
16	S	32.06		0	0	0	0
17	Cl	35.45		0	0	0	0
18	Ca	40.08		0	0	0	0
19	Ti	47.90		0	0	0	0
20	Cu	63.54		0	0	0	0
21	Au	197.97		0	0	0	0

Formula Set I Prototype Cell is D7	Formula Set III Prototype Cell is F7
=100*C7/**B1**	=E7/**B2**

Formula Set II Prototype Cell is E7	Formula Set IV Prototype Cell is G7
=D7/B7	=**B3***E7/**B2**

CHAPTER 3

GASES

Chemists think about gases in two different ways. On the one hand, a gas can be treated as a black box that is defined in terms of the properties: pressure (P) and temperature (T). These properties are defined operationally with no underlying mechanisms for the behavior of these properties either sought or implied. The behavior of a mercury-filled manometer, for example, can provide an operational definition for pressure. Likewise, the temperature of a gas can be defined in terms of the behavior of a thermometer.

The other way scientists think about gases is in terms of a theoretical model built up from elementary physical principles — motion, work, energy, momentum, etc. This view (the kinetic theory of gases) treats a gas as a collection of independent, randomly moving particles that obey Newton's laws of motion. The kinetic theory of gases constitutes an attempt to understand the mechanisms supporting the behavior of properties such as temperature and pressure.

In this chapter, you have the opportunity to explore the behavior of gases in both contexts. In Chapter 8 (Thermodynamics), you will return to these issues in greater detail. You may, if you wish, jump ahead to Chapter 8 after completing Chapter 3.

EQUATIONS OF STATE

An *ideal gas* is a gas that consists of molecules that occupy no space and are subject to no intermolecular forces. Many gases approximate the behavior of an ideal gas under conditions of high temperature and low pressure.

Suppose that an ideal gas is confined to a chamber that is closed at one end by a moveable piston (Figure 3.1). The relationships between pressure, volume, and temperature that exist within the container are described by an *equation of state*. The equation of state for an ideal gas already should be familiar (Equation 3.1)

$$PV = nRT \tag{3.1}$$

where P is the pressure exerted on the walls of the chamber (often in atmospheres, atm), V is the volume of the chamber at the current setting of the moveable piston (often in liters, L), n is the amount of gas within the chamber (often in moles, mol), R is the universal gas constant (0.0820575 when expressed in L atm/mol K [K is the temperature in kelvins]), and T is the temperature within the chamber in kelvins.

The equation of state for an ideal gas provides excellent insight into the behavior of noninteracting, dimensionless particles, and gases often behave as though their individual atoms or molecules have such

TO GET THE MOST OUT OF THIS EXERCISE, YOU SHOULD ALREADY KNOW

The definition of the term: equation of state

How to build a simple model with your electronic spreadsheet program by writing labels, values, and formulas directly into cells and by copying the contents of one cell or a range of cells into other cells.

How to write formulas with both absolute and relative variables and how such formulas behave when copied using your spreadsheet program's COPY command.

Figure 3.1

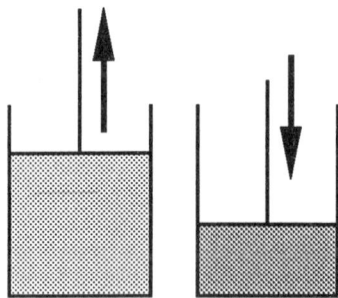

Gases can be described in terms of their temperature (T), pressure (P), and volume (V). If a gas is enclosed in a container with a moveable piston at one end, the relationships between T, P, and V can be explored. Physical experiments show that

- If the temperature of the gas is held constant and the piston is pushed in to reduce the volume of the container, pressure increases (Boyle's law).

- If the pressure of the gas is held constant and its temperature is raised, the piston is pushed out and volume increases (Charles' law).

When precise measurements are taken of pressure, temperature, and volume under "ideal" conditions, the relationships between P, V, and T are accurately predicted by the ideal gas law (Equation 3.1).

properties. Nevertheless, atoms *do* occupy space and *do* exert forces on their neighbors. An equation of state often used to approximate the behavior of a real gas is van der Waal's equation (Equation 3.2)

$$\left(P + \frac{n^2a}{V^2}\right)(V - nb) = nRT \tag{3.2}$$

where a and b are constants that vary with specific gases.[1]

LESSONS

1. Build and verify Model 3.1. Model 3.1 represents the equation of state of an ideal gas (Equation 3.1) solved for pressure (Equation 3.3) and is organized to take advantage of the built-in graphics capability of your electronic spreadsheet program.

$$P = \frac{nRT}{V} \tag{3.3}$$

Recall that you build your model by using the information in the formula and formula set blocks of the model description and that you verify your model by comparing the values in the value block with those that appear on your screen. Formula Set I is constructed by entering the formula of the formula set into the prototype cell (in this case, cell B9) and copying it across the entire range of the formula set (cells B9–B18). Don't forget that boldfaced variables in the formula set should be specified as absolute addresses and plain-faced variables should be specified as relative addresses.

Although the formula contained in cells A10–A18 could also have been specified by a formula set, we wrote it out because there was room. You can still use the COPY command on this formula by placing =A9+B6 into cell A10 (specifying B6 as an absolute address) and copying this cell into cells A11–A18.[2]

Examine the structure of the model until you understand its major features:

a. Cells A1–B6 constitute a parameter table that drives the remainder of the model. Rows 1–3 specify the physical parameters of the model (e.g., moles, gas constant, temperature), while rows 5 and 6 control the range of volumes contained in rows 9–18.

b. Rows 8–18 constitute the table of data you wish to explore. What does the formula in cell A9 do? What do the formulas in cells A10–A18 do?

c. Verify that the formulas in cells B9–B18 do indeed represent the equation of state of an ideal gas.

2. Although your model is now complete and ready to be explored (you may wish to save it onto a diskette before proceeding), you must still build *xy*-plots of the data generated in the table of your model.

Build an on-screen graphic corresponding to the description found in Figure 3.2. You may want to review Exercise 1.3 if you are uncertain about what to do. In brief, Figure 3.2 represents an *xy*-plot (called scatter plot by some programs) in which the values along the *x*-axis (volume) are determined by the contents of cells A9–A18, and the contents of the *y*-axis (pressure) are determined by contents of cells B9–B18.

In this book, *xy*-plots such as in Figure 3.2 will usually be represented by smooth lines. You, of course, can format your plots to your own preference (recall, for example, Figure 1.16).

Generate this graphic on your computer screen now and use it to answer the following questions:

a. Consider a container of gas such as that shown in Figure 3.1. If 1 mol of gas is present in the container at room temperature (20°C = 293.15 K), what happens to the pressure exerted by the gas against the container's walls as the piston is inserted and the volume gradually increases from 1 L to 10 L? From 0.1 L to 1 L? From 0.01 L to 0.1 L?

An ideal gas consists of molecules that occupy no space and are subject to no intermolecular forces, and so one can imagine that an ideal gas can occupy an arbitrarily small volume. What happens to the pressure on the walls of the container as its volume approaches zero?

b. Returning to realistic laboratory conditions where the volume of the container varies between 1 L and 10 L, what happens to the plot of *P* vs *V* as the amount of gas in the container gradually increases from 1 to 5 mol? As the temperature gradually increases from 20°C (293.15 K) to 100°C (373.15 K)?

c. Lesson 2b should have shown you that the plot of *P* vs *V* retains a constant shape as the amount of gas or the temperature increases. It should also have shown you, however, that pressure

Figure 3.2

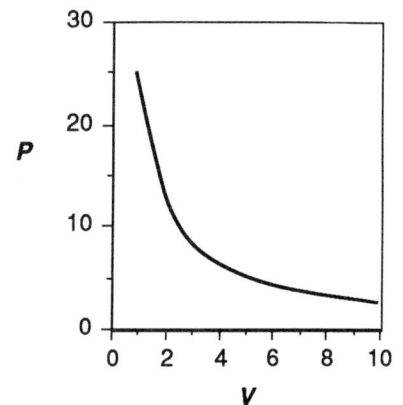

XY-PLOT

Data
Horizontal (*x*) axis: cells A9- A18 (volume)
Vertical (*y*) axis: cells B9–B18 (pressure)

Figure 3.3

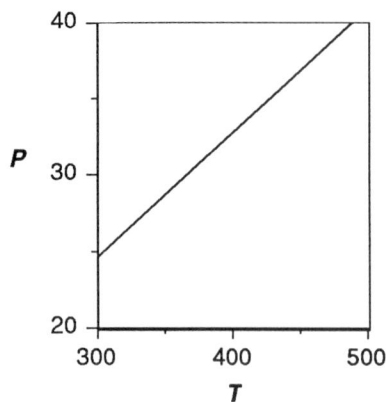

XY-PLOT

Data
Horizontal (x) axis: cells A9- A18
(temperature)
Vertical (y) axis: cells B9–B18
(pressure)

increases with increases in either gas amount or temperature if volume is held constant.

The design of the model does not, however, instantly reveal the precise shape of the curve of, say, P vs T or P vs n. For this you need a slightly different model.

3. Build and verify Model 3.2. Notice that Model 3.2 also uses Equation 3.3 but is arranged so that temperature rather than volume is plotted against pressure.

Construct an on-screen graphic similar to Figure 3.3.

a. Consider a 1-L vessel containing 1 mol of gas at room temperature (20°C = 293.15 K). What happens to the pressure exerted by the gas against the container's walls as the temperature increases from 300 K to 480 K?

b. Notice that your plot of P vs T, unlike P vs V, is a straight line.

4. Imagine that the vessel described in Figure 3.1 is constructed so that the freely moving piston exerts a constant force against the confined gas and that, as a consequence, the pressure of the gas within the container is also a constant. Build, verify, and explore a model in which the volume of the container can be seen as a function of temperature.

5. Build, verify, and explore a model in which the volume of the container can be seen as a function of the amount of confined gas.

6. Thus far, you have built models capable of exploring the behavior of an ideal gas consisting of molecules that occupy no space and that are subject to no intermolecular forces. Now let's explore the behavior of a nonideal gas represented by the van der Waal's equation (Equation 3.2).

Build and verify Model 3.3 as well as its supporting on-screen graphic (Figure 3.4).

a. When your model is initialized with the values displayed in Model 3.3, the ideal gas law and the van der Waal's equation give comparable (but not identical) results. Under these conditions of volume and temperature, the real gas defined by the constants a and b behaves almost "ideally."

b. Push the piston into the chamber of gas and explore the range of chamber volumes between 0.2 L and 2.0 L by setting V_0 and ΔV to 0.2 L (cells B8 and B9 of your model). As the volume decreases (and consequently, pressure increases), does the gas described by the van der Waal's equation continue to obey the ideal gas law? If not, what happens? Explain the behavior of the van der Waal's gas by invoking either or both of the properties of real gases that are ignored by model for an ideal gas (i.e., molecules occupy space, and molecules are subject to intermolecular forces).

c. Return the piston to its settings of $V_0 = \Delta V = 1$ L. Now withdraw the piston from the chamber of gas and increase chamber volume by setting V_0 and ΔV to 10 L. What happens to the correspondence between "real" and ideal gas behavior? Once again, explain the correspondance of the ideal gas and the van der Waal's gas in terms of the properties of real gases that are ignored by the equation of state of an ideal gas.

d. In general, would you expect a real gas in a real laboratory to approach ideal behavior as its confining volume decreased? Increased?

7. Reinitialize your model with the values displayed in Model 3.3.

a. Gradually reduce the temperature in the chamber of gas from 302.15 K to 150 K. When the volume of 1 mol of gas is varied from 1 to 10 L, at which temperature would you expect real gases to conform most closely to the ideal gas law?

b. Increase the temperature in the chamber of gas from 302.15 K to 1000 K. Over the volume range of 1 to 10 L, at which temperature would you expect gases to conform most closely to the ideal gas law?

c. All other things being equal, would you expect a real gas to approach or diverge from ideal behavior as temperature dropped?

8. Recall the introduction to this exercise and the statement that "many gases approximate the behavior of an ideal gas under conditions of high temperature and low pressure." Write a brief "laboratory" report of your "experiments" and defend or refute this statement. Before writing your report, you may wish to explore your models under a wider range of conditions. Keep in mind, however, that

Figure 3.4

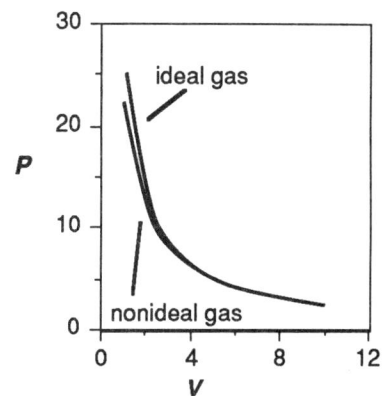

XY-PLOT

Data
Horizontal (*x*) axis: cells A14–A23 (volume)
Vertical (*y*) axis: cells B14–B23 (ideal), and C14–C23 (van der Waal's)

Figure 3.5

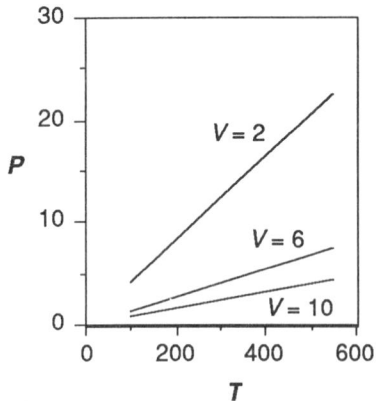

XY PLOT

Data
Horizontal (x) axis: cells B7–K7
(temperature)
Vertical (y) axis: cells B9–K9,
B13–K13, and B17–K17 (pressure at
three different volumes)

under extreme conditions your models will give erroneous results. At very low temperatures and volumes, for example, the van der Waal's calculations of Model 3.3 produce anomolous negative pressures.

Van der Waal's equation is only a simple approximation of the behavior of real gases. For example, the correction factor a is not a constant at all temperatures and pressures but in fact overcorrects at low temperatures and high pressures. The first term of van der Waal's equation for pressure decreases proportionally with T so that at low temperatures the second factor, n^2a/V^2, becomes relatively larger. When V is also small, that factor can become larger than the first term and give negative pressures.

PROBLEMS

1. Build and verify Model 3.4. Notice that Formula Set III has an unusual structure. The variable corresponding to cell address B7 has a plain-faced B and a boldfaced 7 (B**7**). The variable corresponding to cell address A8 has a boldfaced A and a plain-faced 8 (**A**8). This means that

    *   Variable B7 should be treated as an absolute variable when copied down the rows and as a relative variable when copied across the columns because we wish to use the range of values in all cells of row 7 in other rows.

    *   Variable A8 should be treated as an absolute variable when copied across the columns and as a relative variable when copied down the rows because we wish to use the range of values in all cells of column A in other columns.

    (If these instructions make no sense, it's time to return to the instruction manual that came with your spreadsheet program and study the various features of the COPY command.)

    The beauty of models organized like Model 3.4 is that you can monitor the effect of two variables simultaneously (in this case, the effect of volume and temperature on pressure).

    a. Build an on-screen graphic that plots P vs T for several different container volumes (Figure 3.5). Review your understanding of the behavior of an ideal gas by examining this plot. Does the plot behave as expected?

b. Plot P vs V at several different temperatures (Figure 3.6). Review your understanding of the behavior of an ideal gas by examining this plot. Does the plot behave as expected?

2. Recently, some electronic spreadsheet manufacturers have begun to incorporate three-dimensional graphing capability into their products. Such plots have been sadly underused in many disciplines because of the difficulties involved in their construction. If your program has this capability, you will want to play with it (Figure 3.7). Even if your program does not have this capability, you may be able to find commercial or public-domain software that can generate three-dimensional graphics from tables of data exported from spreadsheets organized like Model 3.4.

a. Examine Figure 3.7 and notice that pressure increases linearly with temperature.

b. Examine Figure 3.7 and notice that pressure is not a linear function of volume.

Figure 3.6

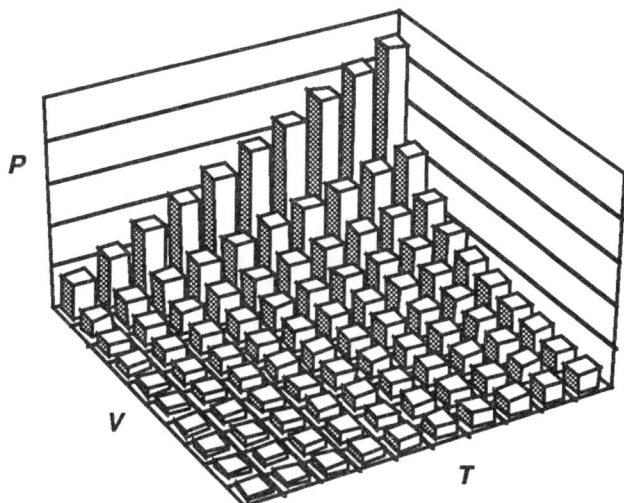

XY PLOT

Data
Horizontal (x) axis: cells A8–A17 (volume)
Vertical (y) axis: cells B8–B17, F8–B17, and J8–B17 (pressure at three different temperatures)

Figure 3.7

3D PLOT

As we complete this book, spreadsheet software is just beginning to appear that is capable of plotting *surfaces* as well as *lines*. This graph was created by the Macintosh spreadsheet program Wingz.

Data
Horizontal (x) axis: cells B7–K7 (temperature)
Horizontal (y) axis: cells A8–A17 (volume)
Vertical (z) axis: cells B8–K17 (pressure)

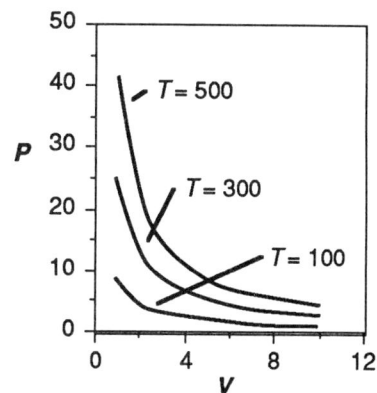

MODELS

Model 3.1

BUILD and verify the model described in Model 3.1.

ENTER data into cells B1–B3 and B5 and B6. Cells B1–B3 contain the number of moles of gas in the container, the universal gas constant, and the temperature in K, respectively. Cells B5 and B6 contain the starting volume and the volume increment used to determine the range of volumes for which pressure is calculated.

READ output from the model at cells A9–A18 and B9–B18. Cells A9–A18 contain a range of container volumes for which pressure will be calculated. Cells B9–B18 contain the pressure of the containment vessel calculated by the ideal gas law.

NOTICE that smoother curves can be achieved by extending the model beyond row 18 and reducing the size of ΔV (cell B6).

FORMULAS

	A	B
1	n (mol) =	1
2	R (L atm/mol K) =	0.0820575
3	T (K) =	302.15
4		
5	Vo =	1
6	ΔV =	1
7		
8	Vol (L)	Press (atm)
9	=B5	
10	=A9+B6	
11	=A10+B6	Formula
12	=A11+B6	Set
13	=A12+B6	I
14	=A13+B6	
15	=A14+B6	
16	=A15+B6	
17	=A16+B6	
18	=A17+B6	

VALUES

	A	B
1	n (mol) =	1
2	R (L atm/mol K) =	0.0820575
3	T (K) =	302.15
4		
5	Vo =	1
6	ΔV =	1
7		
8	Vol (L)	Press (atm)
9	1	24.79367363
10	2	12.39683681
11	3	8.264557875
12	4	6.198418406
13	5	4.958734725
14	6	4.132278938
15	7	3.541953375
16	8	3.099209203
17	9	2.754852625
18	10	2.479367363

| Formula Set I
Prototype Cell is B9 |
|---|
| =(**B1** ***B2** ***B3**)/A9 |

FORMULAS

	A	B
1	n (mol) =	1
2	R (L atm/mol K) =	0.0820575
3	V (L) =	1
4		
5	To =	302.15
6	ΔT =	20
7		
8	T (K)	Press (atm)
9	=B5	
10	=A9+B6	
11	=A10+B6	Formula
12	=A11+B6	Set
13	=A12+B6	I
14	=A13+B6	
15	=A14+B6	
16	=A15+B6	
17	=A16+B6	
18	=A17+B6	

VALUES

	A	B
1	n (mol) =	1
2	R (L atm/mol K) =	0.0820575
3	V (L) =	1
4		
5	To =	302.15
6	ΔT =	20
7		
8	T (K)	Press (atm)
9	302.15	24.79367363
10	322.15	26.43482363
11	342.15	28.07597363
12	362.15	29.71712363
13	382.15	31.35827363
14	402.15	32.99942363
15	422.15	34.64057363
16	442.15	36.28172363
17	462.15	37.92287363
18	482.15	39.56402363

Formula Set I
Prototype Cell is A1
$=B1*B2*A9/B3$

Model 3.2

BUILD and verify the model described in Model 3.2.

ENTER data into cells B1–B3 and B5 and B6. Cells B1–B3 contain the number of moles of gas in the container, the universal gas constant, and the volume of the containment vessel, respectively. Cells B5 and B6 contain the starting temperature and the temperature increment used to determine the range of volumes for which pressure is calculated.

READ output from the model at cells A9–A18 and B9–B18. Cells A9–A18 contain a range of container temperatures for which pressure will be calculated. Cells B9–B18 contain the pressure of the containment vessel calculated by the ideal gas law.

Model 3.3

BUILD and verify the model described in Model 3.3.

ENTER data into cells B1–B3, B5 and B6, and B8 and B9. Cells B1–B3 contain the number of moles of gas in the container, the universal gas constant, and the temperature in K, respectively. Cells B5 and B6 are the van der Waal's constants that specify the behavior of a specific, "real" gas. Cells B8 and B9 contain the starting volume and the volume increment used to determine the range of volumes for which pressure is calculated.

READ output from the model at cells A9–A18, B9–B18, and C14–C23. Cells A14–A23 contain a range of container volumes for which pressure will be calculated. Cells B14–B23 contain the pressure of the containment vessel calculated by the ideal gas law. Cells C14–C23 contain the pressure of the containment vessel calculated by the van der Waal's equation.

NOTICE that the formula used to calculate the van der Waal's equation uses A14*A14 rather than A14^2 to determine the value of V^2. Electronic spreadsheets use the same algorithm to calculate integer powers (e.g., 10^2) that they use to calculate fractional powers (e.g., $10^{2.3}$). This algorithm can lead to small errors in the last decimal position (e.g., 1.999999999 vs 2). For this reason, many spreadsheet modelers prefer to modify formulas containing small integer exponents by multiplying them out explicitly (e.g., A14*A14 rather than A14^2).

FORMULAS

	A	B	C
1	n (mol) =	1	
2	R (L atm/mol K) =	0.0820575	
3	T (K) =	302.15	
4			
5	a (L^2 atm/mol^2) =	3.592	
6	b (L/mol) =	0.04267	
7			
8	Vo =	1	
9	ΔV =	1	
10			
11		Ideal	van der Waal's
12		Gas Law	Equation
13	Volume	Pressure	Pressure
14	=B8	=(B1*B2*B3)/A14	
15	=A14+B9	=(B1*B2*B3)/A15	
16	=A15+B9	=(B1*B2*B3)/A16	Formula
17	=A16+B9	=(B1*B2*B3)/A17	Set
18	=A17+B9	=(B1*B2*B3)/A18	I
19	=A18+B9	=(B1*B2*B3)/A19	
20	=A19+B9	=(B1*B2*B3)/A20	
21	=A20+B9	=(B1*B2*B3)/A21	
22	=A21+B9	=(B1*B2*B3)/A22	
23	=A22+B9	=(B1*B2*B3)/A23	

VALUES

	A	B	C
1	n (mol) =	1	
2	R (L atm/mol K) =	0.0820575	
3	T (K) =	302.15	
4			
5	a (L^2 atm/mol^2) =	3.592	
6	b (L/mol) =	0.04267	
7			
8	Vo =	1	
9	ΔV =	1	
10			
11		Ideal	van der Waal's
12		Gas Law	Equation
13	Volume	Pressure	Pressure
14	1	24.7937	22.3068
15	2	12.3968	11.7691
16	3	8.2646	7.9847
17	4	6.1984	6.0408
18	5	4.9587	4.8577
19	6	4.1323	4.0621
20	7	3.5420	3.4904
21	8	3.0992	3.0597
22	9	2.7549	2.7236
23	10	2.4794	2.4541

Formula Set I
Prototype Cell is C14
=((**B1*****B2*****B3**)/(A14-(**B1*****B6**)))-((**B1*****B1*****B5**)/(A14*A14))

FORMULAS

	A	B	C	D	E	F	G	H	I	J	K
1	n (mol) =	1									
2	R (L atm/mol K) =	0.082									
3											
4	Vo =	1	To =	100							
5	ΔV =	1	ΔT =	50							
6											
7	V(L)\T(°K)	=D4				Formula Set I					
8		=B4									
9											
10											
11	Formula					Formula					
12	Set					Set					
13	II					III					
14											
15											
16											
17											

VALUES

	A	B	C	D	E	F	G	H	I	J	K
1	n (mol) =	1									
2	R (L atm/mol K) =	0.082									
3											
4	Vo =	1	To =	100							
5	ΔV =	1	ΔT =	50							
6											
7	V(L)\T(°K)	100	150	200	250	300	350	400	450	500	550
8	1	8.206	12.31	16.41	20.51	24.62	28.72	32.82	36.93	41.03	45.13
9	2	4.103	6.154	8.206	10.26	12.31	14.36	16.41	18.46	20.51	22.57
10	3	2.735	4.103	5.471	6.838	8.206	9.573	10.94	12.31	13.68	15.04
11	4	2.051	3.077	4.103	5.129	6.154	7.18	8.206	9.231	10.26	11.28
12	5	1.641	2.462	3.282	4.103	4.923	5.744	6.565	7.385	8.206	9.026
13	6	1.368	2.051	2.735	3.419	4.103	4.787	5.471	6.154	6.838	7.522
14	7	1.172	1.758	2.345	2.931	3.517	4.103	4.689	5.275	5.861	6.447
15	8	1.026	1.539	2.051	2.564	3.077	3.59	4.103	4.616	5.129	5.641
16	9	0.912	1.368	1.824	2.279	2.735	3.191	3.647	4.103	4.559	5.015
17	10	0.821	1.231	1.641	2.051	2.462	2.872	3.282	3.693	4.103	4.513

Formula Set I Prototype Cell is C7	Formula Set II Prototype Cell is A9	Formula Set III Prototype Cell is B8
=D5 +B7	=B5 +A8	=(B1 *B2 *B7)/A8

Model 3.4

BUILD and verify the model described in Model 3.4.

ENTER data using cells B1 and B2, B4 and B5, D4 and D5. Cells B1 and B2 hold the number of moles of gas and the universal gas constant, respectively. Cells B4 and B5, D4 and D5 determine the ranges of volume and temperature for which pressure will be calculated.

READ output from the model at cells B8–K17. Each cell in this array contains a value for pressure, given the volume and temperature specified at the edges of the array.

NOTICE that the cell array, B8–K17, can be specified by a single Formula Set (III). The formula in cell B8, =(**B1*****B2***B7)/A8, contains two absolute variables (**B1**, **B2**) and two *hybrid* variables (**B7**, **A8**). B7 behaves as if it were an absolute variable when it is replicated down the rows (**7** is boldfaced) and as if it were a relative variable when it is replicated across the columns (B is plain-faced). What rules govern the hybrid variable **A**8?

PARTIAL PRESSURES

Consider a sample of ideal gas that exerts a pressure P_{tot} on the walls of its closed container (Figure 3.8a) and recall that this pressure can be calculated from its equation of state (Equation 3.4)

$$P_{tot} = \frac{nRT}{V} \tag{3.4}$$

where n is the number of moles of gas within the chamber, R is the universal gas constant, T is the chamber temperature in K, and V is the volume of the chamber.

The critical point to notice in Equation 3.4 is that the total pressure exerted by the gas on the walls of the container is a function of the total number of gas particles and is independent of the type of particles present.[1] If the enclosed chamber contains two or more different gases, P_{tot} still depends on only the total number of gas particles present (Figure 3.8b). Furthermore, each gas exerts a pressure of its own that depends on the number of particles of that specific gas (Equation 3.5, Figure 3.9)

$$P_{tot} = P_1 + P_2 + \ldots + P_n = \frac{n_1 RT}{V} + \frac{n_2 RT}{V} + \ldots + \frac{n_n RT}{V} \tag{3.5}$$

where P_1, P_2, P_n, n_1, n_2, and n_n are the pressures and amounts (in moles) of each of the contributing gases in the mixture.

(a)   (b)

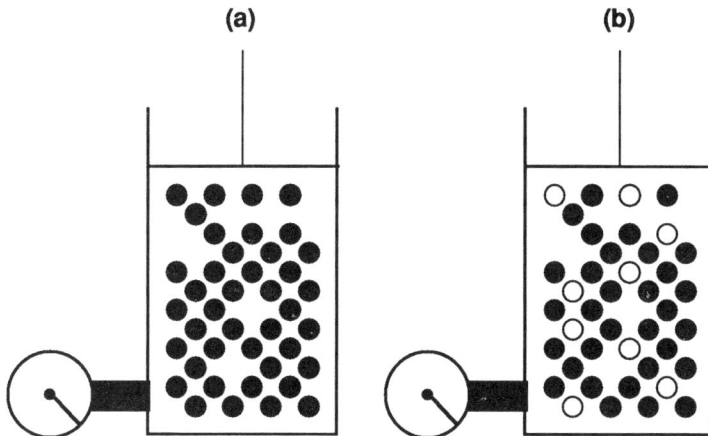

Figure 3.8

a. The pressure on the walls of a container of ideal gas depends on the container's volume and temperature and on the number of gas molecules present in the container.

b. The pressure on the walls of a container of a mixture of two ideal gases (totaling *n* mol of gas) is the same as the pressure on the walls of a container of *n* mol of one ideal gas provided that the container's volume and temperature are identical in both cases.

Equation 3.5 is known as *Dalton's law of partial pressures*, and it provides you with numerous opportunities for exploring important gas behaviors.

Before beginning your explorations, note that traditional partial pressure problems often ask you to calculate the partial pressures P_1, P_2, ... , P_n from a knowledge of the total pressure in the container and the amount of each gas in the container. In this case, Equation 3.5 is not directly usable because it requires different arguments (*n*, *T*, and *V*). However, because

$$\frac{P_i}{P_{tot}} = \frac{P_i V}{P_{tot} V} = \frac{n_i RT}{n_{tot} RT} \qquad (3.6)$$

where P_i and n_i are the partial pressure and number of moles of the *i*th type of gas, it follows that

$$\frac{P_i}{P_{tot}} = \frac{n_i}{n_{tot}} \qquad (3.7)$$

and so

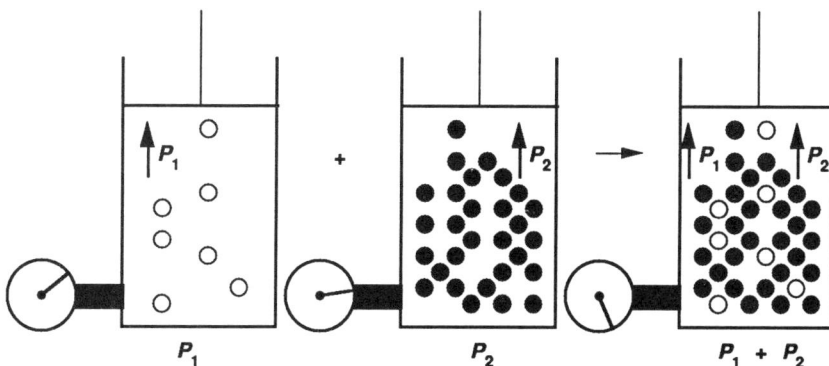

Figure 3.9

Dalton's law of partial pressures states that the total pressure exerted by a mixture of gases on the walls of a confining container is equal to the sum of the partial pressures contributed by each individual gas.

Figure 3.10

A container consisting of a cylinder with a stopcock at one end (to permit the introduction or removal of different kinds of gas) and a moveable piston at the other (to control the volume of the container) is an ideal device for performing the thought experiments implicit in the simulations of this exercise.

$$P_i = \left(\frac{n_i}{n_{tot}}\right) P_{tot} \qquad (3.8)$$

LESSONS

Consider a laboratory device consisting of a cylinder that is closed at the top by a moveable piston into which numerous different gases can be introduced via a stopcock in the bottom of the container (Figure 3.10).

1. Suppose that the chamber shown in Figure 3.10 is charged with 0.709 g of Cl_2 and 0.156 g of H_2. Figure 3.11 shows the output of an electronic spreadsheet model that can simulate the behavior of these gases (plus two others as yet unspecified). You control the temperature and volume of the container and the amount and molecular weight of the gases.

    a. Examine Figure 3.11 and notice that cells B1–B3 specify the gas constant, temperature, and volume of the container, whereas cells B6–E7 specify the molecular weight and amount (in grams) of four gases labeled A, B, C, and D.

      Build the model implied by Figure 3.11 but leave cells B9–F10 blank. Notice that gas A represents Cl_2 (molecular weight of 70.9) and gas B represents H_2 (molecular weight of 2.016). Also notice that the unused slots for gases C and D have been assigned a molecular weight of 1. Some nonzero value must be placed in these cells in order to keep division by zero errors from occurring later in the model.

    b. Cells B9–E9 convert the amount of each gas in grams to the amount of each gas in moles. What formulas should be placed in these cells? Verify you answer by comparing your computer's calculated output with the output in Figure 3.11.

Figure 3.11

Electronic spreadsheets can be used to calculate the partial and total pressures of a mixture of gases confined within an enclosed container when the volume and temperature of the container and the amounts of each gas are known.

	A	B	C	D	E	F
1	R (L atm/mol K) =	0.08206				
2	T (K) =	293.15				
3	V (K) =	4.55957				
4						
5	Gas =	A	B	C	D	Total
6	Mol. Wt. =	70.9	2.016	1	1	
7	Amount (g) =	0.709	0.156	0	0	
8						
9	n (mol) =	0.01	0.07738	0	0	0.08738
10	Partial pressure (atm) =	0.05276	0.40824	0	0	0.461

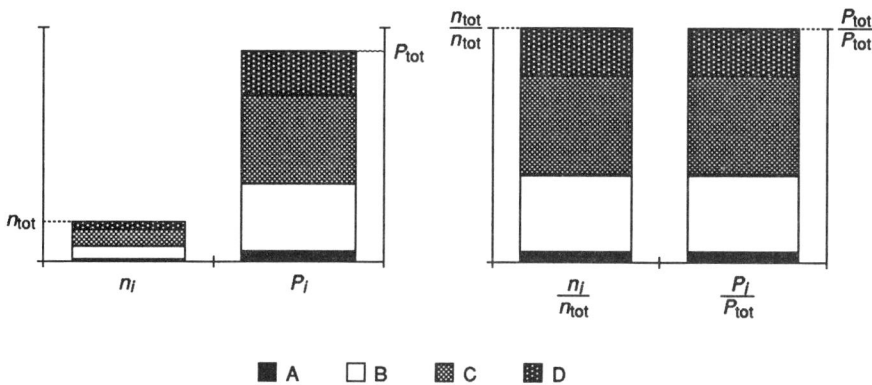

A B C D

Figure 3.12

Stacked-bar charts can be used to represent Dalton's law of partial pressures.

a. The height of the bar can represent the total number of moles n_{tot} of gas (or the total pressure P_{tot} on the walls of the confining container), while individual segments within the bar can be stacked one on the other to represent the number of moles of each type of gas n_i (or the partial pressure P_i associated with that gas).

b. The height of the bar can represent the normalized total number of moles of gas n_{tot}/n_{tot} (or normalized total pressure P_{tot}/P_{tot}), while the individual segments within the bar can be stacked one on the other to represent the mole fraction of each type of gas n_i/n_{tot} (or the normalized partial pressure P_i/P_{tot}).

c. Cell F9 calculates the total amount of gas in moles present in the container. What formula should be placed in this cell? Verify your answer by checking it against Figure 3.11.

d. Cells B10–E10 calculate the partial pressure of each gas in the container by using Equation 3.5. What formulas should be placed in these cells? Verify your answer.

e. Cell F10 calculates the total pressure in the container and can be calculated in one of two ways. On the one hand, the total pressure in the container is the sum of the partial pressures of each gas; on the other hand, the total pressure in the container can be calculated using the equation of state of an ideal gas $P_{tot} = n_{tot}RT/V$.

Verify that the same correct answer is obtained when either expression is entered into cell F10.

f. Model 3.5 fully describes the model you have just built. Take a moment to confirm that you have entered all formulas correctly (especially the unverified formulas related to gases C and D).

2. One of the wonderful properties of computer simulations is their capacity for allowing you to *see* what you already *know*. You can verify some of the obvious properties of gases with your model by using two graphic displays that are rarely used in this book but that are uniquely suited to exploring partial pressures of gases. One is a simple stacked-bar chart (Figure 3.12a), and the other is a stacked-bar chart normalized to correspond to values as percentages of the whole (Figure 3.12b).

If you are using Microsoft Excel, build these on-screen graphics as follows:

Figure 3.13

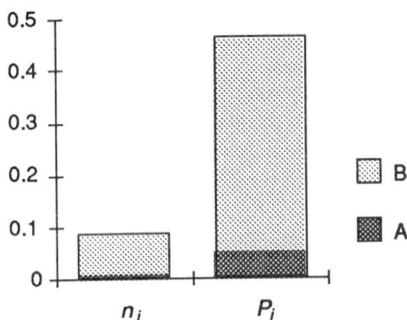

STACKED-BAR PLOT

<u>Data</u>
Horizontal (*x*) axis: n/a
Vertical (*y*) axis: cells B9 and B10
(Gas A), C9 and C10 (Gas B), D9 and
D10 (Gas C), and E9 and E10 (Gas
D)

Notice that no *x*-axis is specified.
Write labels into this region using the
options of your spreadsheet graphics
package, if you wish.

Figure 3.14

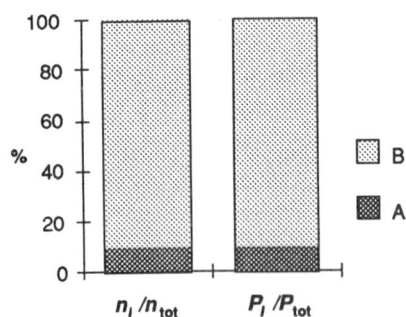

STACKED-BAR PLOT

<u>Data</u>
Horizontal (*x*) axis: n/a
Vertical (*y*) axis: cells B9 and B10
(Gas A), C9 and C10 (Gas B), D9 and
D10 (Gas C), and E9 and E10 (Gas
D)

Notice that no *x*-axis is specified.
Write labels into this region using the
options of your spreadsheet graphics
package, if you wish.

- *Stacked-Bar Chart* (*Actual Values*), (Figure 3.13): Select the block of cells B9–E10 (the number of moles and the partial pressures for each of the four possible gases) and place it on your clipboard with the COPY command. Now select the NEW option from the FILE menu and click CHART, followed by OK. When the blank chart window appears on your screen, select the PASTE SPECIAL option from the EDIT menu, tell your software that data are organized in columns rather than rows, and then click the OK button. Next select COLUMN from the GALLERY menu and tell your software that you want a stacked-bar chart of the actual values (as opposed to percentages of total) in cells B9–E10. In Version 2.2 of Microsoft Excel, this is Option 3 and is identified by the fact that each of the stacked bars in the graphic is a variable height. You will probably also want to select ADD LEGEND from the CHART menu in order to identify the region of the stacked-bar chart that is associated with each gas.

You can add other refinements to your chart, if you wish. Be sure to save your chart by selecting SAVE from the FILE menu before proceeding.

- *Stacked-Bar Chart* (*Values as Percentage of Total*), (Figure 3.14): After you have saved your first stacked-bar chart, it is an easy matter to create the next one. First, make a copy of your saved chart file by selecting SAVE AS from the FILE menu and save the chart under a new name. Then select COLUMN from the GALLERY menu and change the type of chart to a stacked-bar chart in which values are plotted as percentages of the whole. This is Option 5 in Version 2.2 of Microsoft Excel and can be identified as the stacked-bar chart in which all bars are the same height.

SAVE your file and add other refinements, if you wish.

If you are using Lotus 1-2-3 or one of its closest clones, build these on-screen graphics as follows:

- *Stacked-Bar Chart* (*Actual Values*), (Figure 3.13): Place your cursor at cell B9 of your spreadsheet and execute the command sequence /G T S (COMMAND GRAPH TYPE STACK-BAR). Next press X (for *x*-range), type in A9.A10 (or, better, select this range by pointing with your cursor), and press <ENTER>. Press A (for the first *y*-range) and enter the range B9–B10. Use the B, C, and D keys to enter the ranges C9 and C10, D9 and D10, and

E9 and E10, respectively. View your graph by pressing V (VIEW), and then save the memory of this graphic with your spreadsheet file by using N (NAME).

You can add other refinements to your graph, if you wish.

• *Stacked-Bar Chart* (*Values as Percentage of Total*), (Figure 3.14): Within the TYPE menu, there is an option called FEATURES. From FEATURES you can use the command 100% to modify your stacked-bar chart to plot normalized data.

Save the memory of this graphic by using N (NAME).

You can flip back and forth between your two different kinds of stacked-bar charts by using / G N U (COMMAND GRAPH NAME USE).

If you are using spreadsheet software other than Lotus 1-2-3 or Microsoft Excel, you should look for the appropriate command sequence in your user's manual.

a. Examine your plot reflecting the amounts (in moles) and partial pressures (in atmospheres) of the gases (Figure 3.13).

The stacked bar on the left represents the moles of each gas. The total height of the bar reflects the total number of moles of gas. Each bar segment represents the number of moles associated with each of the gases in the mixture.

The stacked bar on the right reflects the partitioning of the partial pressures of each gas in the mixture. The total height of the bar represents the total pressure of the gas mixture. Each partition within the bar identifies the partial pressure of each of the gases in the mixture.

b. After you understand what your stacked-bar chart represents evacuate your simulated chamber of all gas by setting the gas amounts to zero (cells B7–E7). Your bar height should drop to zero. (For this exercise, you may want to fix the range of values along the y-axis of your data so that it always displays values between 0 and 7. Otherwise, you must keep watching the numbers along this axis to get a sense of proportion.)

Now gradually increase the amount of Cl_2 gas in the container until a total of 8 g of Cl_2 is present.

Next gradually increase the amount of H_2 gas in the container until a total of 2 g of H_2 is present. Why does 2 g of H_2 make a larger contribution to the total pressure in the container than 8 g of Cl_2?

c. What would you expect to happen to the partial and total pressures in the container as the temperature within the container drops? Would you expect the ratio of the partial pressure of Cl_2 to the partial pressure of H_2 to change?

Verify your intuition with your model.

d. What would you expect to happen to the partial and total pressures in the container as the volume of the container increases? Would you expect the ratio of the partial pressure of Cl_2 to the partial pressure of H_2 to change?

Verify your intuition with your model.

3. You can gain a slightly different insight into gas behavior by exploring your simulation using Figure 3.14, where all values are presented as a percentage of a total.

a. Examine your plot representing Figure 3.14. The stacked bar on the left represents the mole fraction of each gas and the total number of moles of gas present. The stacked bar on the right represents the partitioning of the partial pressures of each gas in the mixture in terms of their percentage of total pressure.

b. Evacuate your simulated chamber of all gas by setting the gas amounts to zero (cells B7–E7) and then gradually increase the amount of Cl_2 gas in the container until a total of 8 g of Cl_2 is present.

Next, gradually increase the amount of H_2 gas in the container until a total of 2 g of H_2 is present.

Explain in simple English why the two stacked bars are identical throughout the time your container is being charged with gas.

c. What would you expect to happen to these plots as the temperature of the container decreases?

Verify your intuition with your model.

d. What would you expect to happen to these plots as the volume of the container increases?

Verify your intuition with your model.

4. Write a brief essay (two or three paragraphs) summarizing the explorations with your model. Focus on the manner in which the number of molecules of each individual gas contributes to total pressure and on the influence of how container temperature and volume effect total pressure and the partial-pressure partitions.

5. Near the end of the introduction to this exercise, it was noted that partial pressures are often calculated from a knowledge of the total pressure in the container and the amount of each gas in the container. Calculating partial pressures in this way offers a slightly different perspective.

a. Figure 3.15 represents the output of an electronic spreadsheet model that calculates the partial pressures in a container of gas from a knowledge of the amount of each gas and the total pressure on the container walls. Notice that row 2 has been left blank. You will fill this in later.

Cells B1 and B3 specify the gas constant and total pressure within the container. Cells B6–E7 specify the molecular weight and amount (in grams) of four gases labeled A, B, C, and D.

Build the model implied by Figure 3.15 but leave cells B9–F11 blank. Notice that gas A represents Cl_2 (molecular weight of 70.9) and gas B represents H_2 (molecular weight of 2.016). Also notice that the unused slots for gases C and D have been assigned a molecular weight of 1. Some nonzero value must be placed in these cells in order to keep division by zero errors from occurring later in the model.

	A	B	C	D	E	F
1	R (L atm/mol K) =	0.08206				
2						
3	Pressure (atm) =	0.461				
4						
5	Gas =	A	B	C	D	Total
6	Mol. Wt. =	70.9	2.016	1	1	
7	Amount (g) =	0.709	0.156	0	0	
8						
9	n (mol) =	0.01	0.07738	0	0	0.08738
10	Mole fraction =	0.11444	0.88556	0	0	1
11	Partial pressure (atm) =	0.05276	0.40824	0	0	0.461

Figure 3.15

Electronic spreadsheets can calculate the partial pressures of a mixture of gases confined within an enclosed container when the total pressure and the amounts of each gas are known.

b. Cells B9–E9 convert the amount of each gas in grams to the amount of each gas in moles. What formulas should be placed in these cells? Verify you answer by comparing your computer's calculated output with the output in Figure 3.15.

c. Cell F9 calculates the total amount of gas in moles present in the container. What formula should be placed in this cell? Verify your answer by checking it against Figure 3.15.

d. Cells B10–E10 calculate the mole fraction of each gas in the container. What formulas should be placed in these cells? Verify your answer.

e. Cell F10 sums the mole fractions of each gas in the mixture. This number should always total to 1.

f. Cells B11–E11 calculate the partial pressure of each gas in the container. What formulas should be placed in these cells? Verify your answer.

g. Cell F11 sums the partial pressures of each gas to calculate the total pressure in the container. This value should be the same as that entered into cell B3.

6. The model you built using Figure 3.15 gives the same results as Model 3.5, but temperature and container volume are not specified (why not?). If temperature were specified, however, then volume could be calculated.

a. Enter additional information into your Figure 3.15 model by writing the label T(K) = into cell A2 and the value 293.15 (room temperature in K) into cell B2.

If the label Volume (L) = is placed in cell A13, what formulas should be entered into cells B13–F13?

b. Model 3.6 is our answer. Take a moment to verify that you have entered all formulas correctly (especially the unverified formulas related to gases C and D).

c. Build appropriate on-screen graphics for this model and explore it much as you explored Model 3.5.

PROBLEMS

1. Design, build, verify, and explore a model that determines the mole fraction, number of moles, and number of grams of up to four different gases in a mixture from a knowledge of temperature, volume, and partial pressure. Our answer is presented as Model 3.7.

2. An evacuated steel reaction cylinder (called a *bomb*) is filled to a pressure of 6.0 atm with SO_3. SF_6 gas is then introduced until the total pressure is 9.0 atm. Modify one of your models to answer the following questions

   a. There are an infinite number of temperature–volume combinations that satisfy the conditions above. Plot T vs V for these allowable combinations.

   b. Given the following balanced equation,

   $$2SO_3 + SF_6 \rightarrow 3SO_2F_2$$

   Which of the starting gases, if any, is in excess?

MODELS

FORMULAS

	A	B	C	D	E	F
1	R (L atm/mol K) =	0.08206				
2	T (K) =	293.15				
3	V (l) =	4.55957				
4						
5	Gas –	A	B	C	D	Total
6	Mol. Wt. =	70.9	2.016	1	1	
7	Amount (g) =	0.709	0.156	0	0	
8						
9	n (mol) =	Formula Set I				=SUM(B9:E9)
10	Partial pressure (atm) =	Formula Set II				=SUM(B10:E10)

VALUES

	A	B	C	D	E	F
1	R (L atm/mol K) =	0.08206				
2	T (K) =	293.15				
3	V (K) =	4.55957				
4						
5	Gas =	A	B	C	D	Total
6	Mol. Wt. =	70.9	2.016	1	1	
7	Amount (g) =	0.709	0.156	0	0	
8						
9	n (mol) =	0.01	0.07738	0	0	0.08738
10	Partial pressure (atm) =	0.05276	0.40824	0	0	0.461

Formula Set I Prototype Cell is B9	Formula Set II Prototype Cell is B10
=B7/B6	=B9*B1*B2/B3

Model 3.5

BUILD and verify Model 3.5.

ENTER data using cells B1–B3 (gas constant, temperature, volume), B6–E6 (molecular weights of gases A through D, respectively), and B7–E7 (grams of gases A through D, respectively).

READ output from the model at cells B9–E9 (number of moles of each gas), F9 (total moles of all gases), B10–E10 (partial pressure of each gas), and F10 (total pressure of all gases).

Model 3.6

BUILD and verify Model 3.6.

ENTER data using cells B1–B3 (gas constant, temperature, pressure), B6–E6 (molecular weights of gases A through D, respectively), and B7–E7 (grams of gases A through D, respectively).

READ output from the model at cells B9–E9 (number of moles of each gas), F9 (total moles of all gases), B10–E10 (mole fraction of each gas), B11–E11 (partial pressure of each gas), F10 (total pressure of all gases), and B13–F13 (volume of container).

FORMULAS

	A	B	C	D	E	F
1	R (L atm/mol K) =	0.08206				
2	T (K) =	293.15				
3	Pressure (atm) =	0.461				
4						
5	Gas =	A	B	C	D	Total
6	Mol. Wt. =	70.9	2.016	1	1	
7	Amount (g) =	0.709	0.156	0	0	
8						
9	n (mol) =		Formula Set I			=SUM(B9:E9)
10	Mole fraction =		Formula Set II			=SUM(B10:E10)
11	Partial pressure (atm) =		Formula Set III			=SUM(B11:E11)
12						
13	Volume (L) =		Formula Set IV			=F9*B1*B2/B3

VALUES

	A	B	C	D	E	F
1	R (L atm/mol K) =	0.08206				
2	T (K) =	293.15				
3	Pressure (atm) =	0.461				
4						
5	Gas =	A	B	C	D	Total
6	Mol. Wt. =	70.9	2.016	1	1	
7	Amount (g) =	0.709	0.156	0	0	
8						
9	n (mol) =	0.01	0.07738	0	0	0.08738
10	Mole fraction =	0.11444	0.88556	0	0	1
11	Partial pressure (atm) =	0.05276	0.40824	0	0	0.461
12						
13	Volume (L) =	4.55957	4.55957	#NUM!	#NUM!	4.55957

Formula Set I Prototype Cell is B9
=B7/B6

Formula Set III Prototype Cell is B11
=B10*B3

Formula Set II Prototype Cell is B10
=B9/SUM(B9:E9)

Formula Set IV Prototype Cell is B13
=B9*B1*B2/B11

FORMULAS

	A	B	C	D	E	F
1	R (L atm/mol K) =	0.08206				
2	T (K) =	293.15				
3	V (L) =	4.55957				
4						
5	Gas =	A	B	C	D	Total
6	Mol. Wt. =	70.9	2.016			
7	Partial pressure (atm) =	0.05276	0.40824			=SUM(B7:E7)
8						
9	Mole fraction =	Formula Set I				=SUM(B9:E9)
10	n (mol) =	Formula Set I				=SUM(B10:E10)
11	Amount (g) =	Formula Set I				=SUM(B11:E11)

VALUES

	A	B	C	D	E	F
1	R (L atm/mol K) =	0.08206				
2	T (K) =	293.15				
3	V (L) =	4.55957				
4						
5	Gas =	A	B	C	D	Total
6	Mol. Wt. =	70.9	2.016			
7	Partial pressure (atm) =	0.05276	0.40824			0.461
8						
9	Mole fraction =	0.11445	0.88555	0	0	1
10	n (mol) =	0.01	0.07738	0	0	0.08738
11	Amount (g) =	0.70903	0.156	0	0	0.86503

Formula Set I
Prototype Cell is B9

=B7/ **F7**

Formula Set II
Prototype Cell is B10

=B7* **B3**/(**B1** * **B2**)

Formula Set III
Prototype Cell is B11

=B10*B6

Model 3.7

BUILD and verify Model 3.7.

ENTER data using cells B1–B3 (gas constant, temperature, volume), B6–E6 (molecular weights of gases A through D, respectively), and B7–E7 (partial pressure of gases A through D, respectively).

READ output from the model at cells B9–E9 (mole fraction of each gas), F9 (total of all mole fractions should always equal 1), B10–E10 (number of moles of each gas), F10 (total moles of all gases), B11–E11 (amount of each gas in grams), and F11 (total grams of gas).

KINETIC THEORY AND
THE IDEAL GAS LAW

TO GET THE MOST OUT OF THIS
EXERCISE, YOU SHOULD
ALREADY KNOW

The definition of the terms:

force
energy

How gases behave under the ideal
gas law.

How gases are viewed in the kinetic
molecular theory of gases.

In Exercises 3.1 and 3.2, you explored the behavior of certain measure-able properties of gases — temperature and pressure. You treated these properties as "something you measure" without pausing to consider how these properties might be interpreted in terms of modern atomic theory.

One of the great successes of modern chemistry was the development of a kinetic molecular theory of gases in the nineteenth century. This theory explained the behavior of gases by applying such physical concepts as motion, work, energy, and momentum to the view that gases are collections of randomly moving particles of definable mass and negligible size that do not interact except by making perfectly elastic collisions with one another.[1]

When gases are viewed in this light, temperature is a measure of the speed at which the gas particles are moving, and pressure is a measure of the force of collision of the gas particles on a unit area of the container wall.

In this exercise, you just briefly glimpse at the kinetic molecular theory of gases and how it is reflected in the observable behavior of gases.

Perhaps you remember from your elementary physics that Newton's second law of motion defines a force F as

$$F = ma = \frac{dp}{dt} \qquad (3.9)$$

where m is the mass of the moving object, a is the acceleration of the object, and dp/dt is the change of momentum of the object with time (i.e., the time derivative of momentum p).

Equation 3.9 can be used to show that a system of randomly moving particles of negligible size (interacting only through elastic collisions) exerts a force per unit area P on the walls of its confining container of

$$P = \frac{nmc^2}{3V} \qquad (3.10)$$

where n is the number of moles of particles in the gas, m is the molecular weight of the particles, c is the root mean square speed of the particles, and V is the volume of the container that confines the gas.

Not all particles in a gas move at the same speed, and c is one statistical measure of the "average" speed of the particles and can be calculated from a knowledge of the temperature T and mass m of a gas (Equation 3.11)

$$c = \sqrt{\frac{3RT}{m}} \qquad (3.11)$$

where R is the universal gas constant and T must be measured in K.

The Maxwell–Boltzmann speed-distribution equation (Equation 3.12) can be used to find the number of gas molecules traveling at any specific speed.

$$f(v) = 4\pi \left(\frac{m}{2\pi RT}\right)^{3/2} v^2 \exp\left(\frac{-mv^2}{2RT}\right) \qquad (3.12)$$

where $f(v)$ is a measure of the relative probability that a molecule is traveling at speed v.

In this exercise, you build simple models of gas behavior using Equations 3.10–3.12.

LESSONS

1. The Maxwell–Boltzmann speed-distribution equation (Equation 3.12) can be verified experimentally using the device shown in Figure 3.16. A stream of low-pressure gas is delivered to a set of

Figure 3.16

One experimental device frequently used to measure molecular speeds uses a series of rotating disks attached to a single shaft. Each disk contains a narrow slot, arranged at different angles on different disks. When a stream of low-pressure gas is directed toward a detector beyond the array of disks, only gas molecules in a narrow range of molecular speeds can successfully pass each slot in turn on its journey toward the detector. The specific range of molecular speeds that reach the detector is a function of the speed of rotation of the disks. Note that the the pressure of the gas must be so low that few molecules collide with each other in transit.

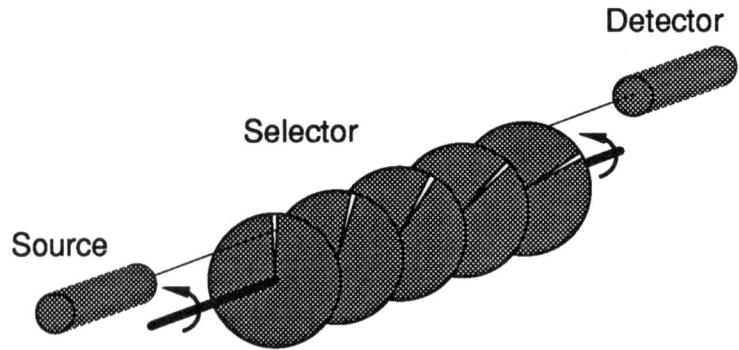

rotating shutters, and a detector positioned beyond the shutters measures the number of gas molecules that successfully run the gauntlet. Different shutter rotation speeds pass molecules traveling at different speeds. A typical experiment consists of a series of measurements at different shutter rotation speeds and different temperatures.

a. Build and verify Model 3.8. Use Figure 3.17 to build its related on-screen graphic.

b. Examine the structure of Model 3.8 until its design is clear.

2. Examine the distribution of speeds in a sample of oxygen molecules (O_2) at 100 K, 300 K, and 500 K.

It probably doesn't surprise you that only a few molecules lie at the extreme low and high ends of the spectrum of speeds, but did you expect the distribution of speeds to be asymmetric? Unlike the familiar, normal distribution curve (the bell-shaped curve) the distribution of fraction of molecules vs speed is asymmetric. The highest point on the curve, the *most probable speed*, and the "average" speed (the root mean square speed c) are not the same.

3. Plot the distribution of molecular speeds in a sample of UF_6 at 100 K, 300 K, and 500 K. Compare your results with those you obtained for O_2. How do they differ? Why do they differ?

4. Build and verify Model 3.9. Also build the on-screen graphics described in Figures 3.18 and 3.19. Notice that in this model different dimensions are used. Molecular weight, for example, is expressed as kg mol^{-1}, so that R can be expressed as J K^{-1} mol^{-1} (1 Joule = 1 kg m s^{-2}).

Figure 3.17

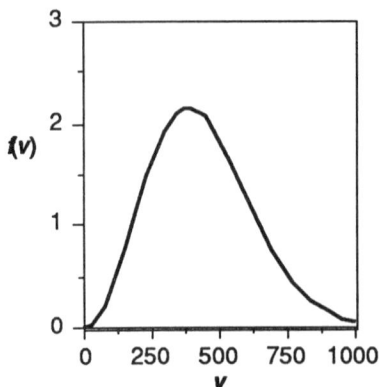

XY–PLOT

Data
Horizontal (*x*) axis: cells A15–A55 (*v*)
Vertical (*y*) axis: cells B15–B55 [*f*(*v*)]

a. Review the structure of Model 3.9 and note especially that although column B uses the ideal gas law to calculate pressure, column E calculates pressure using the kinetic molecular theory of gases (Equation 3.10).

b. Explore the pressure and root mean square speed of 1 mol of O_2 gas confined to a volume of 1 m³ in a range of temperatures between 0 K and 1000 K. Note particularly that the ideal gas law and the molecular kinetic theory of gases give identical results. (However, they are *both* wrong at very low temperatures where the gas molecules crowd together and cannot be treated as dimensionless particles!)

c. Suppose the volume of gas is increased to 2 m³. What happens to pressure? What happens to the speed of the gas molecules? Explain.

(NOTE: You will discover in Chapter 8 that an increase in the volume of a gas involves quite subtle issues. Does the gas increase its volume by performing work against the walls of its container? Can the gas exchange heat with its surroundings? Assume in this case that the volume increase occurs without work on the part of the gas.)

d. If 1 mol of UF_6 replaces 1 mol of O_2 in an enclosed volume, how do pressure and molecular speed change? Explain this result.

PROBLEMS

1. Design, build, verify, and explore a model in which the pressure and molecular speed of a mixture of gases can be displayed across a range of temperatures. After exploring your model, write a brief essay describing Dalton's law of partial pressures in terms of the kinetic molecular theory of gases.

2. The following gas-phase reaction has an activation energy of 103 kJ mol⁻¹:

$$N_2O_5 \rightarrow NO_3 + NO_2$$

On average, could a collision between a molecule of N_2 (present in the air contained in the reaction vessel) and a molecule N_2O_5 transfer enough energy to permit this reaction to occur at 500 K? Recall that the kinetic energy of an object is $mv^2/2$, where v is its speed. How, then, can such reactions occur?

Figure 3.18

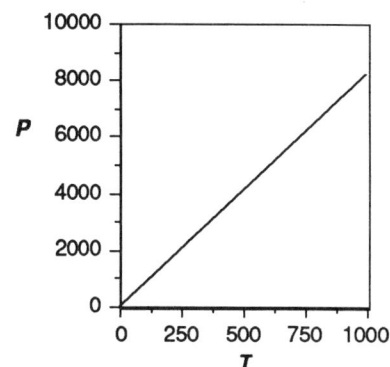

XY–PLOT

Data
Horizontal (x) axis: cells A11–A31 (T)
Vertical (y) axis: cells E11–E31 (P)

Figure 3.19

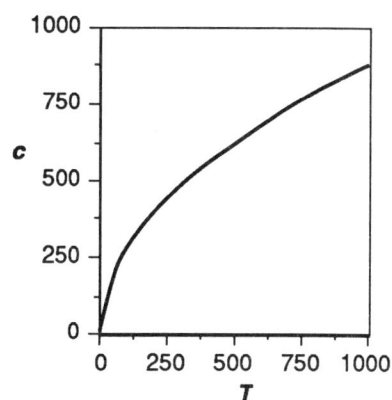

XY–PLOT

Data
Horizontal (x) axis: cells A11–A31 (T)
Vertical (y) axis: cells D11–D31 (c)

Note that this plot implies an ideal gas. At very low temperatures, *no* gas is ideal.

(HINT: Calculate the average kinetic energy of a molecule of N_2 at 500 K, and assume that all of this energy could be transferred to N_2O_5.)

MODELS

Model 3.8

BUILD and verify Model 3.8.

ENTER data using cells B1–B3 (moles and molecular weight of gas, respectively; universal gas constant), B5 (temperature), and B7 and B8 (starting and incremental gas-particle velocity).

READ output from the model at cells A14–A54 (gas-particle velocity), B14–B54 [a measure of probability, $f(v)$, that a gas particle will have velocity v], and C14–C54 (moles of gas at a specific velocity range).

NOTICE that the values in cells C14–C54 are valid only if most of the gas in the sample is described as the velocities in cells A14–A54.

FORMULAS

	A	B	C
1	n (mol) =	2	
2	M (kg/mol) =	0.032	
3	R (J/K mol) =	8.314	
4			
5	T (°K) =	290	
6			
7	v0 (m/sec) =	0	
8	Δv (m/sec) =	25	
9			
10	c =	=SQRT((3*B3*B5)/B2)	
11			
12		1000 x	
13	v (m/sec)	f(v) (sec/m)	Moles
14	=B7		
15–17	Formula Set I	Formula Set II	Formula Set III
53–55			
56	TOTAL =	=SUM(B14:B54)	=SUM(C14:C54)

VALUES

	A	B	C
1	n (mol) =	2	
2	M (kg/mol) =	0.032	
3	R (J/K mol) =	8.314	
4			
5	T (°K) =	290	
6			
7	v0 (m/sec) =	0	
8	Δv (m/sec) =	25	
9			
10	c =	475.433355	
11			
12		1000 x	
13	v (m/sec)	f(v) (sec/m)	Moles
14	0	0	0
15	25	0.02401219	0.0012048
16	50	0.09486106	0.00475962
17	75	0.20905675	0.01048934
53	975	0.06678001	0.00335066
54	1000	0.05062182	0.00253993
55			
56	TOTAL =	39.8607828	2

Formula Set I
Prototype Cell is A15

=A14+ **B8**

Formula Set III
Prototype Cell is C14

=**B1***B14/ **B56**

Formula Set II
Prototype Cell is B14

=1000*4*PI()*((**B2**/(2*PI()***B3*****B5**))^(1.5))*A14*A14 *EXP((-**B2***A14*A14)/(2* **B3*****B5**))

FORMULAS

	A	B	C	D	E
1	n (mol) =	1			
2	M (kg/mol) =	0.032			
3	R (J/K mol) =	8.314			
4	Vol (m^3) =	1			
5					
6	To =	0			
7	ΔT =	50			
8					
9	Ideal Gas Law			Kinetic Theory	
10	T	P		c	P
11	=B6				
12	Formula	Formula		Formula	Formula
13	Set	Set		Set	Set
14	I	II		III	IV
15					

VALUES

	A	B	C	D	E
1	n (mol) =	1			
2	M (kg/mol) =	0.032			
3	R (J/K mol) =	8.314			
4	Vol (m^3) =	1			
5					
6	To =	0			
7	ΔT =	50			
8					
9	Ideal Gas Law			Kinetic Theory	
10	T	P		c	P
11	0	0		0	0
12	50	415.7		197.412956	415.7
13	100	831.4		279.184079	831.4
14	150	1247.1		341.929269	1247.1
15	200	1662.8		394.825911	1662.8

Formula Set I Prototype Cell is A12
=A11+ **B7**

Formula Set III Prototype Cell is D11
=SQRT(3* **B3** *A11/ **B2**)

Formula Set II Prototype Cell is B11
=(**B1** * **B3** *A11)/ **B4**

Formula Set IV Prototype Cell is E11
=**B1** * **B2** *D11*D11/(3* **B4**)

Model 3.9

BUILD and verify Model 3.9.

ENTER data using cells B1–B4 (moles of gas, molecular weight of gas, universal gas constant, container volume) and B6 and B7 (temperature).

READ output from the model at cells cells A11–A31 (temperature), B11–B31 (pressure according to the ideal gas law), D11–D31 (root mean square speed of the gas particles), and E11–E31 (pressure according to the kinetic molecular theory of gases),

NOTICE that the dimensions of this model are different than those of Exercise 3.2. This is because the kinetic theory of gases is a physical theory and is usually phrased in the mks system.

CHAPTER 4

LIQUIDS, SOLUTIONS, AND PHASE EQUILIBRIA

Liquids and solutions are distinguished from gases by the presence of *cohesive forces* between the molecules.[1] Liquids do not expand indefinitely to fill their containers but rather are contained within a measurable volume that changes only with changes in temperature, amount, etc. Just as the theory of gases benefits from the concept of an ideal gas and its equation of state, the theory of liquids and solutions benefits from the concept of an ideal solution and Raoult's law.

Consider a closed, evacuated container into which a single liquid substance is introduced (Figure 4.1). Initially, all molecules of the substance are in the liquid state. With time, however, a vapor (gas) forms above the liquid as molecules at the surface of the liquid leave the liquid and enter the vapor (Figure 4.1a). As the number of molecules in the vapor increases, some molecules leave the gas and return to the liquid (Figure 4.1b). At the beginning, more molecules escape the liquid than return to it, but as the amount of gas increases, so do the number of molecules returning to the liquid. Eventually, the rate of escape and return become equal, and the vapor pressure above the liquid assumes a constant value (Figure 4.1c).

Notice that even after the vapor pressure assumes a constant value, molecules continue to move to and from the liquid phase at constant, equal rates. The system is said to be in *dynamic equilibrium*. The concept of dynamic equilibrium is central to many processes in chemistry and will be explored in depth in Chapters 5 and 7. In Chapter 7, you will build working models of systems in which their approach to equilibrium can be explicitly monitored.

Figure 4.1

(a) Molecules near the surface of a liquid can escape the liquid and and enter the gas phase.

(b) Once a vapor exists, molecules in the gas can be recaptured by the liquid.

(c) As the amount of gas increases, the rate at which gas molecules are recaptured by the liquid increases. Eventually the rate at which the molecules of a substance leave the liquid equals the rate at which they return to the liquid from the gas. In this state of *dynamic equilibrium*, the vapor pressure above the liquid never changes.

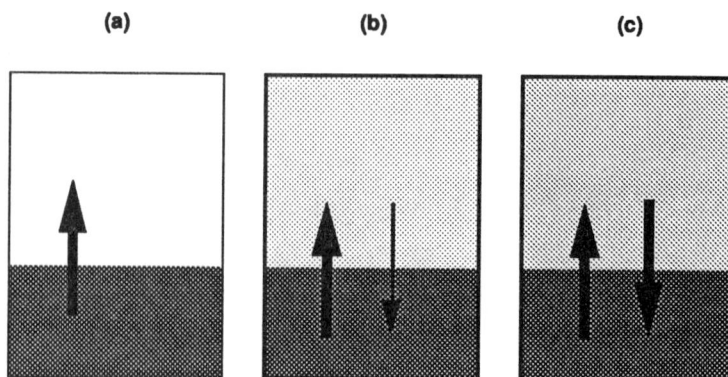

(a)   (b)   (c)

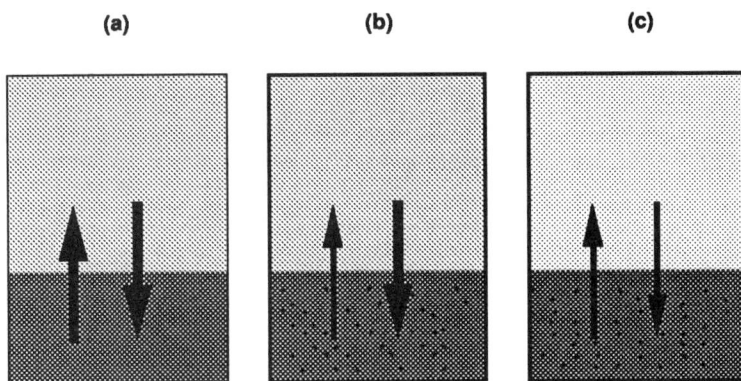

Figure 4.2

The equilibrium vapor pressure of a solution depends on its composition.

(a) A pure substance exhibits a characteristic vapor pressure at a specific temperature.

(b) As the number of nonvolatile molecules (solute) added to a liquid substance (solvent) increases, the number of solvent molecules escaping the liquid decreases. The rate at which gas molecules return to the liquid remains the same initially but gradually drops as the amount of gas is reduced.

(c) Ultimately, a new equilibrium vapor pressure is established in which the (reduced) rate at which molecules enter and depart the liquid is again the same.

Raoult's law predicts the value of the equilibrium vapor pressure of a solution as its composition changes. For example, suppose that a solution consists of a volatile substance A and a nonvolatile substance B such that the mole fraction of A in the solution is X_A (Figure 4.2). According to Raoult's law, the equilibrium vapor pressure of substance A, P_A is

$$P_A = X_A P°_A \qquad (4.1)$$

where $P°_A$ is the vapor pressure of substance A in its pure form.

If substance B is also volatile, its vapor pressure is calculated in similar fashion (Equation 4.2):

$$P_B = X_B P°_B \qquad (4.2)$$

The total vapor pressure above the solution can be calculated using Dalton's law of partial pressures (Equation 4.3):

$$P_{tot} = P_A + P_B = X_A P°_A + X_B P°_B \qquad (4.3)$$

These deceptively simple equations have some quite delightful consequences. In the exercises that follow, you will calculate the molecular weight of a dissolved solute by the extent to which it reduces the vapor pressure of a solvent, and you will build a computer simulation of a distillation device that is capable of separating a solution into its component substances.

At the end of the chapter, you will return to the properties of pure substances and build a computer model of the pressure–temperature diagram of a one-component system and predict the conditions under which a substance is a solid, liquid, or gas and the conditions under which one or more of these phases can coexist.

IDEAL SOLUTIONS: RAOULT'S LAW

TO GET THE MOST OUT OF THIS
EXERCISE, YOU SHOULD
ALREADY KNOW

The definition of the terms:
ideal gas
mole
mole fraction
liquid
solution

How to build a simple model with your
electronic spreadsheet program by
writing labels, values, and formulas
directly into cells and by copying the
contents of one cell or a range of cells
into other cells.

How to build *xy*-plots using your
electronic spreadsheet program.

Raoult's law describes some of the behavior of an ideal solution. It states that the vapor pressure P_A of a substance in a solution is proportional to the mole fraction of that substance in the solution (Equation 4.4):

$$P_A = X_A P°_A \tag{4.4}$$

where X_A is the mole fraction of substance A in the solution and $P°_A$ is the vapor pressure of substance A in its pure form.

If there is more than one volatile substance in the solution, each substance obeys Raoult's law, and, according to Dalton's law of partial pressures, the total vapor pressure of the solution equals the sum of the vapor pressures of each substance (Equation 4.5):

$$P_{tot} = P_A + P_B + ... + P_z \tag{4.5}$$

where P_{tot} is the total vapor pressure of the solution and $P_A, P_B, ..., P_z$ are the vapor pressures of each volatile component of the solution.

Although few solutions behave ideally over their full range of possible compositions, many solutions approximate ideal behavior when, for example, the amount of solvent greatly exceeds the amount of solute.

In the lessons and problems that follow, you will see how Raoult's law can be used to find the molecular weight of an unknown solute, and you will explore the behavior of both ideal and real solutions that contain multiple volatile components. In the problems, you will also explore other so-called colligative properties of solutions (i.e., properties that depend on the *number* of solute molecules present).

LESSONS

1. Figure 4.3 represents the output of an electronic spreadsheet model designed to calculate the vapor pressure of a binary solution of benzene (a volatile substance) and napthalene (an essentially non-volatile substance). Cells B2 and B3 contain values that represent the molecular weights of benzene and napthalene (78 and 128, respectively). Cell B5 contains a value representing the vapor pressure of pure benzene at 25°C (i.e., 0.1252 atm). Cells A9–A19 represent the grams of benzene solvent in the solution (the initial values of the model assume a constant amount of 50 g of benzene). Cells B9–B19 represent the grams of napthalene solute in the solution (the initial values of the model assume a gradually increasing amount of napthalene).

    a. Cells C9–C19 and D9–D19 calculate the moles of solvent and solute in each solution. What formulas belong in these cells?

    b. Cells E9–E19 and F9–F19 calculate the mole fractions of solvent and solute in each solution. What formulas belong in these cells?

    c. Cells G9–G19 calculate the vapor pressure of each solution using Raoult's law. What formulas belong in these cells?

    Our answer to the design of the model shown in Figure 4.3 is contained in Model 4.1 at the end of this exercise.

	A	B	C	D	E	F	G
1	MOLECULAR WEIGHTS						
2	Solvent =	78	[benzene]				
3	Solute =	128	[naphthalene]				
4							
5	P0 (atm) =	0.1252	[benzene]				
6							
7		grams		moles		mole fraction		Vapor
8	Solvent	Solute	Solvent	Solute	Solvent	Solute	Pressure
9	50	0	0.641026	0	1	0	0.1252
10	50	10	0.641026	0.078125	0.891365	0.108635	0.111599
11	50	20	0.641026	0.15625	0.80402	0.19598	0.100663
12	50	30	0.641026	0.234375	0.732265	0.267735	0.09168
13	50	40	0.641026	0.3125	0.672269	0.327731	0.084168
14	50	50	0.641026	0.390625	0.621359	0.378641	0.077794
15	50	60	0.641026	0.46875	0.577617	0.422383	0.072318
16	50	70	0.641026	0.546875	0.539629	0.460371	0.067562
17	50	80	0.641026	0.625	0.506329	0.493671	0.063392
18	50	90	0.641026	0.703125	0.4769	0.5231	0.059708
19	50	100	0.641026	0.78125	0.450704	0.549296	0.056428

Figure 4.3

This electronic spreadsheet output calculates the vapor pressure of a solution of volatile solvent and nonvolatile solute. Can you build it?

Figure 4.4

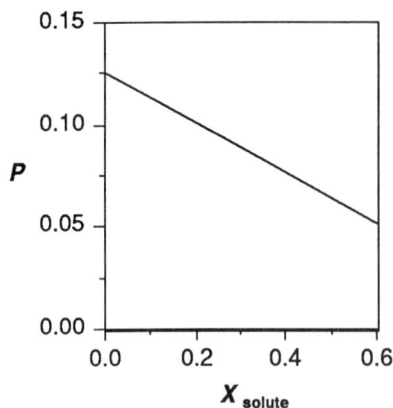

XY–PLOT

Data
Horizontal (x) axis: cells F9–F19
(mole fraction of solute)
Vertical (y) axis: cells G9–G19 (vapor
pressure)

2. Build an on-screen graphic for the plot of vapor pressure vs mole fraction of solute (Figure 4.4).

   a. As initialized, your model plots the vapor-pressure reduction that occurs when 0 to 100 g of napthalene are dissolved in 50 g of benzene (assuming ideal behavior). Contrast the changes you see in the vapor pressures of these solutions with those that occur when equivalent amounts of napthalene are dissolved in 100 g of benzene; 200 g of benzene.

   b. Suppose that your ideal benzene–napthalene solution has a constant mass of 100 g and its constituents vary from 100 g of benzene (0 g of napthalene) to 0 g of benzene (100 g of napthalene). Change the values in cells A9–B19 to reflect this situation and comment on your results. What is the vapor pressure of pure napthalene? Does this make sense given the assumptions used to build the model?

3. Restore your model to its initial values.

   a. What is the vapor pressure of a solution consisting of 50 g of benzene and 50 g of napthalene?

   b. Suppose 50 g of benzene form a solution with 50 g of an unknown, nonvolatile substance and the measured vapor pressure of the solution is 0.0925 atm. Is the molecular weight of this unknown substance greater than, equal to, or less than that of napthalene?

      By trial and error, substitute new values for the molecular weight of the unknown solute into cell B3 of your model until your ideal solution reflects a vapor pressure of 0.0925 atm for the conditions noted above.

   c. Write a brief paragraph on how the the molecular weight of a solute influences the vapor pressure of a solution.

   d. Describe an experimental procedure that you could use to determine the molecular weight of a substance by dissolving it in a known amount of solvent of known molecular weight.

4. Finding the molecular weight of a substance from its effect on the vapor pressure of a solvent is probably a problem for your pocket calculator. To verify your understanding, however, you may want to examine Figure 4.5.

	A	B	C	D
1	P0 (atm) =	0.125		
2	P (atm) =	0.119		
3				
4	X (solv) =	0.952		
5				
6	SOLVENT		SOLUTE	
7	M.W. =	78		
8	grams =	7.8	grams =	0.625
9	moles =	0.1	moles =	0.00504202
10			M.W. =	123.958333

Figure 4.5

This electronic spreadsheet model can calculate the molecular weight of a nonvolatile solute from its effect on the vapor pressure of a solution. Can you build it?

a. Figure 4.5 is a small electronic spreadsheet model that calculates the molecular weight of a solute from its effect on solvent vapor pressure. Cell B1 contains the value for the vapor pressure of pure solvent (benzene is used in this example). Cell B2 contains the value for the vapor pressure of the binary solution of solvent and solute. Cells B7 and B8 contain the molecular weight and grams of solvent in the solution, respectively. Cell D8 contains the grams of solute in the solution.

b. Cell B4 calculates the mole fraction of solvent in the solution from a knowledge of the vapor pressure of pure solvent and the vapor pressure of the solution. What formula belongs in this cell?

c. Cell B9 calculates the number of moles of solvent in the solution from a knowledge of solvent molecular weight and grams of solvent in the solution. What formula belongs in this cell?

d. Cell D9 calculates the number of moles of solute in the solution from a knowledge of the grams of solute in the solution, the mole fraction of solvent in the solution, and the definition of mole fraction. After performing a bit of algebra, what formula belongs in cell D9?

e. Cell D10 calculates the presumed molecular weight of the solute from its effect on solution vapor pressure. What formula belongs in cell D10?

Model 4.2 at the end of this exercise contains our answer to the preceding questions.

f. Determining molecular weight by solution vapor–pressure reduction is not very accurate. Your model was initialized with data obtained from a binary solution of benzene and napthalene.

Other, more accurate measurements of solute molecular weight can be obtained from other, more sensitive colligative properties. One of these, measurements of the osmotic pressure of a solution, is discussed in Problem 3.

5. Lessons 1 through 4 explored the behavior of binary, ideal solutions with one volatile and one nonvolatile component. Recall, however, that Raoult's law applies even when there are two or more volatile components in a solution.

   a. Build and verify Model 4.3. Model 4.3 calculates the vapor pressures of two volatile substances in solution (columns E and F). It then uses Dalton's law of partial pressures to calculate the total vapor pressure of the solution (column G).

   Examine the structure of the model until you understand how it has been built.

   b. Build an on-screen graphic that plots the vapor pressures of substances A and B as well as the total vapor pressure vs the mole fraction of substance B in solution (Figure 4.6).

   c. Model 4.3 has been initialized with values appropriate for solutions of benzene and methylbenzene. Examine the plots of each vapor pressure vs the mole fraction of methylbenzene as the amount of methylbenzene in solution increases.

   When equal numbers of molecules of benzene and methylbenzene are present in solution, are there equal numbers of benzene and methylbenzene molecules in the vapor? (Recall that the pressure of an ideal gas depends on the number and not the kind of molecules present.) Do you think that this might form the basis of a laboratory technique for separating benzene from methylbenzene?

   d. Explore the behavior of solutions in which methylbenzene is replaced by progressively less volatile substances.

   What does your model predict for the behavior of solutions of benzene and napthalene? Does it make sense?

   e. Suppose that two substances of identical vapor pressure (in their pure form) are mixed in solution. When equal numbers of molecules of the two substances are present in solution, what is the composition of the vapor?

Figure 4.6

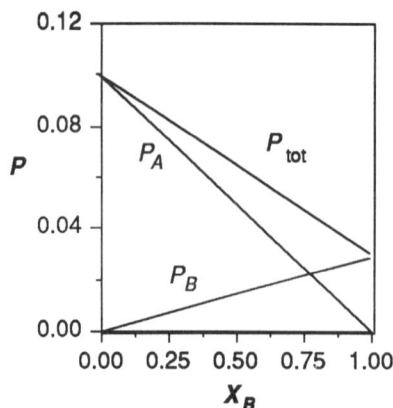

XY–PLOT

Data
Horizontal (x) axis: cells D6–D16 (mole fraction of solvent B)
Vertical (y) axis: cells E6–E16 (vapor pressure of A), F6–F16 (vapor pressure of B), and G6–G16 (total vapor pressure)

When unequal numbers of molecules of the two substances are present in solution, what is the composition of the vapor?

Do you think that these two substances could be separated from each other on the basis of their vapor pressures?

6. Not many solutions behave ideally over the full range of possible compositions. Below is a table of data from a laboratory experiment in which the vapor pressures of chloroform and acetone were measured over a full range of compositions.

Vapor Pressure in Atmospheres

X (Acetone)	Chloroform	Acetone	Total
0.0	0.3855	0	0.3855
0.2	0.2921	0.0526	0.3447
0.4	0.1908	0.1342	0.325
0.6	0.1079	0.2421	0.35
0.8	0.0461	0.3553	0.4013
1.0	0	0.4566	0.4566

Modify Model 4.3 to reflect both the ideal behavior and the actual laboratory behavior of chloroform–acetone solutions. (Notice that you can obtain the vapor pressure of both pure substances from the table.) Our answer is shown in Model 4.4 and Figure 4.7.

a. How does the actual behavior of acetone correspond to its predicted ideal behavior?

b. How does the actual behavior of chloroform correspond to its predicted ideal behavior?

c. Notice that the actual behavior of the total vapor pressure of the solution can be predicted from Dalton's law of partial pressures once the actual behavior of each component is known.

7. Solutions of acetone and carbon disulfide show a different kind of deviation from ideal behavior. Build a model to compare the ideal and actual laboratory behavior of acetone–carbon disulfide solutions.

Figure 4.7

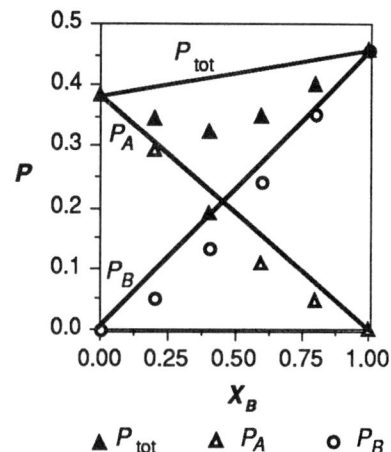

XY–PLOT

Data
Horizontal (x) axis: cells D6–D16 (mole fraction of solvent B)
Vertical (y) axis: cells E6–E16 (calculated vapor pressure of A), F6–F16 (calculated vapor pressure of B), G6–G16 (calculated total vapor pressure), H6–H16 (actual vapor pressure of A), I6–I16 (actual vapor pressure of B), and J6–J16 (actual total vapor pressure)

NOTE: You may find it useful to plot calculated data using lines and actual data using plotting symbols, as we have done here.

Vapor Pressure in Atmospheres

X (Acetone)	CS$_2$	Acetone	Total
0.0	0.6737	0	0.6737
0.2	0.6079	0.0526	0.6605
0.4	0.5566	0.1342	0.6908
0.6	0.4974	0.2421	0.7395
0.8	0.3684	0.3553	0.7237
1.0	0	0.4566	0.4566

Compare and contrast the behavior of carbon disulfide–acetone solutions with solutions of chloroform and acetone.

PROBLEMS

1. A nonvolatile reagent used to determine the iron level in water has the empirical formula C$_3$H$_2$N. A 0.150-g sample is dissolved in 8.500 g of dichloromethane (CH$_2$Cl$_2$), and the vapor pressure of the solution is measured and found to be 398.1 mm Hg at 24°C. (At 24°C, pure dichloromethane has a vapor pressure of 400.0 mm Hg.)

    What is the molecular weight of the reagent? What is its molecular formula?

    Use your electronic spreadsheet to prepare a plot of the ideal behavior of such solutions over a wide range of compositions.

2. Anthraquinone is an important building block for natural and artificial dyes. Analysis of a sample of pure anthraquinone gives the following results by weight:

C	80.75%
H	3.87%
O	15.37%

    A 0.223-g sample is dissolved in 7.800 g of benzene, and the vapor pressure of the solution is measured and found to be 268.8 mm Hg at 50°C. (At 50°C, pure benzene has a vapor pressure of 271.7 mm Hg.)

    What is the molecular weight of the anthraquinone? What is its molecular formula?

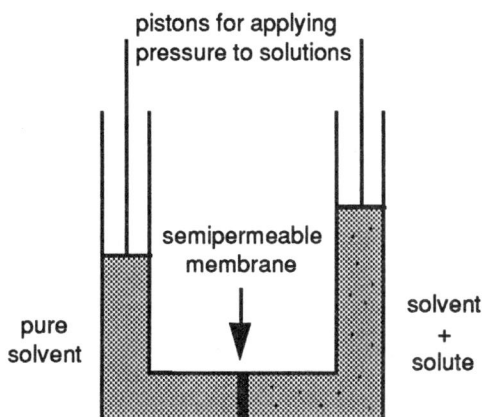

Figure 4.8

If pure solvent is separated from its solution by a membrane permeable to solvent but not solute, molecules of solvent cross the membrane to enter the solution. The pressure that must be applied to the solution to prevent the entry of solvent is the osmotic pressure of the solution. Osmotic pressure is a colligative property of the solution and is very sensitive to even small changes in solute concentration.

Use your electronic spreadsheet to prepare a plot of the ideal behavior of anthraquinone–benzene solutions over a wide range of compositions.

3. Vapor pressure is a *colligative property* of solutions. Colligative properties depend on only the number and not the type of molecules of dissolved solute. There are many colligative properties in addition to vapor pressure. The boiling point of a solution is raised, for example, as the number of dissolved molecules of solute increases. Also, the freezing point of a solution is decreased as solute concentration increases.

One particularly useful colligative property of solutions is osmotic pressure. If pure solvent is separated from a solution by a membrane permeable to solvent but not solute, pure solvent moves across the membrane into the solution (Figure 4.8). Solvent can be prevented from moving across the membrane by applying pressure on the solution side of the membrane. The amount of pressure required to prevent net movement of solvent is the osmotic pressure π of the solution. For dilute solutions,

$$\pi = \frac{nRT}{V} \tag{4.6}$$

where π is osmotic pressure (in atm), n is the number of moles of dissolved solute, R is the universal gas constant (0.0820575 L atm mol^{-1} K), T is temperature (in K), and V is the volume (in L) of the solution.

The changes in osmotic pressure resulting from dissolved solute are quite large compared to changes in vapor pressure. Polymer chemists and biochemists use osmotic-pressure changes to find the molecular weights of large molecules (macromolecules). Macromolecules can have molecular weights of 100,000 or more and often only very dilute solutions can be prepared. (How many grams of NaCl yield a mole fraction of 0.1 when combined with 1000 g of H_2O? How many grams of a synthetic polymer with a molecular weight of 112,000 yield a mole fraction of 0.1 when combined with 1000 g of H_2O?)

a. Design, build, and verify a model that compares the effect of solute on vapor pressure with its effect on osmotic pressure. Start with Model 4.1, if you wish.

b. Plot the vapor pressure and osmotic pressure vs grams of dissolved solute for a variety of solutions with substances of different molecular weight. Include several macromolecules in your survey.

MODELS

Model 4.1

BUILD and verify Model 4.1.

ENTER data using cells B2 and B3 (molecular weights of solvent and solute), cell B5 (vapor pressure of pure solvent), and cells A9–A19 and B9–B19 (grams of solvent and solute in solution).

READ output from the model at cells G9–G19 (vapor pressure of solution).

FORMULAS

	A	B	C	D	E	F	G
1	MOLECULAR WEIGHTS						
2	Solvent =	78	[benzene]				
3	Solute =	128	[naphthalene]				
4							
5	P0 (atm) =	0.1252	[benzene]				
6							
7	—grams—		—moles—		—mole fraction—		Vapor
8	Solvent	Solute	Solvent	Solute	Solvent	Solute	Pressure
9	50	0					
10	50	10					
11	50	20					
12	50	30	Formula Set I	Formula Set II	Formula Set III	Formula Set IV	Formula Set V
13	50	40					
14	50	50					
15	50	60					
16	50	70					
17	50	80					
18	50	90					
19	50	100					

VALUES

	A	B	C	D	E	F	G
1	MOLECULAR WEIGHTS						
2	Solvent =	78	[benzene]				
3	Solute =	128	[naphthalene]				
4							
5	P0 (atm) =	0.1252	[benzene]				
6							
7	—grams—		—moles—		—mole fraction—		Vapor
8	Solvent	Solute	Solvent	Solute	Solvent	Solute	Pressure
9	50	0	0.641026	0	1	0	0.1252
10	50	10	0.641026	0.078125	0.891365	0.108635	0.111599
11	50	20	0.641026	0.15625	0.80402	0.19598	0.100663
12	50	30	0.641026	0.234375	0.732265	0.267735	0.09168
13	50	40	0.641026	0.3125	0.672269	0.327731	0.084168
14	50	50	0.641026	0.390625	0.621359	0.378641	0.077794
15	50	60	0.641026	0.46875	0.577617	0.422383	0.072318
16	50	70	0.641026	0.546875	0.539629	0.460371	0.067562
17	50	80	0.641026	0.625	0.506329	0.493671	0.063392
18	50	90	0.641026	0.703125	0.4769	0.5231	0.059708
19	50	100	0.641026	0.78125	0.450704	0.549296	0.056428

Formula Set I Prototype Cell is C9	Formula Set II Prototype Cell is D9	Formula Set III Prototype Cell is E9
=A9/**B2**	=B9/**B3**	=C9/(C9+D9)

Formula Set IV Prototype Cell is F9	Formula Set V Prototype Cell is G9
=D9/(C9+D9)	=E9***B5**

FORMULAS

	A	B	C	D
1	P0 (atm) =	0.125		
2	P (atm) =	0.119		
3				
4	X (solv) =	=B2/B1		
5				
6	SOLVENT		SOLUTE	
7	M.W. =	78		
8	grams =	7.8	grams =	0.625
9	moles =	=B8/B7	moles =	=(B9-B4*B9)/B4
10			M.W. =	=D8/D9

VALUES

	A	B	C	D
1	P0 (atm) =	0.125		
2	P (atm) =	0.119		
3				
4	X (solv) =	0.952		
5				
6	SOLVENT		SOLUTE	
7	M.W. =	78		
8	grams =	7.8	grams =	0.625
9	moles =	0.1	moles =	0.00504202
10			M.W. =	123.958333

Model 4.2

BUILD and verify Model 4.2.

ENTER data using cells B1 and B2 (vapor pressure of solvent and solution), B8 and B9 (molecular weight and grams of solvent), and D8 (grams of solute).

READ output from the model at cell D10 (molecular weight of solute).

Model 4.3

BUILD and verify Model 4.3.

ENTER data using cells B1 and B2 (vapor pressure of pure solvents) and A6–A16 and B6–B16 (grams of solvent A and solvent B).

READ output from the model at cells E6–E16, F6–F16, and G6–G16 (vapor pressure of solvent A, solvent B, and total solution).

FORMULAS

	A	B	C	D	E	F	G
1	P0.A (atm) =	0.1	[benzene]				
2	P0.B (atm) =	0.03	[methylbenzene]				
3							
4	|------moles------|		|------mole fraction------|		|------vapor pressure------|		
5	Solvent A	Solvent B	Solvent A	Solvent B	Solvent A	Solvent B	Total
6	0.5	0					
7	0.45	0.05					
8	0.4	0.1					
9	0.35	0.15	Formula Set I	Formula Set II	Formula Set III	Formula Set IV	Formula Set V
10	0.3	0.2					
11	0.25	0.25					
12	0.2	0.3					
13	0.15	0.35					
14	0.1	0.4					
15	0.05	0.45					
16	0	0.5					

VALUES

	A	B	C	D	E	F	G
1	P0.A (atm) =	0.1	[benzene]				
2	P0.B (atm) =	0.03	[methylbenzene]				
3							
4	|------moles------|		|------mole fraction------|		|------vapor pressure------|		
5	Solvent A	Solvent B	Solvent A	Solvent B	Solvent A	Solvent B	Total
6	0.5	0	1	0	0.1	0	0.1
7	0.45	0.05	0.9	0.1	0.09	0.003	0.093
8	0.4	0.1	0.8	0.2	0.08	0.006	0.086
9	0.35	0.15	0.7	0.3	0.07	0.009	0.079
10	0.3	0.2	0.6	0.4	0.06	0.012	0.072
11	0.25	0.25	0.5	0.5	0.05	0.015	0.065
12	0.2	0.3	0.4	0.6	0.04	0.018	0.058
13	0.15	0.35	0.3	0.7	0.03	0.021	0.051
14	0.1	0.4	0.2	0.8	0.02	0.024	0.044
15	0.05	0.45	0.1	0.9	0.01	0.027	0.037
16	0	0.5	0	1	0	0.03	0.03

Formula Set I
Prototype Cell is C6

=A6/(A6+B6)

Formula Set II
Prototype Cell is D6

=B6/(A6+B6)

Formula Set III
Prototype Cell is E6

=C6*B1

Formula Set IV
Prototype Cell is F6

=B2*D6

Formula Set V
Prototype Cell is G6

=E6+F6

FORMULAS

	A	B	C	D	E	F	G	H	I	J
1	P0.A (atm) =	0.3855	[chloroform]							
2	P0.B (atm) =	0.4566	[acetone]							
3										
4	|------------moles------------|		|--mole fraction--|	|----calc. vapor pressure---|		|--actual vapor pressure--|	
5	Solvent A	Solvent B	Solv. A	Solv. B	Solv. A	Solv. B	Total	Solv. A	Solv. B	Total
6	0.5	0	Form Set I	Form Set II	Form Set III	Form Set IV	Form Set V	0.3855	0	0.3855
7	0.45	0.05								
8	0.4	0.1						0.2921	0.0526	0.3447
9	0.35	0.15								
10	0.3	0.2						0.1908	0.1342	0.325
11	0.25	0.25								
12	0.2	0.3						0.1079	0.2421	0.35
13	0.15	0.35								
14	0.1	0.4						0.0461	0.3553	0.4013
15	0.05	0.45								
16	0	0.5						0	0.4566	0.4566

VALUES

	A	B	C	D	E	F	G	H	I	J
1	P0.A (atm) =	0.3855	[chloroform]							
2	P0.B (atm) =	0.4566	[acetone]							
3										
4	|------------moles------------|		|--mole fraction--|	|----calc. vapor pressure---|		|--actual vapor pressure--|	
5	Solvent A	Solvent B	Solv. A	Solv. B	Solv. A	Solv. B	Total	Solv. A	Solv. B	Total
6	0.5	0	1	0	0.3855	0	0.3855	0.3855	0	0.3855
7	0.45	0.05	0.9	0.1	0.347	0.0457	0.3926			
8	0.4	0.1	0.8	0.2	0.3084	0.0913	0.3997	0.2921	0.0526	0.3447
9	0.35	0.15	0.7	0.3	0.2699	0.137	0.4068			
10	0.3	0.2	0.6	0.4	0.2313	0.1826	0.4139	0.1908	0.1342	0.325
11	0.25	0.25	0.5	0.5	0.1928	0.2283	0.4211			
12	0.2	0.3	0.4	0.6	0.1542	0.2739	0.4282	0.1079	0.2421	0.35
13	0.15	0.35	0.3	0.7	0.1157	0.3196	0.4353			
14	0.1	0.4	0.2	0.8	0.0771	0.3653	0.4424	0.0461	0.3553	0.4013
15	0.05	0.45	0.1	0.9	0.0386	0.4109	0.4495			
16	0	0.5	0	1	0	0.4566	0.4566	0	0.4566	0.4566

Formula Set I Prototype Cell is C6	Formula Set II Prototype Cell is D6	Formula Set III Prototype Cell is E6
=A6/(A6+B6)	=B6/(A6+B6)	=C6*B1

Formula Set IV Prototype Cell is F6	Formula Set V Prototype Cell is G6
=B2*D6	=E6+F6

Model 4.4

BUILD and verify Model 4.4.

ENTER data using cells B1 and B2 (vapor pressure of pure solvents) and A6–A16 and B6–B16 (grams of solvent A and solvent B). Use cells H6–H16, I6–I16, and J6–J16 to enter experimental data for comparsion with ideal solution behavior.

READ output from the model at cells E6–E16, F6–F16, and G6–G16 (vapor pressure of solvent A, solvent B, and total solution).

NOTICE that xy-plots can be used effectively to compare and contrast ideal and actual solution behavior (e.g., Figure 4.7).

SIMPLE AND FRACTIONAL DISTILLATION

If several pure liquids have different vapor pressures at the same temperature, it is a consequence of Raoult's law that solutions of these liquids will have different compositions than their vapors.

Consider two pure liquids A and B that exhibit vapor pressures of P°_A and P°_B at 50°C. If these two liquids are combined into a solution such that each has a mole fraction of 0.5, the vapor pressure of A and B of this solution will be $0.5P^\circ_A$ and $0.5P^\circ_B$, respectively. From the ideal gas law, it should be clear that if P°_A has a value that is twice that of P°_B, then the number of molecules of A in the vapor will be twice the number of molecules of B.

In general, if X_A and X_B are mole fractions of A and B in a binary solution, then mole fractions of A and B in the vapor will be

$$X_A \text{ (vapor)} = \frac{P_A}{P_A + P_B} = \frac{P^\circ_A X_A}{P^\circ_A X_A + P^\circ_B X_B} \qquad (4.7)$$

$$X_B \text{ (vapor)} = \frac{P_B}{P_A + P_B} = \frac{P^\circ_B X_A}{P^\circ_A X_A + P^\circ_B X_B} \qquad (4.8)$$

where X_A (vapor) and X_B (vapor) are mole fractions of A and B in the vapor and P_A and P_B are vapor pressures of A and B above the solution. [There are many solutions that deviate significantly from ideal behavior (e.g., ethanol–water) and where this simple treatment would be inappropriate.]

As the temperature of the binary solution increases, the vapor pressures of the two substances A and B also increase until the total vapor pressure of the solution equals the pressure of the surrounding atmosphere (i.e., the solution boils). The differences in the vapor pressures of these substances at the boiling point of the solution can be used to separate the substances in a process called *distillation* (Figure 4.9):

- The solution to be purified is placed in a container and heated to its boiling point.

- The vapor is cooled in a condenser and collected. The condensed vapor is enriched for some components; the remaining solution is enriched for others.

- This process can be repeated as many times as desired.

If the boiling point of a solution is so high that some components can decompose, separations can be performed at lower temperatures by reducing the external pressure with a vacuum pump.[1] Vacuum distillations are performed routinely at pressures of 0.01 atm or less.

In this exercise, use your electronic spreadsheet to explore the consequences of both simple and fractional distillations. Simple distillations occur when the vapor above a boiling liquid is condensed once (Figure

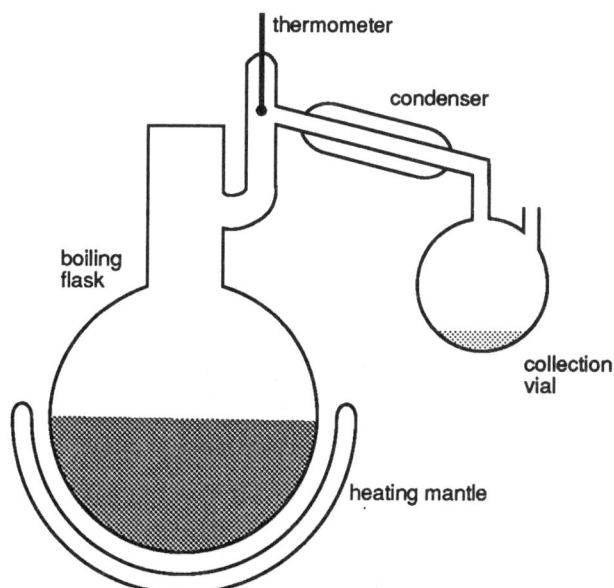

Figure 4.9

Distillation can separate solutions that are mixtures of liquids of different boiling points. In simple distillation, the solution is placed in a boiling flask and heated to its boiling point. The vapor passes up into a condensor, condensed to liquid, and collected in a collection flask. If the boiling points of the liquids in the original solution are very different, a single evaporation–condensation cycle can effect almost total separation. If the boiling points of the liquids in the original solution are similar but not identical, multiple evaporation–condensation cycles can still accomplish almost total separation. Many devices exist that can perform multiple evaporation–condensation cycles in a single experiment.

Figure 4.10

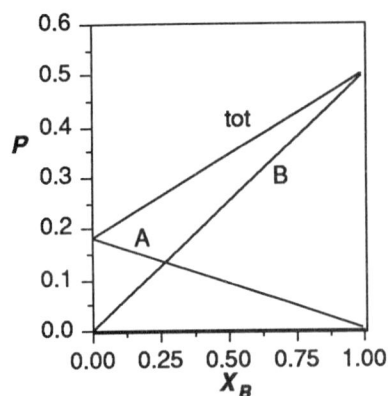

XY–PLOT

Data
Horizontal (x) axis: cells B7–B57
(mole fraction of B in solution)
Vertical (y) axis: cells C7–C57,
D7–D57, and E7–E57 (vapor pressures)

Figure 4.11

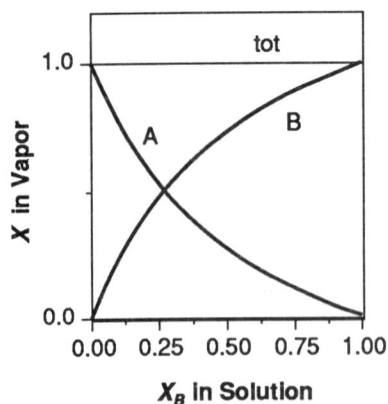

XY–PLOT

Data
Horizontal (x) axis: cells B7–B57
(mole fraction of B in solution)
Vertical (y) axis: cells F7–F57,
G7–G57, and H7–H57 (mole fractions
and sum of mole fractions in vapor)

4.9). Fractional distillations occur when the vapor above a boiling liquid is condensed repeatedly (by placing glass beads in a fractionation column, for example, so that hundreds of tiny condensations and vaporizations occur as the vapor rises through the column). Fractional distillation techniques can be used to cleanly separate liquids that differ in their boiling points by only a few degrees Celsius.

LESSONS

1. Build and verify Model 4.5. Model 4.5 is similar to Model 4.3 in that it calculates the equilibrium vapor pressures of two substances above a binary solution (columns C–E). It also, however, calculates the mole fraction of the two substances in the vapor (columns F–H).

    Because you will use this model to explore the behavior of a *boiling* solution, cells B2 and B3 of the model are used together to specify the *relative* vapor pressures of the two substances at the boiling point of the solution. (As the solution boils, its composition changes because its components depart at different rates. This changing composition changes the boiling point of the solution. It would also change the values in cells B2 and B3 if the dimensions of pressure were specified. Model 4.5, however, relies only on the ratio of the values in B2 and B3 — a values that changes more slowly with changes in temperature).[2]

    a. Explore the structure of Model 4.5 until you understand how it has been constructed.

    b. Build two on-screen graphics: A plot of vapor pressures vs the mole fraction of substance B in the solution (Figure 4.10) and a plot of mole fractions in the vapor vs the mole fraction of substance B in the solution (Figure 4.11).

2. Model 4.5 is initialized to represent a binary solution of toluene and benzene. (Toluene is represented by substance A, benzene by substance B.)

    a. If this solution is prepared such that it contains equal numbers of each molecule, which substance has the higher vapor pressure (Figure 4.10)? Which substance has the greater number of molecules in the vapor (Figure 4.11)? Why? (Use the ideal gas law in your explanation.)

b. Consider the simple distillation of binary solutions of toluene and benzene at several different compositions. Use your on-screen graphic of Figure 4.11 to determine the relationship between the mole fraction of benzene in the boiling solution and the mole fraction of benzene in the vapor.

3. Plot vapor pressure vs mole fraction of B for a boiling binary solution in which substance A is very much less volatile than substance B. Gradually increase the volatility of A until A and B have equivalent vapor pressures in their pure form. Discuss your results in terms of the suitability of simple distillation as a method of separating the components of binary solutions.

4. Build the on-screen graphic represented by Figure 4.12. This figure provides a different perspective on the distillation process and will make it easier for you to understand the nature of fractional distillation. The top (straight) line plots total vapor pressure (at equilibrium) vs X_B (mole fraction of B) in the *liquid*. The bottom (curved) line plots total vapor pressure (at equilibrium) vs X_B in the *vapor*. Use the guidance below to explore this graphic, using the values with which your model was initialized. Figure 4.13 gives you our answers.

a. Suppose that the mole fraction of substance B, X_B, in a binary solution of A and B originally has a value of 0.2. Print your on-screen graphic and mark the point on the liquid curve corresponding to $X_B = 0.2$ as number 1.

b. Extend a horizontal line from point 1 to where it intersects the curved (vapor) line of your plot. This is the mole fraction of B in the vapor under the same conditions. [The horizontal line moves from the composition of the liquid (at a specific total pressure) to the composition of the vapor (at the same total pressure).] Mark this point as number 2.

c. Extend a vertical line upward from point 2 to where it intersects the straight (liquid) line of your plot. Mark this point as number 3. If the vapor at point 2 were condensed into a liquid, this point would identify the mole fraction of B in this liquid.

d. If the liquid specified by point 3 is at equilibrium with its vapor phase, identify the composition of the vapor with the number 4.

Figure 4.12

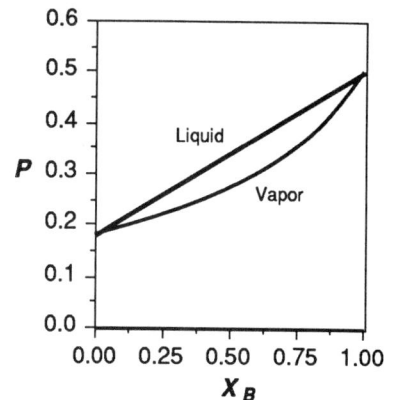

XY–PLOT

Data
Horizontal (x) axis: cells B7–B57 (mole fraction of B in solution)
Vertical (y) axis: cells E7–E57 (total vapor pressure of the solution)

Horizontal (x) axis: cells G7–G57 (mole fraction of B in vapor)
Vertical (y) axis: cells E7–E57 (total vapor pressure of the solution)

Note that P in this diagram is in arbitrary units (only relative vapor pressures are specified).

If your spreadsheet doesn't permit you to specify to different x-ranges, use total vapor pressure as your x-axis.

Figure 4.13

The pressure-composition diagram of a solution represents the conditions under which a liquid is in equilibrium with its vapor. The liquid line constitutes a plot of pressure vs composition of the liquid. The vapor line constitutes a plot of pressure vs composition of the vapor. Point 1, for example, represents a pressure and composition (mole fraction of B) at which the liquid is in equilibrium with its vapor. If a horizontal line is drawn to point 2, the composition of the vapor at the same pressure can be identified. If the vapor at point 2 is fully condensed into a liquid, point 3 is achieved (the vapor of which has the composition noted at point 4).

Notice that the composition of the solution changes as it "steps" across the pressure-composition diagram. Although this kind of separation is similar to what happens in fractional distillation, it is not the same. In fractional distillation, pressure is held constant and temperature changes (see Figure 4.14). Furthermore, if a device were built that actually fractionated materials in this fashion, the composition of the starting material would also gradually change.

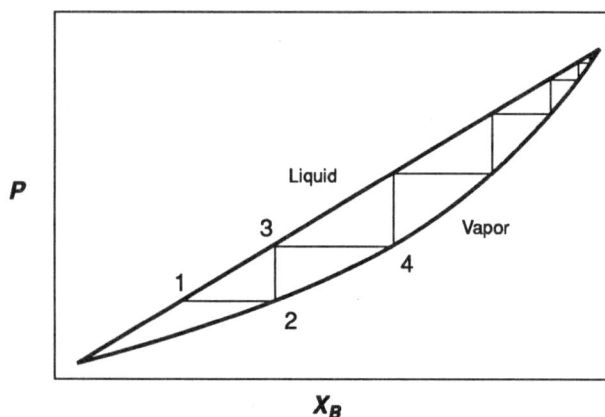

You can proceed in this fashion several times to follow the separation of the components of your original solution to any arbitrary degree of purity. Although a device could be built that would separate your binary solution into its components according to the preceding principles, such a device would not represent fractional distillation. At each step in this purification, the total vapor pressure of the binary solution changes (recall Figure 4.10). *Whether you are working at 1 atm or in a vacuum, distillations occur at constant pressure. In fractional distillation, it is the boiling point of the solution that changes, not the vapor pressure.*

Figure 4.12 represents a *pressure-composition diagram* of your binary solution. To follow the separation that occurs in fractional distillation, you must build a *temperature-composition diagram.*

5. To plot the temperature of a boiling liquid and its vapor as a function of composition, you must calculate vapor pressures of its components as a function of temperature (Equation 4.9):

$$P_1^\circ = P_0^\circ \exp\left[-\left(\frac{\Delta H_{vap}}{R}\right)\left(\frac{1}{T_1} - \frac{1}{T_0}\right)\right] \qquad (4.9)$$

where P_0° is the known vapor pressure at temperature T_0 and P_1° is the pressure calculated at temperature T_1. R is the universal gas constant, and ΔH_{vap} is the heat that must be absorbed by 1 mol of substance to vaporize. If T_0 is the boiling point of the pure substance in an open flask, then P_0° has a value of 1 atm. (You will learn more about Equation 4.9 in Exercise 4.3 and about ΔH in Chapter 8).

Once you know vapor pressure as a function of temperature, use Raoult's law and Dalton's law of partial pressures to determine the relationship between the boiling point of a solution and its composition. If a binary solution is brought to boiling in an environment exhibiting an external pressure P_{ext}, then

$$P_{ext} = P^\circ_A X_A + P^\circ_B X_B = P^\circ_A X_A + P^\circ_B (1 - X_A) \qquad (4.10)$$

It follows that

$$X_A = \frac{P_{ext} - P^\circ_B}{P^\circ_A - P^\circ_B} \qquad (4.11)$$

where X_A and X_B are the mole fractions of A and B in the vapor phase.

a. Build and verify Model 4.6. Model 4.6 calculates the temperature-composition diagram of a binary solution at its boiling point. The model has been built and initialized to represent two imaginary substances exposed to 1 atm of pressure such that substances A and B have boiling points of 150.1 and 80.1°C, respectively. Notice that columns C and D represent Equation 4.9 under the conditions specified and that column E represents Equation 4.11. Explore the model until you understand its structure.

b. Build the on-screen graphic represented by Figure 4.14. This figure is similar to Figure 4.13, except that temperature rather than pressure is now represented as a function of composition.

c. Suppose that the mole fraction of substance A, X_A, in a binary solution of A and B originally has a value of 0.2. Print your on-screen graphic and mark this point on your hardcopy as number 1.

d. Extend a horizontal line from point 1 to where it intersects the vapor line of your plot. This is the mole fraction of A in the vapor at the same temperature. Mark this intersection point as number 2.

e. Extend a vertical line down from point 2 to where it intersects the liquid line of your plot. Mark this point as number 3. If the vapor at point 2 were condensed into a liquid, this point would identify the mole fraction of A in this liquid. *Notice that to accomplish this the temperature of the vapor is lowered. Fractional distillation occurs because of a temperature gradient that exists along a fractionating column.*

f. As the liquid specified by point 3 again heats and vaporizes, identify the composition of this vapor with the number 4.

Figure 4.14

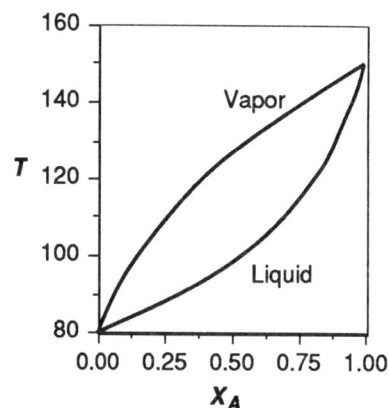

XY–PLOT

Data
Horizontal (x) axis: cells E10–E24 (mole fraction of A in the liquid)
Vertical (y) axis: cells A10–A24 (temperature)

Horizontal (x) axis: cells G10–G24 (mole fraction of A in the vapor)
Vertical (y) axis: cells A10–A24 (temperature)

Proceed in this fashion several times to follow the separation of the components of your original solution to any arbitrary degree of purity. These steps represent the separations that would occur at each step in a fractional distillation.

6. Explore Model 4.6 with different, imaginary binary solutions and determine the number of steps required to achieve 90% purity of one of the substances in the vapor. Observe especially the effect of the difference in boiling points on this process.

PROBLEM

You are now in a position to build a reasonably complete simulation of the simple distillation of a binary solution in which the contents of a boiling flask are transferred incrementally to a condensation flask (Figure 4.9). We propose a model built on the structure of Model 4.5, which assumes that the ratio of the vapor pressures of the two pure components of the solution does not change with temperature.[3] Using Model 4.5 as your starting point, attempt to build a model in which

1. At each step in the distillation, 1/100 of the liquid remaining in the boiling flask is transferred via distillation to a flask in which the vapor of the boiling solution is condensed (i.e., subtract the appropriate amounts of A and B from the boiling flask and add them to the condensation flask).

2. A running inventory of the contents of both flask is maintained.

You will also find it useful to calculate the temperature of the vapor just as it departs the surface of the boiling liquid. This temperature can be used to identify the composition of the remaining liquid and indicate when the distillation should be stopped. (In actual distillations, the temperature of the vapor as it enters the condensor provides a much sharper indicator, but we couldn't think of a way to make that calculation here.) Use your knowledge of stoichiometry and Raoult's law to first build a model that calculates the mole fractions of the components in the boiling and collection flasks as the distillation proceeds. Then, if you wish, add a column to your model that calculates the boiling point of the liquid in the boiling flask, using the Clapeyron equation.

Our answer to this problem is contained in Model 4.7 and it can be used to make a number of interesting observations:

a. When your boiling solution contains two substances with similar but not identical vapor pressures (say, 0.5 and 0.2 in arbitrary pressure units), plot the amounts and mole fractions of A and B as they change in both the boiling and reaction flasks during distillation. You should notice a relatively smooth progression in which the contents of the boiling flask are gradually enriched for the least volatile component and the contents of the condensate start out enriched for the most volatile component but gradually approach the contents of the original solution prior to distillation.

b. When your boiling solution contains two substances with very different vapor pressures (say, 0.5 and 0.02 in arbitrary pressure units) plot the amounts and mole fractions of A and B as they change in both the boiling flask and reaction flask during distillation. You should notice a specific point at which the composition of the boiling flask and condensation flask begins to change dramatically. If you were to monitor an actual distillation with a thermometer suspended in the vapor above the boiling liquid, how would you know when to stop the distillation?

    (The temperature change displayed by your model is not as good an endpoint indicator as the temperature change of the vapor farther away from the boiling liquid — say, at the point where the vapor enters the condensor. Do you know why?)

c. Write a report on your results and describe a procedure in which two substances of very different vapor pressures can be separated.

MODELS

Model 4.5

BUILD and verify Model 4.5.

ENTER data using cells B2 and B3 (relative vapor pressures of pure liquids).

READ output from the model at cells C7–C57, D7–D57, and E7–E57 (vapor pressures of components and solution) and F7–F57, G7–G57, and H7–H57 (mole fractions of components in vapor and sum of mole fractions).

FORMULAS

	A	B	C	D	E	F	G	H
1	Relative Vapor Pressures:							
2	P0(A) =	0.182						
3	P0(B) =	0.502						
4								
5	|---mole fraction---|		|---------vap. press.---------|			|---------mole fraction---------|		
6	Liq. A	Liq. B	Liq. A	Liq. B	Total	Vapor A	Vapor B	Total
7	1	0	Form Set I	Form Set II	Form Set III	Form Set IV	Form Set V	Form Set VI
8	0.98	0.02						
9	0.96	0.04						

	A	B
56	0.02	0.98
57	0	1

VALUES

	A	B	C	D	E	F	G	H
1	Relative Vapor Pressures:							
2	P0(A) =	0.182						
3	P0(B) =	0.502						
4								
5	|---mole fraction---|		|---------vap. press.---------|			|---------mole fraction---------|		
6	Liq. A	Liq. B	Liq. A	Liq. B	Total	Vapor A	Vapor B	Total
7	1	0	0.182	0	0.182	1	0	1
8	0.98	0.02	0.17836	0.01004	0.1884	0.94671	0.05329	1
9	0.96	0.04	0.17472	0.02008	0.1948	0.89692	0.10308	1

	A	B	C	D	E	F	G	H
56	0.02	0.98	0.00364	0.49196	0.4956	0.00734	0.99266	1
57	0	1	0	0.502	0.502	0	1	1

Formula Set I — Prototype Cell is C7: =A7*B2

Formula Set IV — Prototype Cell is F7: =C7/E7

Formula Set II — Prototype Cell is D7: =B7*B3

Formula Set V — Prototype Cell is G7: =D7/E7

Formula Set III — Prototype Cell is E7: =C7+D7

Formula Set VI — Prototype Cell is H7: =F7+G7

FORMULAS

	A	B	C	D	E	F	G	H
1	R [J/(mol K)]=	8.314						
2	T0 (°C) =	80.1						
3	ΔT (°C) =	5						
4			A	B				
5	b.p. =		150.1	80.1				
6	ΔH(J/mol) =		34000	34000				
7					Solution -->		Vapor ----->	
8	T	T	P(A)	P(B)	X(A)	X(B)	X(A)	X(B)
9	(°C)	K	(atm)	(atm)				
10	=B2							
11								
12								
13	Formula	Formula	Formula	Formula	Formula	Formula	Formula	Formula
14	Set	Set	Set	Set	Set	Set	Set	Set
15	I	II	III	IV	V	VI	VII	VIII
16								
17								
18								
19								
20								
21								
22								
23								
24								

VALUES

	A	B	C	D	E	F	G	H
1	R [J/(mol K)]=	8.314						
2	T0 (°C) =	80.1						
3	ΔT (°C) =	5						
4			A	B				
5	b.p. =		150.1	80.1				
6	ΔH(J/mol) =		34000	34000				
7					Solution -->		Vapor ----->	
8	T	T	P(A)	P(B)	X(A)	X(B)	X(A)	X(B)
9	(°C)	K	(atm)	(atm)				
10	80.1	353.25	0.14739	1	0	1	0	1
11	85.1	358.25	0.17324	1.17536	0.17499	0.82501	0.03032	0.96968
12	90.1	363.25	0.20272	1.37534	0.32008	0.67992	0.06489	0.93511
13	95.1	368.25	0.2362	1.60249	0.44097	0.55903	0.10416	0.89584
14	100.1	373.25	0.27408	1.85953	0.54214	0.45786	0.14859	0.85141
15	105.1	378.25	0.3168	2.14932	0.62718	0.37282	0.19869	0.80131
16	110.1	383.25	0.36479	2.47491	0.69897	0.30103	0.25498	0.74502
17	115.1	388.25	0.41852	2.83948	0.75982	0.24018	0.318	0.682
18	120.1	393.25	0.4785	3.2464	0.81159	0.18841	0.38835	0.61165
19	125.1	398.25	0.54524	3.69917	0.85581	0.14419	0.46662	0.53338
20	130.1	403.25	0.61927	4.20146	0.89372	0.10628	0.55345	0.44655
21	135.1	408.25	0.70117	4.75709	0.92632	0.07368	0.64951	0.35049
22	140.1	413.25	0.79151	5.37004	0.95446	0.04554	0.75547	0.24453
23	145.1	418.25	0.89091	6.04443	0.97883	0.02117	0.87206	0.12794
24	150.1	423.25	1	6.78452	1	0	1	0

Formula Set I Prototype Cell is A11	**Formula Set V** Prototype Cell is E10
=A10+ **B3**	=(1-D10)/(C10-D10)

Formula Set II Prototype Cell is B10	**Formula Set VI** Prototype Cell is F10
=A10+273.15	=1-E10

Formula Set III Prototype Cell is C10	**Formula Set VII** Prototype Cell is G10
=EXP(-(**C6** /**B1**)*(1/B10-1/(**C5** +273.15)))	=E10*C10

Formula Set IV Prototype Cell is D10	**Formula Set VIII** Prototype Cell is H10
=EXP(-(**D6** /**B1**)*(1/B10-1/(**D5** +273.15)))	=F10*D10

Model 4.6

BUILD and verify Model 4.6.

ENTER data using cells B1–B3 (gas constant, starting and incremental temperatures), C5 and D5 (boiling points of A and B), and C6 and D6 (enthalpy of vaporization of substances A and B).

READ output from columns A–H (below row 9). Columns A and B contain temperature in °C and K, C and D are the vapor pressures of each component of the solution at the specified temperature, E and F are the mole fractions of the solution components, and G and H are the mole fractions of the vapor components.

Model 4.7

BUILD and verify Model 4.7.

ENTER data using cells B2 and B3 (starting amounts of liquids A and B), E2 and E3 (relative vapor pressures of A and B), E4 (temperature at which the vapor pressures are specified), and I2 (enthalpy of vaporization for the two components of the binary solution). Recall that the model assumes that ΔH is identical for both components.

READ output from columns A–T (below row 7). The contents of these columns, respectively, are fraction distilled, amounts of A and B remaining in the boiling flask, mole fractions of A and B in the boiling flask, vapor pressures of A and B above the boiling solution, mole fractions of A and B in the vapor, fraction distilled (repeated with the one row offset required by the model), amounts condensed (total, A, and B) in this step of the distillation, total amounts condensed (total, A, and B), and mole fractions of A and B in the condensate, temperature of the vapor immediately above the boiling liquid.

NOTICE that the model assumes that the distillation proceeds in steps and that each step transfers 1/100 of the remaining contents of the boiling flask to the condensation flask.

FORMULAS

	A	B	C	D	E	F	G	H	I	J
1	Initial Amounts			Relative Vapor Pressures:						
2	Liq. A =	0.2	mol	P0(A) =	0.02			ΔH =	7355	
3	Liq. B =	0.2	mol	P0(B) =	0.5			ΔH/R =	3700.1	
4				at T =	50	°C				
5										
6	Fract.	|------------boiling flask------------|				|------------rising vapor------------|				
7	Dist.	|------amt------|		|-mole fraction-|		|-vap. press.--|		|-mole fraction-|		
8		Liq. A	Liq. B	Liq. A	Liq. B	Liq. A	Liq. B	Vap. A	Vap. B	
9	0	=B2	=B3	Form Set IV	Form Set V	Form Set VI	Form Set VII	Form Set VIII	Form Set IX	
10	Form Set I	Form Set II	Form Set III							
11										

FORMULAS (continued)

	K	L	M	N	O	P	Q	R	S	T
1										
2										
3										
4										
5										
6	Fract.	|----------------------------condensation flask----------------------------|								
7	Dist.	|---amount condensed---|		|------total amounts-------|		|-mole fraction--|			
8		Tot	Liq. A	Liq. B	Tot	Liq. A	Liq. B	Liq. A	Liq. B	Temp
9	Form Set X	Form Set XI	Form Set XII	Form Set XIII	=L9	=M9	=N9	Form Set XVII	Form Set XVIII	Form Set XIX
10					Form Set XIV	Form Set XV	Form Set XVI			
11										

VALUES

	A	B	C	D	E	F	G	H	I	J
1	Initial Amounts			Relative Vapor Pressures:						
2	Liq. A =	0.2	mol	P0(A) =	0.02			ΔH =	7355	
3	Liq. B =	0.2	mol	P0(B) =	0.5			ΔH/R =	3700.1	
4				at T =	50	°C				
5										
6	Fract.	|------------boiling flask------------|				|------------rising vapor------------|				
7	Dist.	|------amt------|		|-mole fraction-|		|-vap. press.--|		|-mole fraction-|		
8		Liq. A	Liq. B	Liq. A	Liq. B	Liq. A	Liq. B	Vap. A	Vap. B	
9	0	0.2	0.2	0.5	0.5	0.01	0.25	0.0385	0.9615	
10	0.01	0.1998	0.1962	0.5047	0.4953	0.0101	0.2477	0.0392	0.9608	
11	0.0199	0.1997	0.1923	0.5094	0.4906	0.0102	0.2453	0.0399	0.9601	

VALUES (continued)

	K	L	M	N	O	P	Q	R	S	T
1										
2										
3										
4										
5										
6	Fract.	|----------------------------condensation flask----------------------------|								
7	Dist.	|---amount condensed---|		|------total amounts-------|		|-mole fraction--|			
8		Tot	Liq. A	Liq. B	Tot	Liq. A	Liq. B	Liq. A	Liq. B	Temp
9	0.01	0.004	0.0002	0.0038	0.004	0.0002	0.0038	0.0385	0.9615	93.045
10	0.0199	0.004	0.0002	0.0038	0.008	0.0003	0.0077	0.0388	0.9612	93.358
11	0.0297	0.0039	0.0002	0.0038	0.0119	0.0005	0.0114	0.0392	0.9608	93.677

Formula Set I Prototype Cell is A10
=O9/(**B2**+**B3**)

Formula Set II Prototype Cell is B10
=B9-M9

Formula Set III Prototype Cell is C10
=C9-N9

Formula Set IV Prototype Cell is D9
=B9/(B9+C9)

Formula Set V Prototype Cell is E9
=C9/(B9+C9)

Formula Set VI Prototype Cell is F9
=D9***E2**

Formula Set VII Prototype Cell is G9
=E9***E3**

Formula Set VIII Prototype Cell is H9
=F9/(F9+G9)

Formula Set IX Prototype Cell is I9
=G9/(F9+G9)

Formula Set X Prototype Cell is K9
=O9/(**B2**+**B3**)

Formula Set XI Prototype Cell is L9
=(B9+C9)/100

Formula Set XII Prototype Cell is M9
=H9*L9

Formula Set XIII Prototype Cell is N9
=I9*L9

Formula Set XIV Prototype Cell is O10
=O9+L10

Formula Set XV Prototype Cell is P10
=P9+M10

Formula Set XVI Prototype Cell is Q10
=Q9+N10

Formula Set XVII Prototype Cell is R9
=P9/O9

Formula Set XVIII Prototype Cell is S9
=Q9/O9

Formula Set XIX Prototype Cell is T9
=**I3**/(LN(F9+G9)+(**I3**/(273+**E4**)))-273

Model 4.7 (continued)

PHASE DIAGRAMS

Chapter 3 introduced the behavior of gases, and early exercises in this chapter introduced the behavior of liquids and solutions and the manner in which they interact with gases. Although we do not devote a separate chapter to the properties of solids, this last exercise in Chapter 4 may offer some insights into the manner in which solids, liquids, and gases coexist in the physical world. You will build and explore an electronic spreadsheet model that summarizes in a *phase diagram* what is known about the equilibrium between the solid, liquid, and gas phases of a pure substance.

Pure compounds can exist in one of three states: solid, liquid, or gas.[1] Recall that some molecules in a pure liquid can escape the liquid into the vapor phase and vice versa. The equilibrium that exists between a liquid and its vapor is just one example of a family of equilibria that exists between each of the phases of matter (Figure 4.15). Not all these equilibria, however, can exist simultaneously. Which equilibria exist at a given moment is a function of pressure and temperature. This can be seen clearly in the pressure–temperature or phase diagram of a pure substance (Figure 4.16).

- Consider a solid at "low" temperature. A solid at many "low" temperatures is in equilibrium with its vapor but cannot become a liquid. At equilibrium, the solid exhibits a characteristic vapor pressure that is determined by the rates at which molecules leave and return to the solid. As the temperature of the solid increases, the vapor pressure of the solid also increases (Figure 4.17a). The specific temperature–pressure combinations at which a solid is in equilibrium with its vapor are *sublimation points* (Figures 4.16 and 4.17a).

- A liquid at "high" temperature behaves in a similar fashion. Over a wide temperature range, a liquid exhibits a characteristic vapor pressure (i.e., it is in equilibrium with its vapor phase) but cannot become a solid (Figure 4.17b). The specific temperature–pressure combinations at which a solid is in equilibrium with its vapor are *boiling points* (Figures 4.16 and 4.17b).

- There is a very narrow temperature range over which a solid is in equilibrium with its liquid. Very large changes in pressure are required to effect quite small changes in this equilibrium temperature (Figure 4.17c). The specific temperature–pressure combinations at which a solid is in equilibrium with its liquid are *melting points* (Figures 4.16 and 4.17c).

If the three curves described in Figure 4.17a–c are combined, they yield the phase diagram (Figure 4.16). Notice that there is a specific temperature and pressure at which a solid is in equilibrium with both its liquid and its vapor. This is the same temperature and pressure at which a liquid is in equilibrium with both its solid and its vapor. This remarkable point, the *triple point*, is the only temperature and pressure at which solid, liquid, and vapor coexist in equilibrium.

Figure 4.15

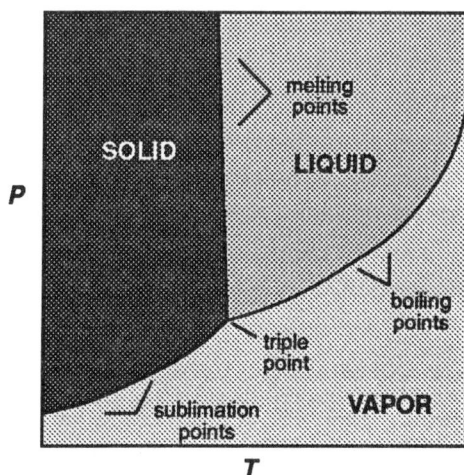

Molecules present in a solid, under some conditions of temperature and pressure, leave the solid phase and enter the liquid phase. Under the same conditions, some molecules of the liquid return to the solid. When the rates at which molecules leave and return to the solid are the same, the system is said to be in *dynamic equilibrium*.

Other equilibria exist between a solid and its vapor and a liquid and its vapor. The temperatures and pressures at which these equilibria are possible are shown in Figure 4.16.

Figure 4.16

The equilibria shown in Figure 4.15 are possible only at certain pressures and temperatures, and *phase diagrams* are used to display these relationships. In this figure, for example, solid and liquid are at equilibrium with each other only at the pressures and temperatures marked "melting points." Solid and vapor are at equilibrium at the pressures and temperatures marked "sublimation points," and liquid and vapor are at equilibrium at "boiling points."

There is a great deal of information to be obtained from phase diagrams, and this exercise offers the opportunity to explore them in some detail.

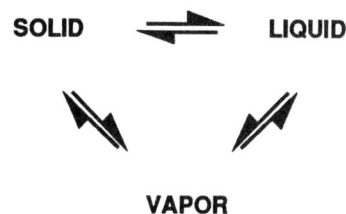

Figure 4.17

Pieces of phase diagrams can be explored separately to gain insight into their structures.

(a) A solid is in equilibrium with its vapor when the rate at which molecules leave the solid is the same as the rate at which they return. For a given temperature, the pressure (P) at which the solid is in equilibrium with its vapor is its *vapor pressure*. Notice that vapor pressure increases as temperature (T) rises.

(b) A liquid is in equilibrium with its vapor when the rate at which molecules leave the liquid is the same as the rate at which they return. For a given temperature, the pressure (P) at which the liquid is in equilibrium with its vapor is called its *vapor pressure*. Vapor pressure increases as temperature (T) rises.

(c) A solid is in equilibrium with its liquid when the rate at which molecules leave the solid is the same as the rate at which they return. Notice that changes in the pressure (P) applied to a solid can change slightly the temperature (T) at which it melts. In the diagram as shown, increases in pressure reduce the melting point. Water behaves in this fashion. Perhaps you know that the pressure of ice skate blades causes the ice to melt and form a thin film of liquid water with reduced frictional resistance.

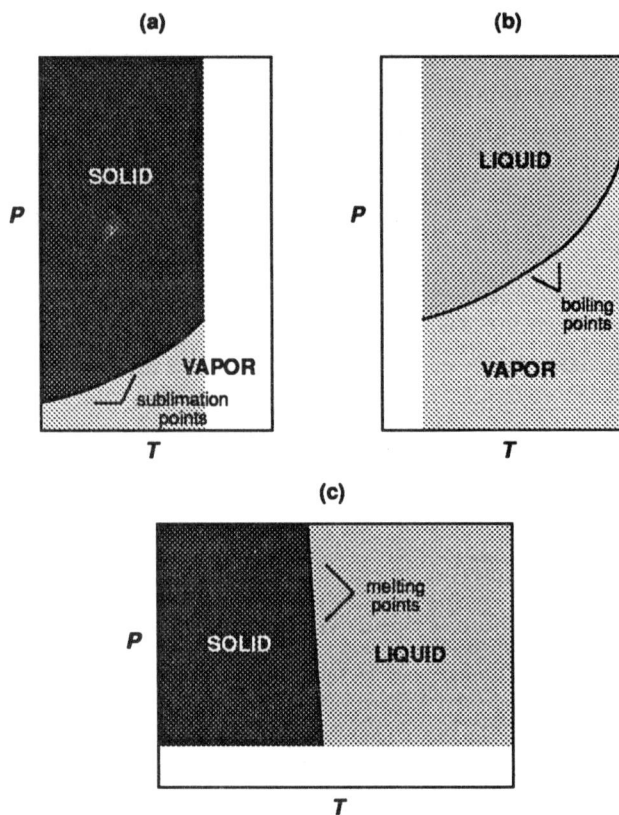

Each equilibrium described above can be calculated by using thermodynamic theory. Clapeyron, in the midnineteenth century, derived an equation that describes the effect of temperature changes on the equilibrium vapor pressure of a one-component system in which two phases are at equilibrium (Equation 4.12).

$$\left(\frac{dP}{dT}\right)_{eq} = \frac{\Delta H}{T\Delta V} \tag{4.12}$$

where $(dP/dT)_{eq}$ is the change in equilibrium pressure that occurs with a change in equilibrium temperature and ΔH, T, and ΔV are the change in enthalpy (heat), temperature (in K), and change in substance volume, respectively.

The derivation of the Clapeyron equation is beyond the scope of this book, but you can find it in any textbook of physical chemistry. Picture the thought processes involved in deriving this equation by considering a liquid in equilibrium with its vapor:

• If the temperature of a pure liquid substance rises by some small amount dT, the rate at which molecules of the substance leave the liquid and enter the vapor also increases. Equilibrium can be restored only when the vapor pressure increases to some point

$(P + dP)$ such that the rate at which molecules entering the liquid equals the rate at which they depart the liquid.

- According to thermodynamic theory, two systems in equilibrium with each other have equivalent values of a quantity known as Gibbs free energy. If the Gibbs free energy of a liquid is changed by some amount dG because of a rise in temperature, then the Gibbs free energy of its vapor must be changed by an equivalent amount. (You will explore the nature of Gibbs free energy in Exercise 8.3).

- The Clapeyron equation is obtained from the necessity of assuring that the free energy of two different phases remain the same.

The methods needed to calculate the changes in enthalpy and volume (ΔH and ΔV) depend on the type of transition:

- The temperature–pressure values that specify the equilibrium between a solid and its liquid (Figure 4.17c) can be obtained from the Clapeyron Equation by rearrangement and integration between limits (Equation 4.13):

$$P_1 = P_0 + \left(\frac{\Delta H_{fus}}{\Delta V_{fus}}\right)\ln\left(\frac{T_1}{T_0}\right) \qquad (4.13)$$

where P_0 and T_0 are the equilibrium pressure and temperature of a known state, P_1 is the calculated equilibrium pressure at a new temperature T_1, and ΔH_{fus}, ΔV_{fus} are the changes in enthalpy (heat) and volume of the substance that accompany melting. Note that temperature T must be stated in some absolute scale such as K.

- The temperature–pressure values that specify the equilibrium between a liquid and its vapor (Figure 4.17b) can be obtained from the Clapeyron equation by first assuming that the vapor is an ideal gas and that the volume of the liquid is negligible compared to the volume of the gas. In this case, ΔV is a consequence only of changes in the volume of the vapor, and, from the ideal gas law,

$$P\Delta V = nR\Delta T \qquad (4.14)$$

or, for 1 mol of substance[2]

$$\Delta V = \frac{R\Delta T}{P} \qquad (4.15)$$

Inserting Equation 4.15 into the Clapeyron equation (Equation 4.12), rearranging, and integrating yield the Clausius–Clapeyron equation for the equilibrium of a liquid and its vapor (Equations 4.16):

$$P_1 = P_0 \exp\left[-\left(\frac{\Delta H_{vap}}{R}\right)\left(\frac{1}{T_1} - \frac{1}{T_0}\right)\right] \tag{4.16}$$

where ΔH_{vap} is the change in heat in a substance that accompanies vaporization.

- The temperature–pressure values that specify the equilibrium between a solid and its vapor (Figure 4.17a) can be obtained from the Clapeyron equation using the same simplifying assumptions described for vaporization. In this case, however, it is the change in the heat of a substance that accompanies sublimation (ΔH_{sub}) that is relevant (Equation 4.17).

$$P_1 = P_0 \exp\left[-\left(\frac{\Delta H_{sub}}{R}\right)\left(\frac{1}{T_1} - \frac{1}{T_0}\right)\right] \tag{4.17}$$

Even if you find the Clapeyron equation to be something of a "black box," perhaps you will find insight into the nature of phase diagrams by using them in the lessons that follow. As you advance in your study of chemistry, this experience will be valuable in helping you understand the underlying thermodynamic theory of phase phenomena.

LESSONS

In the following lessons, slowly build the phase diagram for carbon dioxide (CO_2) piece by piece.

1. Figure 4.18 uses the Clapeyron equation (Equation 4.13) to calculate the values of pressure and temperature at which solid carbon dioxide (dry ice) and liquid CO_2 are in equilibrium.

   a. Cells B1 and B2 represent the values for pressure and temperature, respectively, at which the solid, liquid, and vapor phases of CO_2 can coexist at equilibrium (i.e., the triple point). Cells B4 and E3 represent the change in volume and change in internal energy, respectively, that occur in 1 mol of CO_2 during its transition from the solid to the liquid phase at constant pressure. Cell B8 defines the change in temperature desired for each row of the table in cells A10–B19.

	A	B	C	D	E	F
1	P (trip. pnt) =	5.1	atm			
2	T (trip. pnt) =	216.5	K			
3				ΔH(fus) =	57.6	J/mol
4	ΔV(fus) =	0.0000207	m^3/mol			
5						
6						
7	SOLID-LIQUID					
8	ΔT =	0.001				
9	T (K)	P (atm)				
10	216.5	5.1				
11	216.501	17.9526664				
12	216.502	30.8052734				
13	216.503	43.657821				
14	216.504	56.5103093				
15	216.505	69.3627382				
16	216.506	82.2151078				
17	216.507	95.0674179				
18	216.508	107.919669				
19	216.509	120.77186				
20						
21						

Figure 4.18

Equation 4.13 can be used to build an electronic spreadsheet model that calculates the values of pressure and temperature at which solid carbon dioxide (dry ice) and liquid CO_2 are in equilibrium. Can you build this model?

Cell A10 defines the beginning temperature in a range of temperatures for which the equilibrium pressures of solid–liquid CO_2 are calculated. It is taken as the temperature of the triple point of CO_2 and so has the formula =B2. Cells A11 and below each increment the value of the temperature in the cell above by an amount equal to the value of ΔT in cell B8. The formula in cell B11, for example, is =A10+B8. What are the formulas in the other cells in this column?

Cells B10 and below represent Equation 4.13. What formulas belongs in these cells?

b. Build and verify the model that is implied by Figure 4.18. Our answer is found in a subset of the cells in Model 4.8.

c. Build an on-screen graphic that plots pressure vs temperature for states at which the solid and liquid phases of CO_2 are in equilibrium (Figure 4.19). Notice that, for dry ice, the pressure of the system must increase as the temperature increases. Does this seem reasonable? Can you think of examples in which the opposite occurs?

2. Figure 4.20 uses the Clausius–Clapeyron equation (Equation 4.16) to calculate the values of pressure and temperature at which solid carbon dioxide is in equilibrium with its vapor.

a. Cells B1 and B2 represent the values for pressure and temperature, respectively, at which the solid, liquid, and vapor phases of CO_2 can coexist at equilibrium (i.e., the triple point). Cells E1 and E5 represent the universal gas constant and change in internal energy, respectively, that occur in 1 mol of CO_2 during

Figure 4.19

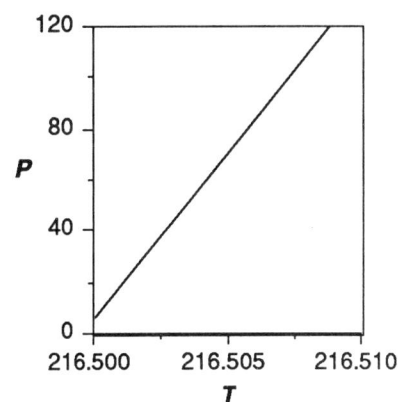

P vs *T* plot; horizontal axis T from 216.500 to 216.510, vertical axis P from 0 to 120.

XY–PLOT

Data
Horizontal (x) axis: cells A10–A19 (temperature)
Vertical (y) axis: cells B10–B19 (pressure)

Figure 4.20

Equation 4.16 can be used to build an electronic spreadsheet model that calculates the values of pressure and temperature at which solid carbon dioxide (dry ice) is in equilibrium with its vapor. Can you build this model?

	A	B	C	D	E	F
1	P (trip. pnt) =	5.1	atm	R =	8.314	J/(mol K)
2	T (trip. pnt) =	216.5	K			
3						
4						
5				ΔH(sub) =	10410	J/mol
6						
7			SOLID-VAPOR			
8			ΔT =	-20		
9			T (K)	P (atm)		
10			216.5	5.1		
11			196.5	2.83091474		
12			176.5	1.37514099		
13			156.5	0.55541746		
14			136.5	0.17199418		
15			116.5	0.03561293		
16			96.5	0.00383899		
17						
18						
19						
20						
21						

its transition from the solid to the vapor phase. Cell D8 defines the change in temperature desired for each row of the table in cells C10–B16.

Cells C10 and below calculate the range of temperatures to be examined. What formulas belong in these cells? (HINT: The same procedure was used in column A.)

Cells D10 and below represent Equation 4.16. What formulas belong in these cells?

b. Build and verify the model that is implied by Figure 4.20. Our answer is found in a subset of the cells in Model 4.8.

c. Build an on-screen graphic that plots pressure vs temperature for states at which the solid and vapor phases of CO_2 are in equilibrium (Figure 4.21). As the temperature of a block of dry ice increases, what happens to its vapor pressure? Use your knowledge of kinetic molecular theory to explain what is happening.

Figure 4.21

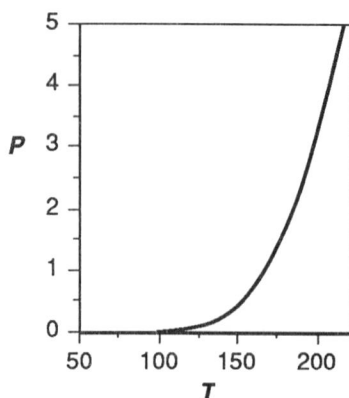

XY–PLOT

Data
Horizontal (x) axis: cells C10–C16 (temperature)
Vertical (y) axis: cells D10–D16 (pressure)

3. Figure 4.22 uses the Clausius–Clapeyron equation (Equation 4.17) to calculate the values of pressure and temperature at which liquid carbon dioxide is in equilibrium with its vapor.

a. Build and verify the model represented by Figure 4.22. Our answer is found in a subset of the cells in Model 4.8.

b. Build an on-screen graphic that plots pressure vs temperature for states at which the liquid and vapor phases of CO_2 are in equilibrium (Figure 4.23). Examine the temperature range at

	A	B	C	D	E	F
1	P (trip. pnt) =	5.1	atm		R =	8.314	J/(mol K)
2	T (trip. pnt) =	216.5	K				
3							
4					ΔH(vap) =	10370	J/mol
5							
6							
7					LIQUID-VAPOR		
8						ΔT =	20
9						T (K)	P (atm)
10						216.5	5.1
11						236.5	8.30155826
12						256.5	12.5242769
13						276.5	17.8036667
14						296.5	24.1356084
15						316.5	31.4851507
16						336.5	39.795134
17						356.5	48.9937308
18						376.5	59.0005835
19						396.5	69.7315425
20						416.5	81.1021645
21						436.5	93.0301825

Figure 4.22

Equation 4.17 can be used to build an electronic spreadsheet model that calculates the values of pressure and temperature at which liquid carbon dioxide is in equilibrium with its vapor. Can you build this model?

which liquid CO_2 is in equilibrium with its vapor phase and compare it to the temperature range at which solid CO_2 is in equilbrium with its vapor phase (Lesson 2). Compare and contrast the conditions under which liquid and solid CO_2 are in equilibrium with their vapor.

4. Model 4.8 is a model that combines the calculations built in Lessons 1–3. Build and verify Model 4.8. (If you are proficient with the editing and exporting features of your spreadsheet program, assemble Model 4.8 from the pieces that you have already built.)

    a. If your spreadsheet program supports *xy*-plots in which each data set can have a different *x*-range (e.g., Microsoft Excel), build the on-screen graphic specified in Figure 4.24.

    b. If your spreadsheet program supports *xy*-plots in which only one *x*-range may be specified for your data sets (e.g., Lotus 1-2-3), edit your model to correspond to the structure shown in Figure 4.25 and then build an on-screen graphic similar to Figure 4.24.

You have now built an electronic spreadsheet representation of the phase diagram of carbon dioxide. Although correct thermodynamic values for CO_2 were used to generate your phase diagram, the results generated by the model are only approximately correct. This is because CO_2 vapor is not an ideal gas and the assumptions used in the derivation of the Clausius–Clapeyron equation are not strictly valid.

Figure 4.23

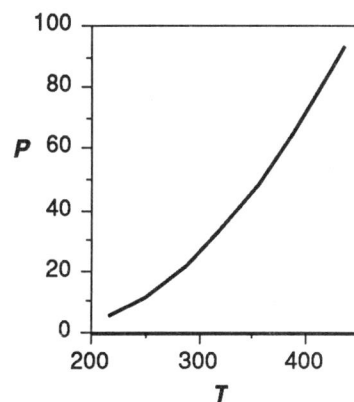

XY–PLOT

Data
Horizontal (*x*) axis: cells E10–E21 (temperature)
Vertical (*y*) axis: cells F10–F21 (pressure)

Figure 4.24

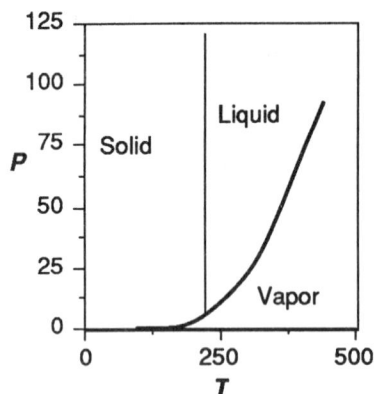

XY–PLOT

Data
Horizontal (*x*) axis: cells A10–A19 (temperature)
Vertical (*y*) axis: cells B10–B19 (pressure)

Data
Horizontal (*x*) axis: cells C10–C16 (temperature)
Vertical (*y*) axis: cells D10–D16 (pressure)

Data
Horizontal (*x*) axis: cells E10–E21 (temperature)
Vertical (*y*) axis: cells F10–F21 (pressure)

This plot can be created in this fashion on spreadsheets that support *xy*-plots capable of multiple *x*-ranges. If your program uses the same *x*-range for all *xy*-data build the model represented by Figure 4.25 and use the plot parameters below.

Data
Horizontal (*x*) axis: cells A9–A37 (temperature)
Vertical (*y*) axis: cells B9–B37 (pressure)

Figure 4.25

Some electronic spreadsheet programs can specify only a single *x*-range for the data in an *xy*-plot. If your program has this feature (e.g., Lotus 1-2-3), edit Model 4.8 by moving all data to be plotted into two single columns as shown.

c. What phase of CO_2 is stable at room temperature and pressure? If you have a printer available, print your phase diagram and mark the location where $P = 1$ atm and $T = 300$ K.

d. At a pressure of 1 atm, what happens to a piece of dry ice as its temperature slowly increases (assuming ideal behavior)? Does the dry ice ever melt? Why or why not?

e. Suppose that your piece of dry ice is placed in an container along with CO_2 vapor at 10 atm of pressure and the temperature is slowly raised from 200 K to 300 K to 500 K. Trace the process on your phase diagram and describe the changes that occur in the container.

f. Water ice melts when the pressure increases over a relatively wide range. What happens to dry ice? Can you explain the difference?

g. Trace other paths through your phase diagram and use kinetic molecular theory to describe what is happening at each step along the path.

	A	B	C	D	E	F
1	P (trip. pnt) =	5.1	atm	R =	8.314	J/(mol K)
2	T (trip. pnt) =	216.5	K			
3				ΔH(fus) =	57.6	J/mol
4	ΔV(fus) =	0.0000207	m^3/mol	ΔH(vap) =	10370	J/mol
5				ΔH(sub) =	10410	J/mol
6						
7	SOLID-LIQUID					
8	T (K)	P (atm)	ΔT =	0.001		
9	216.5	5.1				
10	216.501	17.9526664				
11	216.502	30.8052734				
12	216.503	43.657821				
13	216.504	56.5103093				
14	216.505	69.3627382				
15	216.506	82.2151078				
16	216.507	95.0674179				
17	216.508	107.919669				
18	216.509	120.77186				
19						
20	216.5	5.1	ΔT =	-20	SOLID-VAPOR	
21	196.5	2.83091474				
22	176.5	1.37514099				
23	156.5	0.55541746				
24	136.5	0.17199418				
25	116.5	0.03561293				
26	96.5	0.00383899				
27						
28	216.5	5.1	ΔT =	20	LIQUID-VAPOR	
29	236.5	8.30155826				
30	256.5	12.5242769				
31	276.5	17.8036667				
32	296.5	24.1356084				
33	316.5	31.4851507				
34	336.5	39.795134				
35	356.5	48.9937308				
36	376.5	59.0005835				
37	396.5	69.7315425				
38	416.5	81.1021645				
39	436.5	93.0301825				

MODEL

FORMULAS

	A	B	C	D	E	F
1	P (trip. pnt) =	5.1	atm	R =	8.314	J/(mol K)
2	T (trip. pnt) =	216.5	K			
3				ΔH(fus) =	57.6	J/mol
4	ΔV(fus) =	0.0000207	m^3/mol	ΔH(vap) =	10370	J/mol
5				ΔH(sub) =	10410	J/mol
6						
7	SOLID-LIQUID		SOLID-VAPOR		LIQUID-VAPOR	
8	ΔT =	0.001	ΔT =	-20	ΔT =	20
9	T (K)	P (atm)	T (K)	P (atm)	T (K)	P (atm)
10	=B2		=B2		=B2	
11	=A10+B8		=C10+D8		=E10+F8	
12	=A11+B8	Formula	=C11+D8	Formula	=E11+F8	Formula
13	=A12+B8	Set	=C12+D8	Set	=E12+F8	III
14	=A13+B8	I	=C13+D8	II	=E13+F8	
15	=A14+B8		=C14+D8		=E14+F8	
16	=A15+B8		=C15+D8		=E15+F8	
17	=A16+B8				=E16+F8	
18	=A17+B8				=E17+F8	
19	=A18+B8				=E18+F8	
20					=E19+F8	
21					=E20+F8	

VALUES

	A	B	C	D	E	F
1	P (trip. pnt) =	5.1	atm	R =	8.314	J/(mol K)
2	T (trip. pnt) =	216.5	K			
3				ΔH(fus) =	57.6	J/mol
4	ΔV(fus) =	0.0000207	m^3/mol	ΔH(vap) =	10370	J/mol
5				ΔH(sub) =	10410	J/mol
6						
7	SOLID-LIQUID		SOLID-VAPOR		LIQUID-VAPOR	
8	ΔT =	0.001	ΔT =	-20	ΔT =	20
9	T (K)	P (atm)	T (K)	P (atm)	T (K)	P (atm)
10	216.5	5.1	216.5	5.1	216.5	5.1
11	216.501	17.9526664	196.5	2.83091474	236.5	8.30155826
12	216.502	30.8052734	176.5	1.37514099	256.5	12.5242769
13	216.503	43.657821	156.5	0.55541746	276.5	17.8036667
14	216.504	56.5103093	136.5	0.17199418	296.5	24.1356084
15	216.505	69.3627382	116.5	0.03561293	316.5	31.4851507
16	216.506	82.2151078	96.5	0.00383899	336.5	39.795134
17	216.507	95.0674179			356.5	48.9937308
18	216.508	107.919669			376.5	59.0005835
19	216.509	120.77186			396.5	69.7315425
20					416.5	81.1021645
21					436.5	93.0301825

Formula Set I Prototype Cell is B10	Formula Set II Prototype Cell is D10
=B1+(E3/B4)*LN(A10/B2)	=B1*EXP(-(E5/E1)*(1/C10-1/B2))

Formula Set III Prototype Cell is F10
=B1*EXP(-(E4/E1)*(1/E10-1/B2))

Model 4.8

BUILD and verify Model 4.8.

ENTER data using cells B1 and B2 (pressure and temperature values at the triple point), B4 (the change in volume of 1 mol of substance that occurs on melting), E1 (universal gas constant), E3–E5 (change in heat of 1 mol of substance on melting, vaporization, and sublimation), and B8, D8, and F8 (the temperature change desired in the table below each of these values).

READ output from the model at cells A10–B19 (pressure–temperature values at which solid is in equilibrium with its liquid), C10–D16 (pressure–temperature values at which a solid is in equilibrium with its vapor), and E10–F21 (pressure–temperature values at which liquid is in equilibrium with its vapor).

NOTICE that spreadsheet programs that permit only a single x-range in xy-plots cannot use this model to plot all components of a phase diagram simultaneously. The model, however, can be edited to correspond to Figure 4.25 and then a plot similar to Figure 4.24 can be generated.

CHAPTER 5

CHEMICAL EQUILIBRIUM

The concept of *dynamic equilibrium* is central to much of chemistry and deserves a careful introduction:

- The number of people in your college bookstore is in equilibrium when the number of people entering the store in one minute (the rate of entry) equals the number of people departing the store in one minute (the rate of departure). If you tracked the *names* of the people in the store, you would discover that the names are constantly changing (i.e., the process is *dynamic*). If you tracked the *number* of people in the store, you would discover that this number is unchanging (i.e., the process is in equilibrium).

- A liquid is in equilibrium with its vapor when the rate at which molecules leave the liquid and enter the vapor equals the rate at which molecules leave the vapor and enter the liquid. If you tagged some molecules in the liquid with a radioactive isotope, you could demonstrate dynamic equilibrium by tracking the movement of isotope from liquid to vapor while measuring the unchanging amounts in each phase.

- The reactants and products of a chemical reaction are in equilibrium with each other when the rate at which reactants are converted to products equals the rate at which products are converted back to reactants. How might you determine that a chemical reaction is in dynamic equilibrium?

Each example is characterized by two processes: A *forward process* in which, for example, students enter a bookstore and a *reverse process* in which students leave the bookstore. The designation of a process as forward or reverse is usually arbitrary.

Kinetics, the study of the rates at which processes occur, is the subject of Chapter 7, and you will review some aspects of chemical equilibrium in that chapter. In this chapter, you explore chemical equilibrium as a separate topic in which the rates at which changes occur are irrelevant, provided that the changes are reversible and that the forward and reverse rates of change are identical.

Consider a reversible chemical reaction with two reactants and two products (Equation 5.1):

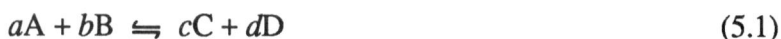

$$a\text{A} + b\text{B} \rightleftharpoons c\text{C} + d\text{D} \qquad\qquad (5.1)$$

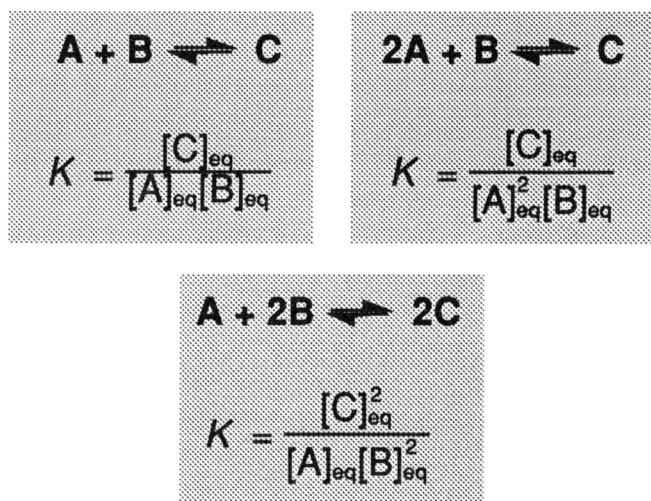

Figure 5.1

The concentrations of reactants and products of a reversible chemical reaction are unchanged while the reaction is in dynamic equilibrium. These concentrations can be calculated from a knowledge of the equilibrium constant K of a reaction (Equation 5.2).

$$A + B \rightleftharpoons C$$

$$K = \frac{[C]_{eq}}{[A]_{eq}[B]_{eq}}$$

$$2A + B \rightleftharpoons C$$

$$K = \frac{[C]_{eq}}{[A]_{eq}^2[B]_{eq}}$$

$$A + 2B \rightleftharpoons 2C$$

$$K = \frac{[C]_{eq}^2}{[A]_{eq}[B]_{eq}^2}$$

where A and B are the reactants and C and D are the products. a, b, c, and d are the number of molecules of each reactant and product that participate in the reaction.

It can be shown that at equilibrium the concentrations of reactants and products, $[A]_{eq}$, $[B]_{eq}$, $[C]_{eq}$, and $[D]_{eq}$ satisfy Equation 5.2:

$$K = \frac{[C]_{eq}^c[D]_{eq}^d}{[A]_{eq}^a[B]_{eq}^b} \tag{5.2}$$

where K is called the equilibrium constant of the reaction. Figure 5.1 shows several examples of reactions and forms of their equilibrium expressions.

A few special cases concerning the setup of Equation 5.2 should be noted:

- Recall from Chapter 4 that dynamic equilibrium between a liquid (or solid) and its vapor does not depend on the amount of liquid (or solid). The vapor pressure of benzene at 25°C, for example, is always 0.125 atm. A similar situation exists when a partially soluble solid is in equilibrium with its solute in solution. In general, the "concentration" of a pure solid or liquid can be regarded as a constant and omitted from expressions such as Equation 5.2. When solid silver chloride, AgCl, dissolves into aqueous solution, for example, the equilibrium established between the partially soluble AgCl and its ions, Ag^+ and Cl^- can be written:

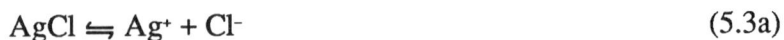

$$AgCl \leftrightharpoons Ag^+ + Cl^- \tag{5.3a}$$

$$K = \frac{[Ag^+][Cl^-]}{[AgCl]} \tag{5.3b}$$

However, because [AgCl] never changes, Equation 5.3b can also be written

$$K[AgCl] = K_{sp} = [Ag^+][Cl^-] \tag{5.4}$$

where K_{sp} is the solubility product constant of AgCl.

- In reactions where one of the reactants is in very large excess, its concentration can be treated as a constant. For example, consider the reaction

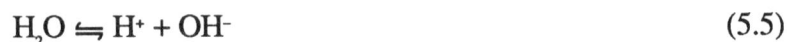

$$H_2O \leftrightharpoons H^+ + OH^- \tag{5.5}$$

At room temperature, the concentration of pure water is essentially unchanged by its dissociation into H^+ and OH^-.[1] In aqueous, acid–base reactions, where this dissociation plays an extremely important role, [H_2O] is usually contained within the dissociation constant such that

$$K[H_2O] = K_w = [H^+][OH^-] = 1 \times 10^{-14} \tag{5.6}$$

at 25°C.

- When dealing with gases, concentrations are usually expressed as partial pressures.

All the described relationships can be powerful tools with which to explore the behavior of system in equilibrium. The exercises that follow show you how.

FINDING EQUILIBRIUM CONCENTRATIONS: A GRAPHICAL METHOD

The equilibrium concentrations of the reactants and products of a chemical reaction can be defined in terms of the equilibrium constant K. Consider, for example, the reaction described by Equation 5.7. At equilibrium, the concentrations of reactants and products must satisfy Equation 5.8:

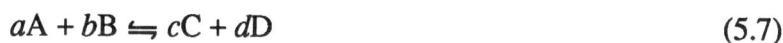

$$aA + bB \rightleftharpoons cC + dD \tag{5.7}$$

$$K = \frac{[C]_{eq}^{c}[D]_{eq}^{d}}{[A]_{eq}^{a}[B]_{eq}^{b}} \tag{5.8}$$

where $[A]_{eq}$, $[B]_{eq}$, $[C]_{eq}$, and $[D]_{eq}$ are the equilibrium concentrations of A, B, C, and D, respectively.

There are, of course, an infinite number of values of $[A]_{eq}$, $[B]_{eq}$, $[C]_{eq}$, and $[D]_{eq}$ that satisfy Equation 5.8. If the starting concentrations of reactants and products are known, however, there is only one set of equilibrium concentrations.

If $[A]_0$, $[B]_0$, $[C]_0$, and $[D]_0$ constitute the starting concentrations of reactants and products and $[A]_{eq}$, $[B]_{eq}$, $[C]_{eq}$, and $[D]_{eq}$ are the equilibrium concentrations, the starting and equilibrium concentrations are related as shown in Figure 5.2.

TO GET THE MOST OUT OF THIS EXERCISE, YOU SHOULD ALREADY KNOW

The definition of the terms:
dynamic equilibrium
chemical reaction
concentration
partial pressure

If the equilibrium concentrations of reactants and products are expressed in terms of starting concentrations and substituted into Equation 5.8, you obtain Equation 5.9:

$$K = \frac{([C]_0 + cx)^c([D]_0 + dx)^d}{([A]_0 - ax)^a([B]_0 - bx)^b} \tag{5.9}$$

where x is a number that is determined by the extent of the reaction.

Equation 5.9 is of enormous general utility for modeling problems in chemical equilibrium. Values of x that satisfy Equation 5.9 can be used to calculate the concentrations of reactants and products at equilibrium (e.g., $[C]_{eq} = [C]_0 + cx$).

One way to find these values of x is to put Equation 5.9 in its rational form (Equation 5.10):

$$K([A]_0 - ax)^a([B]_0 - bx)^b$$
$$- ([C]_0 + cx)^c([D]_0 + dx)^d = 0 \tag{5.10}$$

The roots of Equation 5.10 are, of course, the values of x for which you are searching.

Although there can be more than one solution to Equation 5.10, there is only one solution that yields values of x that correspond to realistic (i.e., nonnegative) values of $[A]_{eq}$, $[B]_{eq}$, $[C]_{eq}$, and $[D]_{eq}$.

Finding the roots of Equation 5.10 can be an arduous task without a computer — sometimes explicit, algebraic expressions (such as the quadratic formula) exist, and sometimes they don't. If an explicit

Figure 5.2

The stoichiometry of a chemical reaction provides a simple method for relating starting and equilibrium concentrations of reactants and products. Let x be a measure of the extent to which a reaction must proceed (to the right) to reach equilibrium. When A is reduced in concentration by ax, then B is reduced in concentration by bx, C is increased in concentration by cx, and D is increased in concentration by dx. Thus, $[A]_{eq} = [A]_0 - ax$, $[B]_{eq} = [B]_0 - bx$, $[C]_{eq} = [C]_0 + cx$, and $[D]_{eq} = [D]_0 + dx$.

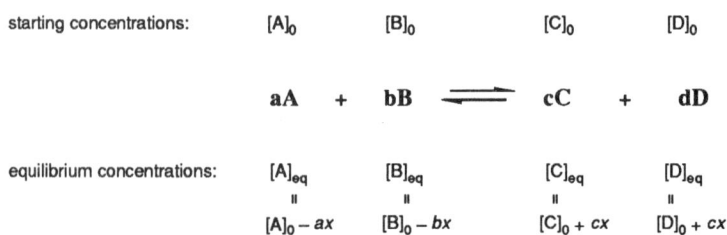

expression doesn't exist, you can always use your electronic spread-sheet to graph the left side of Equation 5.10 and identify the points where the function equals zero (Figure 5.3). Other, more sophisticated computer methods also exist (e.g., Exercise 5.2, Problem 3).

In this exercise, graph Equation 5.10 for a number of chemical reactions and identify the equilibrium concentrations of these reactions from a visual inspection of the graph. In the next exercise, you will study model-building techniques that explicitly calculate equilibrium concentrations. Subsequent exercises will then use these techniques to explore the behavior of chemical systems in equilibrium.

LESSON

NO_2, an airborne pollutant generated by automobiles and some industries, can spontaneously dimerize to form N_2O_4 according to Equation 5.11:

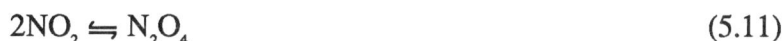

$$2NO_2 \rightleftharpoons N_2O_4 \qquad (5.11)$$

where the equilibrium constant $K = 9$ (when the standard state is chosen to be a partial pressure of 1 atm).

Because NO_2 and N_2O_4 are gases, it is convenient to represent the concentrations $[NO_2]$ and $[N_2O_4]$ in terms of partial pressures (in atm).

If $[NO_2]_0$ and $[NO_2]_0$ are the partial pressures of reactant and product at the beginning of the reaction then, by Equation 5.9, the equilibrium constant K can be written

$$K = \frac{[N_2O_4]_0 + x}{([NO_2]_0 - 2x)^2} \qquad (5.12)$$

The rationalized form of Equation 5.12 is given by Equation 5.13:

$$K([NO_2]_0 - 2x)^2 - [NO_2]_0 - x = 0 \qquad (5.13)$$

a. Build and verify Model 5.1. Model 5.1 evaluates the left side of Equation 5.13 for any set of starting conditions.

    Build and verify the plot shown in Figure 5.4. You will use this plot to visually identify the roots of Equation 5.13.

    Once a root of Equation 5.13 has been identified, its value can be entered manually into cell E1 of the model. Your model uses this value to calculate the partial pressures of NO_2 and N_2O_4 (cells E3 and

Figure 5.3

$f(x)$

roots

0

x

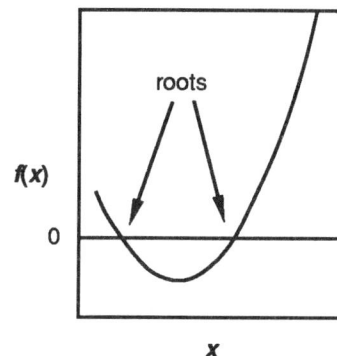

The roots of any function $f(x)$ are the values of x where $f(x) = 0$.

Figure 5.4

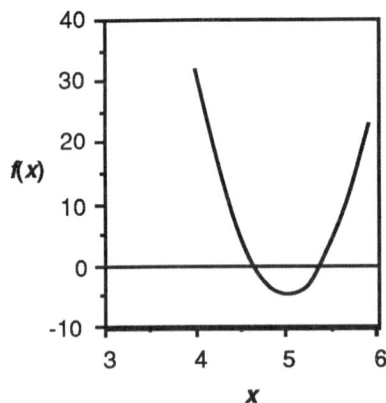

XY–PLOT

<u>Data</u>
Horizontal (x) axis: cells A10–A29 (x)
Vertical (y) axis: cells B10–B29 [f(x)]

E4) and to determine the mole fraction of N_2O_4 in the gas mixture (cell E6). The value of x in cell E1 can be refined by trial and error by checking it against the value of K that it predicts (cell E10).

This process is tedious and can be replaced, in many instances, by the more convenient techniques of Exercise 5.2. It's worthwhile plodding through this process once or twice, however, to acquire a feel for what's really going on.

b. Use your model to find the equilibrium concentrations of NO_2 and N_2O_4 in a chamber initially charged with 10 atm of pure NO_2 (i.e., $[NO_2]_0 = 10$, $[N_2O_4]_0 = 0$). First examine a plot from your model (Figure 5.4) when $x_{start} = -10$ and $x_{incr} = 1$. Then progressively refine your plot by trying values such as those below.

x_{start}	x_{incr}
–2	1
4	0.1
4.5	0.05
Etc.	Etc.

You will soon discover that two roots exist to Equation 5.13 — one value slightly larger than 4.6 and another slightly less than 5.4. Place either of these values into cell E1 and change it by trial and error until the value of K in cell E10 is arbitrarily close to the value of K in cell B1.

Only one of these values of x gives nonnegative partial pressures for both NO_2 and N_2O_4. Which is it? What are the equilibrium partial pressures of NO_2 and N_2O_4 when only 10 atm of NO_2 is present initially?

c. Find the mole fraction of N_2O_4 at equilibrium in a chamber initially charged with the following pressures of pure NO_2:

Initial Pressure NO_2	Equilibrium Mole Fraction N_2O_4
10	_____
5	_____
3	_____
1	_____

If you wish to maximize the dimerization of NO_2 to N_2O_4, would you store NO_2 at high or low pressure?

(In reactions where the number of precursors equals the number of products, K is a pure number without dimensions. However, the dimerization of NO_2 to form N_2O_4 is not such a reaction, and K may be regarded as having the dimensions of atm^{-1} because the numerator of K is atm and the denominator is atm^2.[1] The pressure dependence of this reaction derives from this fact.)

PROBLEMS

1. Another simple reaction that can be studied in your simulated reaction vessel is described by Equation 5.14:

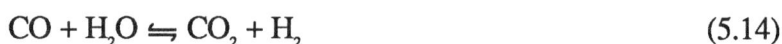

$$CO + H_2O \rightleftharpoons CO_2 + H_2 \qquad (5.14)$$

If $[CO_2]_0$, $[H_2]_0$, $[CO]_0$, and $[H_2O]_0$ are the partial pressures of reactants and products at the beginning of the reaction then, by Equation 5.15, the equilibrium constant K can be written

$$K = \frac{([CO_2]_0 + x)([H_2]_0 + x)}{([CO]_0 - x)([H_2O]_0 - x)} \qquad (5.15)$$

For this reaction, K has a numeric value of 5. The rationalized form of Equation 5.15 is given by Equation 5.16:

$$K([CO]_0 - x)([H_2O]_0 - x) \\ - ([CO_2]_0 + x)([[H_2]_0 + x) = 0 \qquad (5.16)$$

Build a model and on-screen graphic capable of finding the equilibrium partial pressures of the gases in this reaction and then complete the table below. If you have trouble, Model 5.2 offers one solution.

Initial				Equilibrium Mole Fraction	
[CO]	[H$_2$O]	[CO$_2$]	[H$_2$O]	CO	CO$_2$
1.0	1.0	0.0	0.0	____	____
2.0	2.0	0.0	0.0	____	____
3.0	3.0	0.0	0.0	____	____

Is this reaction sensitive to initial pressure? Does the number of precursor molecules equal the number of product molecules in the balanced equation? (See the final comment in the preceding lesson.)

2. The Haber process for the synthesis of ammonia offers an interesting and more complex situation (Equation 5.17):

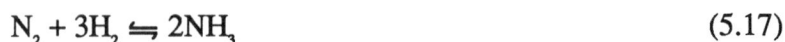

$$N_2 + 3H_2 \rightleftharpoons 2NH_3 \qquad (5.17)$$

In this case, a fourth-order equation (Equation 5.18) must be solved for x:

$$K = \frac{([NH_3]_0 + 2x)^2}{([N_2]_0 - x)([H_2]_0 - 3x)^3} \qquad (5.18)$$

The value of the equilibrium constant K is 0.007 (when the standard state is chosen to be a partial pressure of 1 atm).

Rationalize Equation 5.12 and build a model capable finding the equilibrium partial pressures of each of the gases in the reaction. Our answer is presented as Model 5.3. Complete the table below.

Initial (in atm)			Equilibrium Mole Fraction
$[H_2]$	$[N_2]$	$[NH_3]$	NH_3
1.0	1.0	0.0	_____
2.0	2.0	0.0	_____
3.0	3.0	0.0	_____

Is the Haber reaction sensitive to initial pressure? Does the number of precursor molecules equal the number of product molecules? (See the final comment in the preceding lesson.)

MODELS

FORMULAS

	A	B	C	D	E
1	K =	9		x =	4.64
2					
3	[NO2]0 =	10		[NO2]eq =	=B3-2*E1
4	[N2O4]0 =	0		[N2O4]eq =	=B4+E1
5					
6	x[start] =	4	Mole Fraction [N2O4] =		=E4/SUM(E3:E4)
7	x [incr] =	0.1			
8					
9	x	f(x)		Verification:	
10	=B6			K=	=E4/(E3*E3)
11	=A10+B7	Formula			
12	=A11+B7	Set			
		I			

	A		C	D	E
28	=A27+B7				
29	=A28+B7				

VALUES

	A	B	C	D	E
1	K =	9		x =	4.64
2					
3	[NO2]0 =	10		[NO2]eq =	0.72
4	[N2O4]0 =	0		[N2O4]eq =	4.64
5					
6	x[start] =	4	Mole Fraction [N2O4] =		0.86567164
7	x [incr] =	0.1			
8					
9	x	f(x)		Verification:	
10	4	32		K=	8.95061728
11	4.1	25.06			
12	4.2	18.84			

	A	B	C	D	E
28	5.8	17.24			
29	5.9	23.26			

Formula Set I
Prototype Cell is B10
=**B1***(**B3**-(2*A10))^2-**B4**-A10

Model 5.1

BUILD and verify Model 5.1.

ENTER data using cells B1 (equilibrium constant, K), B3 and B4 (starting pressures), and B6 and B7 (scaling controls for x-values).

PLOT cells A10–A29 (x-axis) vs B10–B29 (y-axis) and identify the values of x where the graph intercepts the x–axis.

ENTER approximate values for roots of x into cell E1 and refine the value by trial and error (until the value of K in cell E10 is arbitrarily close to the value of K in cell B1).

READ output from the model at cells E3 and E4 (equilibrium pressures), and E6 (equilibrium mole fraction of N_2O_4).

Model 5.2

BUILD and verify Model 5.2.

ENTER data using cells B1 (equilibrium constant K), B3–B6 (starting pressures), and B8 and B9 (scaling controls for x-values).

PLOT cells A12–A31 (x-axis) vs B12–B31 (y-axis) and identify the values of x where the graph intercepts the x-axis.

ENTER approximate values for roots of x into cell E1 and refine the value by trial and error (until the value of K in cell E12 is arbitrarily close to the value of K in cell B1).

READ output from the model at cells E3–E6 (equilibrium pressures) and E8 (equilibrium mole fraction of CO_2).

FORMULAS

	A	B	C	D	E
1	K =	5		x =	-2.5
2					
3	[CO]0 =	5		[CO]eq =	=B3-E1
4	[H2O]0 =	10		[H2O]eq =	=B4-E1
5	[CO2]0 =	20		[CO2]eq =	=B5+E1
6	[H2]0 =	30		[H2]eq =	=B6+E1
7					
8	x[start] =	-5		Mole fraction [C02] =	=E5/SUM(E3:E6)
9	x [incr] =	3			
10					
11	x	f(x)		Verification:	
12	=B8			K=	=(E5*E6)/(E3*E4)
13	=A12+B9				
14	=A13+B9				

Formula Set I

	A	B	C	D	E
30	=A29+B9				
31	=A30+B9				

VALUES

	A	B	C	D	E
1	K =	5		x =	-2.5
2					
3	[CO]0 =	5		[CO]eq =	7.5
4	[H2O]0 =	10		[H2O]eq =	12.5
5	[CO2]0 =	20		[CO2]eq =	17.5
6	[H2]0 =	30		[H2]eq =	27.5
7					
8	x[start] =	-5		Mole fraction [C02] =	0.26923077
9	x [incr] =	3			
10					
11	x	f(x)		Verification:	
12	-5	375		K=	5.13333333
13	-2	-84			
14	1	-471			

	A	B	C	D	E
30	49	3129			
31	52	3966			

Formula Set I
Prototype Cell is B12

$$=(B1*(B3-A12)*(B4-A12))-((B5+A12)*(B6+A12))$$

FORMULAS

	A	B	C	D	E
1	K =	0.007		x =	0.155
2					
3	[N2]0 =	1		[N2]eq =	=B3-E1
4	[H2]0 =	3		[H2]eq =	=B4-3*E1
5	[NH3]0 =	0		[NH3]eq =	=B5+2*E1
6					
7	x[start] =	-4		Mole Fraction [NH3] =	=E5/SUM(E3:E5)
8	x [incr] =	0.6			
9					
10	x	f(x)		Verification:	
11	=B7			K=	=E5^2/(E3*E4^3)
12	=A11+B8				
13	=A12+B8				

Formula Set I

	A	B	C	D	E
29	=A28+B8				
30	=A29+B8				

VALUES

	A	B	C	D	E
1	K =	0.007		x =	0.155
2					
3	[N2]0 =	1		[N2]eq =	0.845
4	[H2]0 =	3		[H2]eq =	2.535
5	[NH3]0 =	0		[NH3]eq =	0.31
6					
7	x[start] =	-4		Mole Fraction [NH3] =	0.08401084
8	x [incr] =	0.6			
9					
10	x	f(x)		Verification:	
11	-4	54.125		K=	0.00698124
12	-3.4	24.5990144			
13	-2.8	8.0490704			

	A	B	C	D	E
29	6.8	28.9217744			
30	7.4	98.0493824			

**Formula Set I
Prototype Cell is B11**

$$=B1*(B3-A11)*(B4-(3*A11))^{\wedge}3-(B5+(2*A11))^{\wedge}2$$

Model 5.3

BUILD and verify Model 5.3.

ENTER data using cells B1 (equilibrium constant K), B3–B5 (starting pressures), and B7 and B8 (scaling controls for x-values).

PLOT cells A11–A30 (x-axis) vs B11–B30 (y-axis) and identify the values of x where the graph intercepts the x-axis.

ENTER approximate values for roots of x into cell E1 and refine the value by trial and error (until the value of K in cell E11 is arbitrarily close to the value of K in cell B1).

READ output from the model at cells E3–E5 (equilibrium pressures) and E7 (equilibrium mole fraction of NH_3).

FINDING EQUILIBRIUM CONCENTRATIONS: ANALYTIC AND OTHER METHODS

TO GET THE MOST OUT OF THIS EXERCISE, YOU SHOULD ALREADY KNOW

The definition of the terms:
dynamic equilibrium
chemical reaction
concentration
partial pressure

How to use elementary algebra to find the roots of equations.

Exercise 5.1 introduced the problem of finding the equilibrium concentrations of reactants and products from a set of initial conditions. In summary, the process involves examining a reaction (e.g., Equation 5.19) and expressing the equilibrium constant K first in terms of equilibrium concentrations (Equation 5.20) and then in terms of initial concentrations (Equation 5.21):

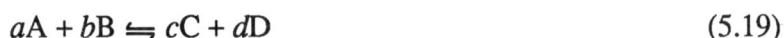

$$aA + bB \rightleftharpoons cC + dD \tag{5.19}$$

$$K = \frac{[C]_{eq}^{c}[D]_{eq}^{d}}{[A]_{eq}^{a}[B]_{eq}^{b}} \tag{5.20}$$

$$K = \frac{([C]_0 + cx)^c([D]_0 + dx)^d}{([A]_0 - ax)^a([B]_0 - bx)^b} \tag{5.21}$$

where $[A]_{eq}$, $[B]_{eq}$, $[C]_{eq}$, and $[D]_{eq}$ are the equilibrium concentrations and $[A]_0$, $[B]_0$, $[C]_0$, and $[D]_0$ are the initial concentrations of reactants and products. x is a measure of the extent to which the reaction must proceed to the right in order to reach equilibrium.

In the case of Equation 5.19, the equilibrium concentrations of reactants and products can be found by

- Solving Equation 5.21 for x.

- Calculating equilibrium concentrations from a knowledge of x and the initial concentrations of reaction components (e.g., $[A]_{eq} = [A]_0 - ax$).

Solving Equation 5.21 for x is usually the most difficult part of this operation. Exercise 5.1 showed you how to use the graphical tools of your electronic spreadsheet program to find the roots of any equation. The technique is instructive but cumbersome.

In this exercise, you explore a number of more convenient alternative methods.

LESSON

Consider a reaction of the general form

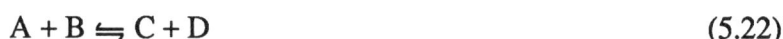

$$A + B \rightleftharpoons C + D \tag{5.22}$$

If $[A]_0$, $[B]_0$, $[C]_0$, and $[D]_0$ are the initial concentrations of reactants and products then, by Equation 5.21, the equilibrium constant K can be written

$$K = \frac{([C]_0 + x)([D]_0 + x)}{([A]_0 - x)([B]_0 - x)} \tag{5.23}$$

or

$$(1 - K)x^2 + ([C]_0 + [D]_0 + K[A]_0 + K[B]_0)x$$
$$+ [C]_0[D]_0 - K[A]_0[B]_0 = 0 \tag{5.24}$$

Equation 5.24 is quadratic and has two roots (Equations 5.25a,b):

$$x = \frac{-b + \sqrt{b^2 - 4ac}}{2a} \tag{5.25a}$$

$$x = \frac{-b - \sqrt{b^2 - 4ac}}{2a} \tag{5.25b}$$

where $a = K - 1$, $b = -[C]_0 - [D]_0 - K[A]_0 - K[B]_0$, and $c = K[A]_0[B]_0 - [C]_0[D]_0$. A quick inspection of Equations 5.25a,b with real data will convince you that only Equation 5.25b has physical meaning.

Equation 5.22 can be used to represent the reaction $CO + H_2O \rightleftharpoons CO_2 + H_2$ (recall Problem 1 of Exercise 5.1). Equation 5.25 therefore offers an opportunity to build a computer model of the behavior of this reaction at equilibrium.

a. Build and verify Model 5.4. Examine the structure of this model until you understand how the model implements Equation 5.25.

b. Determine the equilibrium mole fractions of CO and CO_2 for each of the starting conditions below and graph your results in Figure 5.5.

	Initial Pressures				Equilibrium Mole Fraction	
CO	H_2O	CO_2	H_2		CO	CO2
1.0	1.0	0.0	0.0		_____	_____
2.0	2.0	0.0	0.0		_____	_____
3.0	3.0	0.0	0.0		_____	_____
4.0	4.0	0.0	0.0		_____	_____
5.0	5.0	0.0	0.0		_____	_____
6.0	6.0	0.0	0.0		_____	_____
7.0	7.0	0.0	0.0		_____	_____
8.0	8.0	0.0	0.0		_____	_____

Figure 5.5

Many spreadsheets offer a feature for generating *data tables* or *what-if tables*. Data tables automatically interrogate a model such as Model 5.4 many times and create a table of the results suitable for plotting. If your spreadsheet has such a feature, you need not plot your results here.

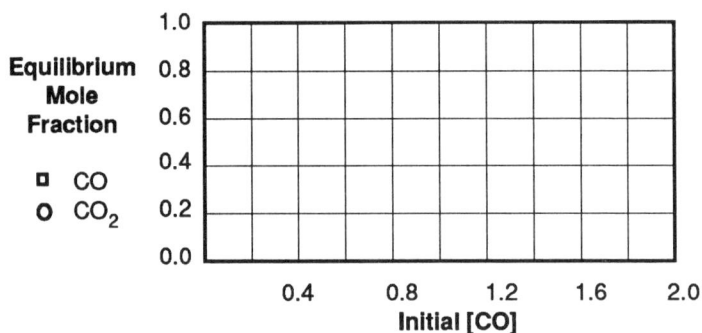

Equilibrium Mole Fraction

□ CO
○ CO_2

Initial [CO]

c. There are many fascinating features to this reaction that you will explore in Exercise 5.3. For now content yourself with the knowledge that you have a computer simulation of a reaction vessel in which you can conveniently explore the outcomes of a large number of thought experiments.

PROBLEMS

1. Consider the dimerization of NO_2 to form N_2O_4 (first discussed in the lesson to Exercise 5.1):

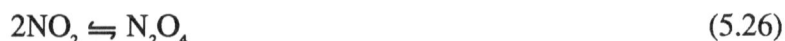

$$2NO_2 \rightleftharpoons N_2O_4 \qquad (5.26)$$

The equilibrium constant for this reaction can be written in terms of initial concentrations by using to Equation 5.21):

$$K = \frac{[N_2O_4]_0 + x}{([NO_2]_0 - 2x)^2} \qquad (5.27)$$

or

$$4Kx^2 - (4K[NO_2]_0 + 1)x + K[NO_2]_0^2 - [N_2O_4]_0 = 0 \qquad (5.28)$$

Equation 5.25b is the physically realistic solution to Equation 5.28 when $a = 4K$, $b = -4K[NO_2]_0 - 1$, and $c = K[NO_2]_0^2 - [N_2O_4]_0$.

a. Build and verify Model 5.5.

b. Find the mole fraction of N_2O_4 at equilibrium in a chamber initially charged with the following pressures of pure NO_2. Plot your results in Figure 5.6

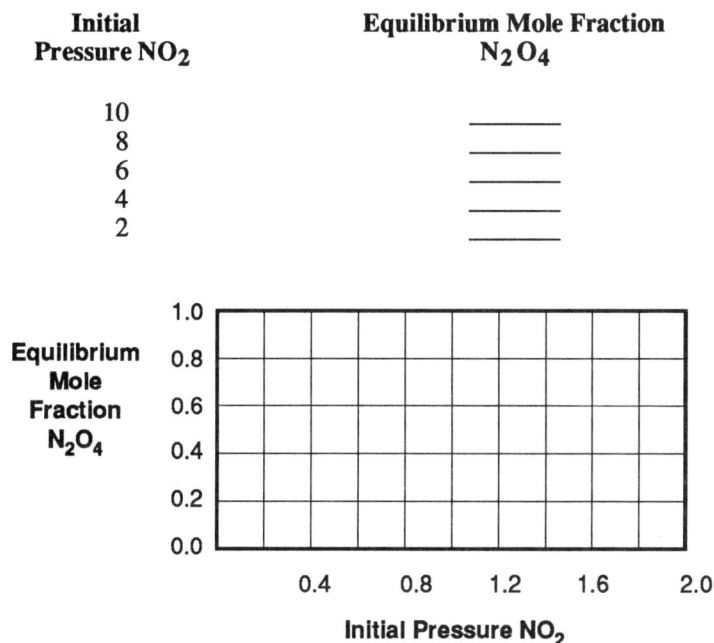

Initial Pressure NO_2	Equilibrium Mole Fraction N_2O_4
10	_____
8	_____
6	_____
4	_____
2	_____

Figure 5.6

You can graph these results on your computer screen if your spreadsheet program supports data tables.

c. If you wish to maximize the dimerization of NO_2 to N_2O_4, would you store NO_2 at high or low pressure?

d. Compare your results with those for the reaction $CO + H_2O \rightleftharpoons CO_2 + H_2$. Recall the discussion in Exercise 5.1. Why is one reaction sensitive to initial total pressure and the other is not?

2. The Haber process for the synthesis of ammonia from N_2 and H_2 offers an interesting and more difficult problem (Equation 5.29):

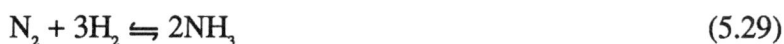

$$N_2 + 3H_2 \rightleftharpoons 2NH_3 \qquad (5.29)$$

In this case, a fourth-order equation (Equation 5.30) must be solved:

$$K = \frac{[NH_3]_{eq}^2}{[N_2]_{eq} \ [H_2]_{eq}^3} \tag{5.30}$$

You can solve this problem using the graphical methods described in Exercise 5.1 or using a numeric method such as Newton's method (Problem 3). Sometimes, however, a clever examination of the equation leads to an alternative approach. Suppose, for example, that you wish to solve only for initial conditions in which NH_3 is absent and N_2 and H_2 are present in the molar ratio 1:3.

A moment's thought will convince you that when initial conditions are so restricted the following relationships hold true:

$$[N_2]_{eq} = [N_2]_0 - \frac{[NH_3]_{eq}}{2} \tag{5.31}$$

$$[H_2]_{eq} = 3[N_2]_0 \tag{5.32}$$

If Equation 5.31 is solved for $[NH_3]_{eq}$ (Equation 5.33),

$$[NH_3]_{eq} = 2([N_2]_0 - [N_2]_{eq}) \tag{5.33}$$

and Equations 5.32 and 5.33 are substituted into Equation 5.30, it follows that

$$[N_2]_{eq} = \sqrt[4]{\frac{4([N_2]_0 - [N_2]_{eq})^2}{27K}} \tag{5.34}$$

At first glance, Equation 5.34 is no easier to solve than Equation 5.24. However, if $[N_2]_{eq}$ on the right side of the equation is presumed to equal zero, then Equation 5.34 can easily be solved for $[N_2]_{eq}$ on the left side of the equation. This new value of $[N_2]_{eq}$ can be fed back into the right side of Equation 5.34 and the process repeated numerous times. After multiple iterations, $[N_2]_{eq}$ converges on the true equilibrium value of N_2 if the initial pressures of the reactants are high enough to avoid the chaos that occasionally accompanies numeric solutions.[1]

a. Build and verify Model 5.6 and study its structure until you understand how the model implements a solution to Equation 5.32.

b. Determine the mole fraction of NH_3 at equilibrium as a function of initial total pressure.

Initial Total Pressure	Equilibrium Mole Fraction NH_3
40	_____
80	_____
100	_____
200	_____
500	_____

c. Compare your results with those obtained using the graphical method in Exercise 5.1.

d. Is the yield of NH_3 in the Haber reaction pressure–sensitive or pressure-insensitive? Why?

3. Some of the lessons and problems in this exercise show how to use simple algebra to find the roots of an equation. When such techniques fail, other methods must be used. In Exercise 5.1, you found the roots of several expressions by examining the graphs of such functions. A field of inquiry known as numeric analysis provides another approach. The numeric methods that evolve out of this inquiry are uniquely suited to computer implementation, and entire textbooks have been written about them. In this problem, you explore one simple and famous method for finding the roots of an equation — Newton's method. You must have some knowledge of elementary calculus to follow this discussion.

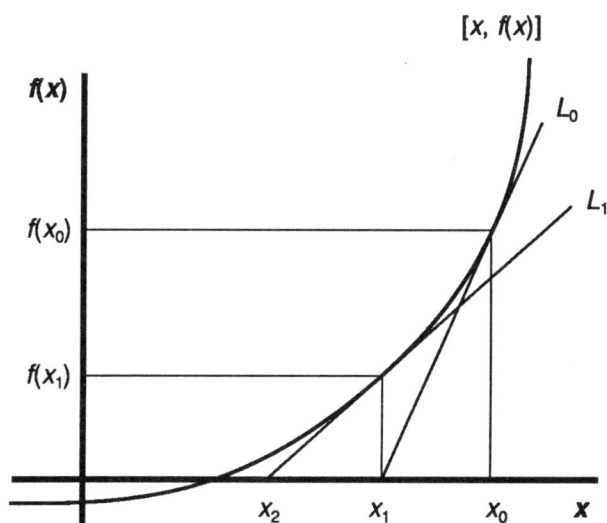

Figure 5.7

In Newton's method, a line drawn tangent to a function $f(x)$ at $[x_0, f(x_0)]$ is extended to intercept with the x-axis (x_1). x_1 is usually closer to a root of $f(x)$ than x_0. A new tangent line drawn at $[x_1, f(x_1)]$ can be used to repeat the process indefinitely. If each new value of x converges on a final common value, a root has been found.

Consider a differentiable function $f(x)$ and a line L_0 that is tangent to that function at $[x_0, f(x_0)]$ (Figure 5.7). It can be shown that the line L_0 intersects the x-axis at x_1 such that

$$x_1 = x_0 - \frac{f(x_0)}{f'(x_0)} \tag{5.35}$$

where $f'(x)$ is the first derivative of $f(x)$ and can be approximated numerically by using the definition of the first derivative and taking Δx equal to a very small number such as 10^{-7} (Equation 5.36):

$$f'(x) \sim \frac{f(x + \Delta x) - f(x)}{\Delta x} \tag{5.36}$$

x_1 is usually a better approximation to a root of $f(x)$ than x_0. Progressively better approximations of the root can be obtained by drawing a new tangent line at $[x_1, f(x_1)]$ and repeating the process (Figure 5.7, Equation 5.37):

$$x_2 = x_1 - \frac{f(x_1)}{f'(x_1)} \tag{5.37}$$

Newton's method is powerful but not infallible. Consult a book on numeric methods to determine the range of cases in which it is successful.

a. Build and verify Model 5.7. Model 5.7 adds Newton's method to a slightly modified version of Model 5.3 (i.e., some cell positions have been relocated for aesthetic reasons). Model 5.7 computes the equilibrium values of reactants and products in the Haber process (Problem 2).

b. Explore Model 5.7 until you see how Newton's method has been implemented. Notice that cell D15 takes a "test" value of a root of $f(x)$ from cell E14 and uses it to calculate $f'(x)$ numerically with $\Delta x = 10^{-7}$.

    Cell D16 then calculates a new "test" value of the root of $f(x)$ (Equation 5.35 or 5.36) and repeats the process. A root has been found if the value of x in each new row converges to a final common value.

c. Use Newton's method to find the roots of $f(x)$. Different test values of x will converge on different roots. Which root is physically correct?

4. Your primary chemistry textbook contains numerous examples of reversible chemical reactions that demonstrate the principles of equilibria. Design, build, verify, and explore computer simulations of at least two examples from your book.

MODELS

FORMULAS

	A	B	C	D	E
1	K =	5			
2	[CO]o =	5		[CO]eq =	=B2-B11
3	[H2O]o =	10		[H2O]eq =	=B3-B11
4	[CO2]o =	20		[CO2]eq =	=B4+B11
5	[H2]o =	30		[H2]eq =	=B5+B11
6					
7	a =	=B1-1		Mole fraction [C02] =	=E4/SUM(E2:E5)
8	b =	=-(B1*B2)-(B1*B3)-B4-B5			
9	c =	=(B1*B2*B3)-(B4*B5)			
10				Verification:	
11	x =	=(-B8-SQRT((B8*B8)-(4*B7*B9)))/(2*B7)		K=	=(E4*E5)/(E2*E3)

VALUES

	A	B	C	D	E
1	K =	5			
2	[CO]o =	5		[CO]eq =	7.58600286
3	[H2O]o =	10		[H2O]eq =	12.5860029
4	[CO2]o =	20		[CO2]eq =	17.4139971
5	[H2]o =	30		[H2]eq=	27.4139971
6					
7	a =	4		Mole fraction [C02] =	0.26790765
8	b =	-125			
9	c =	-350			
10				Verification:	
11	x =	-2.5860029		K=	5

Model 5.4

BUILD and verify Model 5.4.

ENTER data using cells B1 (equilibrium constant) and B2–B5 (initial concentrations of reactants and products).

READ output from the model at cells B11 (extent of reaction), E2–E5 (equilibrium concentrations of reactants and products), and E7 (mole fraction of CO_2).

NOTICE that cell E11 provides an independent verification of the calculations by recalculating the equilibrium constant from the values of the equilibrium concentrations.

Model 5.5

BUILD and verify Model 5.5.

ENTER data using cells B1 (equilibrium constant) and B2 and B3 (initial concentrations of reactants and products).

READ output from the model at cells B9 (extent of reaction), E2–E3 (equilibrium concentrations of reactants and products), and E5 (mole fraction of N_2O_4).

NOTICE that cell E9 provides an independent verification of the calculations by recalculating the equilibrium constant from the values of the equilibrium concentrations.

FORMULAS

	A	B	C	D	E
1	K =	9			
2	[NO2]o =	10		[NO2]eq =	=B2-(2*B9)
3	[N2O4]o =	0		[N2O4]eq =	=B3+B9
4					
5	a =	=4*B1		Mole Fraction [N2O4] =	=E3/(E2+E3)
6	b =	=-(4*B1*B2)-1			
7	c =	=(B1*B2*B2)-B3			
8				Verification:	
9	x =	=(-B6-SQRT((B6*B6) -(4*B5*B7)))/(2*B5)		K=	=E3/(E2*E2)

VALUES

	A	B	C	D	E
1	K =	9			
2	[NO2]o =	10		[NO2]eq =	0.71809564
3	[N2O4]o =	0		[N2O4]eq =	4.64095218
4					
5	a =	36		Mole Fraction [N2O4] =	0.86600313
6	b =	-361			
7	c =	900			
8				Verification:	
9	x =	4.64095218		K=	9

VALUES

	A	B	C	D
1	K =	0.007	[H2]init =	=3*B2
2	[N2]init =	20	[tot]init =	=B2+D1
3				
4		=((4*B2^2)/(27*B1))^0.25		
5			<==	
6			<==	
7			<==	
8			<==	
9		Formula	<==	
10			<==	
11		I	<==	
12			<==	
13			<==	
14			<==	
15			<==	
16			<==	
17			<==	
18			<==	
19			<== Root Solver	
20				
21	[N2] =	=B19		
22	[H2] =	=3*B21		
23	[NH3] =	=2*(B2-B21)		
24				
25	MF(NH3) =	=B23/SUM(B21:B23)		
26				
27	Verification:			
28	K =	=(B23*B23)/(B21*B22*B22*B22)		

VALUES

	A	B	C	D
1	K =	0.007	[H2]init =	60
2	[N2]init =	20	[tot]init =	80
3				
4		9.59212		
5		6.9196	<==	
6		7.75729	<==	
7		7.50479	<==	
8		7.58178	<==	
9		7.55839	<==	
10		7.5655	<==	
11		7.56334	<==	
12		7.564	<==	
13		7.5638	<==	
14		7.56386	<==	
15		7.56384	<==	
16		7.56385	<==	
17		7.56384	<==	
18		7.56384	<==	
19		7.56384	<== Root Solver	
20				
21	[N2] =	7.56384		
22	[H2] =	22.6915		
23	[NH3] =	24.8723		
24				
25	MF(NH3) =	0.45118		
26				
27	Verification:			
28	K =	0.007		

Formula Set I
Prototype Cell is B5

=(4*(**B2**-B4)^2/(27***B1**))^0.25

Model 5.6

BUILD and verify Model 5.6.

ENTER data using cells B1 (equilibrium constant) and B2 (initial concentration of N_2).

READ output from the model at cells B21–E23 (equilibrium concentrations of reactants and products) and B25 (mole fraction of NH_3).

NOTICE that only the partial pressure of N_2 need be specified because all other partial pressures are fixed by the assumptions of the model. Notice also that cell E28 provides an independent verification of the calculations by recalculating the equilibrium constant from the values of the equilibrium concentrations.

Model 5.7

BUILD and verify Model 5.7.

ENTER data using cells B1 (equilibrium constant K), B3–B5 (starting pressures), and B7 and B8 (scaling controls for x-values).

PLOT cells A11–A30 (x-axis) vs B11–B30 (y-axis) and identify the values of x where the graph intercepts the x-axis.

ENTER value for roots of x into cell E1 after examining the plot or the output of Newton's method (cell E24 of this column of data converges).

READ output from the model at cells E3–E5 (equilibrium pressures) and E7 (equilibrium mole fraction of NH_3). These values are correct only after x has been correctly entered into cell E1.

FORMULAS

	A	B	C	D	E	F	G	H
1	K =	0.007		x =	0.155			
2								
3	[N2]0 =	1		[N2]eq =	=B3-E1			
4	[H2]0 =	3		[H2]eq =	=B4-3*E1			
5	[NH3]0 =	0		[NH3]eq =	=B5+2*E1			
6								
7	x[start] =	-4		Mole Fraction [NH3] =	=E5/SUM(E3:E5)			
8	x [incr] =	0.6						
9					Verification:			
10				K =	=E5^2/(E3*E4^3)			
11								
12	x	f(x)		Newton's Method				
13	=B7			Δx =	0.0000001			
14				Test x =	5			
15				x	f(x)	x + Δx	f(x+Δx)	f'(x)
16				=E14				
17								
18								
19	Formula Set I	Formula Set II		Formula Set III	Formula Set IV	Formula Set V	Formula Set VI	Formula Set VII
20								
21								
22								
23								
24								

VALUES

	A	B	C	D	E	F	G	H
1	K =	0.007		x =	0.155			
2								
3	[N2]0 =	1		[N2]eq =	0.845			
4	[H2]0 =	3		[H2]eq =	2.535			
5	[NH3]0 =	0		[NH3]eq =	0.31			
6								
7	x[start] =	-4		Mole Fraction [NH3] =	0.08401084			
8	x [incr] =	0.6						
9					Verification:			
10				K =	0.00698124			
11								
12	x	f(x)		Newton's Method				
13	-4	54.125		Δx =	0.0000001			
14	-3.4	24.5990144		Test x =	5			
15	-2.8	8.0490704		x	f(x)	x + Δx	f(x+Δx)	f'(x)
16	-2.2	0.4580864		5	-51.616	5.0000001	-51.615999	8.3840014
17	-1.6	-1.6031536		11.1564875	1513.24177	11.1564876	1513.24184	702.797777
18	-1	-0.976		9.00331939	451.190604	9.00331949	451.190635	315.527463
19	-0.4	0.0860624		7.57336271	123.444641	7.57336281	123.444656	154.139079
20	0.2	-0.0825856		6.77249741	26.3868925	6.77249751	26.3869016	91.2363072
21	0.8	-2.5596976		6.48328258	2.72185942	6.48328268	2.72186669	72.7697935
22	1.4	-7.8351616		6.44587888	0.04187898	6.44587898	0.04189503	70.5358036
23	2	-15.811		6.44528503	1.0452E-05	6.44528513	1.7502E-05	70.500613
24	2.6	-25.80137		6.44528488	1.1653E-12	6.44528498	7.0501E-06	70.5006039

Formula Set I
Prototype Cell is A14

=A13+B8

Formula Set IV
Prototype Cell is E16

=B1*(B3-D16)*(B4-(3*D16))^3-B5+(2*D16))^2

Formula Set II
Prototype Cell is B13

=B1*(B3-A13)*(B4-(3*A13))^3-B5+(2*A13))^2

Formula Set V
Prototype Cell is F16

=D16+E13

Formula Set III
Prototype Cell is D17

=D16-(E16/H16)

Formula Set VI
Prototype Cell is G16

=B1*(B3-F16)*(B4-(3*F16))^3-B5+(2*F16))^2

Formula Set VII
Prototype Cell is H16

=(G16-E16)/E13

THE EXTENT OF A REACTION

The equilibrium concentrations of reactants and products are often surprising. Recall your studies of reactions of the form

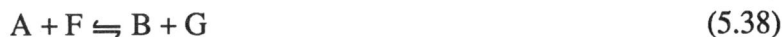

$$A + F \rightleftharpoons B + G \tag{5.38}$$

A chemical reaction such as that described in Equation 5.38 will have reactant and product concentrations that satisfy Equation 5.39:

$$K = \frac{[B]_{eq}[G]_{eq}}{[A]_{eq}[F]_{eq}} \tag{5.39}$$

where $[A]_{eq}$ and $[F]_{eq}$, $[B]_{eq}$, and $[G]_{eq}$ are the concentrations of A, F, B, and G at equilibrium.

Exercises 5.1 and 5.2 showed how to use Equation 5.39 to determine the equilibrium concentrations of a reversible reaction. In this exercise, you exploit your knowledge to explore the behavior of systems in equilibrium.

LESSONS

Imagine that in a summer job you are asked to make product code B from starting material A. Of several methods you find in the library, you select a procedure in which A reacts with F to form the desired product

TO GET THE MOST OUT OF THIS EXERCISE, YOU SHOULD ALREADY KNOW

How to find the equilibrium concentrations of reversible chemical reactions using graphical, analytic, and other techniques.

Figure 5.8

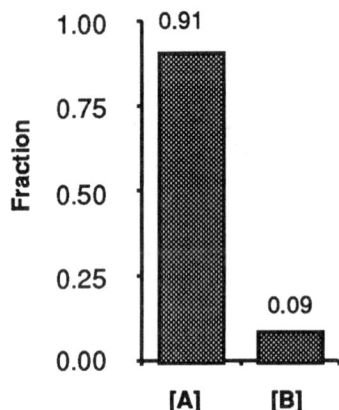

BAR CHART

Data
Cells F6, F7

Options

Scale y-axis 0–1.00
Display value with bar

B and a by-product G (Equation 5.38). You choose this reaction because because F is inexpensive.

A is expensive, and it is important to obtain a high percent conversion of A to B in your reaction vessel. However, because the equilibrium constant K for the reaction is 0.01, you realize that you have a problem.

1. Build and verify Model 5.8. Model 5.8 uses the same principles described in the lesson of Exercise 5.2. It has been reorganized to highlight specific features of the reaction that you will explore. Notice especially that cell F7 calculates the fractional yield of the reaction and that cell F6 shows the amount of reactant that remains at equilibrium. Also build an on-screen graphic (Figure 5.8) to conveniently visualize these results.

   a. Suppose that you perform your reaction under conditions in which both reactant concentrations, [A] and [F], are at 10 mM. Use your model to demonstrate that your maximal theoretical yield is only 9%. (You'll never earn a bonus or be offered a job next summer at this rate.)

   In desperation, you double the concentration of A: $[A]_0 = 20$, $[F]_0 = 10$. What happens to the fractional conversion of A?

   Complete the table below and then write a general rule to predict what happens to the fractional conversion of A as $[A]_0$ increases ($[F]_0$ held constant).

$[A]_0$	$[F]_0$	$[B]_0$	$[G]_0$	% Conversion of A
10	10	0	0	_____
20	10	0	0	_____
40	10	0	0	_____
80	10	0	0	_____

   Can you explain your results?

   b. You gain further insight into your decreasing yields by measuring the equilibrium concentrations of G at the end of the reaction. Complete the table below and then write a general rule to predict $[G]_{eq}$ as $[A]_0$ increases.

$[A]_0$	$[F]_0$	$[B]_0$	$[G]_0$	$[G]_{eq}$
10	10	0	0	_____
20	10	0	0	_____
40	10	0	0	_____
80	10	0	0	_____

Notice that by increasing the amount of your expensive starting material A, you increase the useless conversion of F to G.

c. Your future job prospects continue to look grim until you discover what happens to the percent conversion of A when you increase initial concentrations of F. Complete the table below and write your results.

$[A]_0$	$[F]_0$	$[B]_0$	$[G]_0$	% Conversion of A
10	10	0	0	_____
10	20	0	0	_____
10	40	0	0	_____
10	80	0	0	_____
10	120	0	0	_____

In general, you can always increase the conversion of A by increasing the concentration of F, but can you increase the percent conversion of A to any arbitrary level by increasing $[F]_0$? What real-world limitations might exist on $[F]_0$?

d. Recalling what happened when you tried to increase yields of B by increasing $[A]_0$, you get the clever idea that you might be able to increase the percentage of A that is converted to B by *decreasing* $[A]_0$. Complete the table below.

$[A]_0$	$[F]_0$	$[B]_0$	$[G]_0$	% Conversion of A	$[A]_{eq}$
10	100	0	0	_____	_____
5	100	0	0	_____	_____
3	100	0	0	_____	_____
1	100	0	0	_____	_____
0.1	100	0	0	_____	_____
0.01	100	0	0	_____	_____

Can you increase the percent conversion of A to any arbitrary level by decreasing $[A]_0$?

As percent conversion increases, what happens to $[A]_{eq}$? Is there a practical lower limit to $[A]_0$ if you wish to make significant quantities of B?

2. Your experience with Lesson 1 (and your understanding of the principle of mass action) should suggest that because percent conversion of A increases with both increases in $[F]_0$ and decreases with $[A]_0$, the value of the fraction $[F]_0/[A]_0$ might be an important determinant of reaction extent. Explore this relationship by completing the table below.

$[A]_0$	$[F]_0$	$[B]_0$	$[G]_0$	% Conversion of A	$[F]_0/[A]_0$
10	5	0	0	_____	_____
20	10	0	0	_____	_____
30	15	0	0	_____	_____
50	5	0	0	_____	_____
100	10	0	0	_____	_____
200	20	0	0	_____	_____

Notice that when $[B]_0 = [G]_0 = 0$, the extent of reaction at equilibrium depends on only $[F]_0/[A]_0$.

3. Often, you will want to build a "quick and dirty" model to explore an interesting relationship. The effect of the fraction $[F]_0/[A]_0$ on the percent conversion of A suggests Model 5.9 (and its related on-screen graphics, Figure 5.9) to us. Build, verify, and explore Model 5.9.

PROBLEMS

1. You can always force a greater conversion of one reactant by increasing the concentration of the other. The magnitude of the effect, however, depends on the equilibrium constant K. Use Model 5.8 to complete the table below.

Fraction of A Converted
$[A]_0 = 1.00$, $[B]_0 = [G]_0 = 0.00$

$[F]_0$	$K = 10^{-4}$	$K = 10^{-3}$	$K = 10^{-2}$	$K = 10^{-1}$
1.00	_____	_____	_____	_____
25.00	_____	_____	_____	_____
50.00	_____	_____	_____	_____
75.00	_____	_____	_____	_____
100.00	_____	_____	_____	_____

As $[F]_0$ increases from 1 mM to 100 mM, the extent of the conversion of A to B also increases. How does K influence the extent of the reaction? Why does the value of K have this effect?

Figure 5.9

XY-PLOT

Data
Horizontal (*x*) axis: cells A5–A21
Vertical (*y*) axis: cells F5–F21

Options

Prepare two charts:
(a) Scale *x*-axis from 0 to 150
(b) Scale *x*-axis from 0 to 10

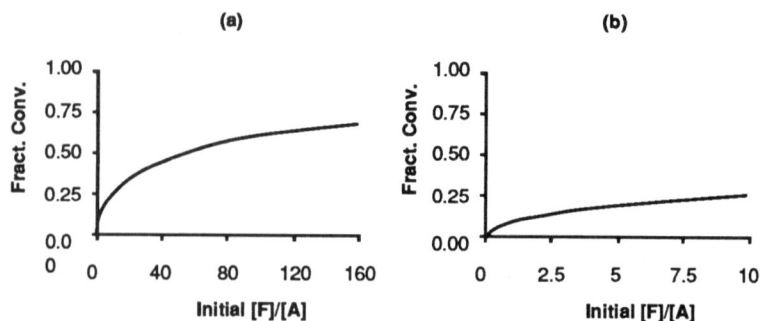

2. Suppose that either or both of the products B and G are present at the beginning of the reaction. Explore the effect of [B] and [G] on equilibrium concentrations of reactants and products. Under these circumstances, does $[F]_0/[A]_0$ solely determine reaction extent?

3. Consider a reaction such as that described by Equation 5.38. Once this reaction has achieved equilibrium, the concentrations of reactants and products remain unchanged. If such a system is "stressed" by increasing or decreasing the concentration of one of its components, the system is no longer in equilibrium, and the concentrations of reactants and products must change for equilibrium to be reestablished. If A is added to a system in equilibrium, this change reduces [A]. If A is removed from a system in equilibrium, this change increases [A]. This effect is sometimes called Lê Chatelier's principle.

   a. Suppose that a system initially exists such that $[A]_0 = [F]_0 = 15$ mM, $[B]_0 = [G]_0 = 2$ mM. What are the equilibrium concentrations of reactants and products?

   b. Suppose that a system initially exists such that the concentrations of reactants and products are the equilibrium conditions discovered in Problem 3a. If a small amount of A is added to the system, is equilibrium re-established by increasing or decreasing [A]?

   c. Explore the effects of changes in other components of the system.

MODELS

Model 5.8

BUILD and verify Model 5.8.

ENTER data using cells B1 (K) and B3, C3, E3, and F3 (initial concentrations of A, F, B, and G).

READ output from the model at cells B4, C4, E4, and F4 (final concentrations of A, F, B, and G) and at F6 and F7 (the fractions $[A]_{eq}/[A]_0$ and $[B]_{eq}/[A]_0$).

NOTICE that cells B6–B10 calculate the various coefficients of the quadratic equation used to solve this problem (Figure 5.8). Cell B11 is a useful device for checking the accuracy of your model. It plugs $[A]_0$, $[F]_0$, $[B]_0$, and $[G]_0$ back into Equation 5.39 to calculate K. Calculated K (cell B11) should equal the K entered into cell B1.

FORMULAS

	A	B	C	D	E	F
1	K =	0.01				
2		[A]	[F]	-->	[B]	[G]
3	Initial	10	10		0	0
4	Final	=B3-B10	=C3-B10		=E3+B10	=F3+B10
5						
6	a	=B1-1			[A]	=B4/B3
7	b	=-((B1*(B3+C3))+(E3+F3))			[B]	=E4/B3
8	c	=(B1*B3*C3)-(E3*F3)				
9	z	=(B7*B7)-(4*B6*B8)				
10	x	=(-B7-SQRT(B9))/(2*B6)				
11	Kcheck	=(E4*F4)/(B4*C4)				

VALUES

	A	B	C	D	E	F
1	K =	0.01				
2		[A]	[F]	-->	[B]	[G]
3	Initial	10	10		0	0
4	Final	9.090909	9.090909		0.909091	0.909091
5						
6	a	-0.99			[A]	0.91
7	b	-0.2			[B]	0.09
8	c	1				
9	z	4				
10	x	0.909091				
11	Kcheck	0.01				

FORMULAS

	A	B	C	D	E	F
1	K =	0.01				
2						Fraction
3	Initial					Convert.
4	[F]/[A]	a	b	c	z	x
5	0.1	=B1-1	=-(B1*(1+A5))	=B1*A5		
6	=A5*(10^0.2)	=B1-1	=-(B1*(1+A6))	=B1*A6		
7	=A6*(10^0.2)	=B1-1	=-(B1*(1+A7))	=B1*A7		
8	=A7*(10^0.2)	=B1-1	=-(B1*(1+A8))	=B1*A8	Formula	Formula
9	=A8*(10^0.2)	=B1-1	=-(B1*(1+A9))	=B1*A9		
10	=A9*(10^0.2)	=B1-1	=-(B1*(1+A10))	=B1*A10	I	II
11	=A10*(10^0.2)	=B1-1	=-(B1*(1+A11))	=B1*A11		
12	=A11*(10^0.2)	=B1-1	=-(B1*(1+A12))	=B1*A12		
13	=A12*(10^0.2)	=B1-1	=-(B1*(1+A13))	=B1*A13		
14	=A13*(10^0.2)	=B1-1	=-(B1*(1+A14))	=B1*A14		
15	=A14*(10^0.2)	=B1-1	=-(B1*(1+A15))	=B1*A15		
16	=A15*(10^0.2)	=B1-1	=-(B1*(1+A16))	=B1*A16		
17	=A16*(10^0.2)	=B1-1	=-(B1*(1+A17))	=B1*A17		
18	=A17*(10^0.2)	=B1-1	=-(B1*(1+A18))	=B1*A18		
19	=A18*(10^0.2)	=B1-1	=-(B1*(1+A19))	=B1*A19		
20	=A19*(10^0.2)	=B1-1	=-(B1*(1+A20))	=B1*A20		
21	=A20*(10^0.2)	=B1-1	=-(B1*(1+A21))	=B1*A21		

Model 5.9

BUILD and verify Model 5.9.

ENTER data using cell B1 (*K*).

READ output from the model at cells F5–F21.

NOTICE that the most effective use of this model involves building the on-screen graphic described in Figure 5.9.

VALUES

	A	B	C	D	E	F
1	K =	0.01				
2						Fraction
3	Initial					Convert.
4	[F]/[A]	a	b	c	z	x
5	0.1	-0.99	-0.011	0.001	0.0041	0.0267
6	0.1585	-0.99	-0.012	0.0016	0.0064	0.0346
7	0.2512	-0.99	-0.013	0.0025	0.0101	0.0444
8	0.3981	-0.99	-0.014	0.004	0.016	0.0567
9	0.631	-0.99	-0.016	0.0063	0.0253	0.072
10	1	-0.99	-0.02	0.01	0.04	0.0909
11	1.5849	-0.99	-0.026	0.0158	0.0634	0.1141
12	2.5119	-0.99	-0.035	0.0251	0.1007	0.1425
13	3.9811	-0.99	-0.05	0.0398	0.1601	0.1769
14	6.3096	-0.99	-0.073	0.0631	0.2552	0.2182
15	10	-0.99	-0.11	0.1	0.4081	0.2671
16	15.849	-0.99	-0.168	0.1585	0.656	0.324
17	25.119	-0.99	-0.261	0.2512	1.0629	0.3888
18	39.811	-0.99	-0.408	0.3981	1.7431	0.4607
19	63.096	-0.99	-0.641	0.631	2.9094	0.5377
20	100	-0.99	-1.01	1	4.9801	0.617
21	158.49	-0.99	-1.595	1.5849	8.8199	0.6944

Formula Set I
Prototype Cell is E5

=((C5*C5)-(4*B5*D5))

Formula Set II
Prototype Cell is F5

=(-C5-SQRT(E5))/(2*B5)

EXERCISE 5.4

EXTRACTIONS

TO GET THE MOST OUT OF THIS
EXERCISE, YOU SHOULD
ALREADY KNOW

How to find the equilibrium concentrations of reversible chemical reactions using graphical, analytic, and other techniques.

When two immiscible liquids are in contact, solute moves between them and establishes a dynamic equilibrium. The equilibrium amounts of solute dissolved in each solvent depend on solvent volume and on the relative solubilities of solute in each solvent.

Suppose, for example, that solid aspirin is dissolved in water, diethyl ether is added, and the mixture is vigorously stirred (Figure 5.10). Ether and water are not miscible, but aspirin is soluble to some extent in both solvents. The dynamic equilibrium established by the movement of aspirin between the two solvents can be described as

$$\text{aspirin in ether} \rightleftharpoons \text{aspirin in water} \qquad (5.40)$$

The equilibrium constant for this reaction is

$$K = \frac{[\text{aspirin.water}]_{eq}}{[\text{aspirin.ether}]_{eq}} \qquad (5.41)$$

where K is the equilibrium constant and $[\text{aspirin.water}]_{eq}$ and $[\text{aspirin.ether}]_{eq}$ are the equilibrium concentrations of aspirin in water and ether, respectively.

Chemical handbooks often tabulate the equilibrium constants for chemical reactions, but they do not list the equilibrium constants for the distribution of solute across two immiscible solvents. This is because such constants can be accurately approximated from the *solubilities* of the solute in the two solvents.

Suppose that solid aspirin is added to two flasks: one containing 100 mL of water, the other containing 100 mL of diethyl ether. In each case, some aspirin dissolves, and an equilibrium is established between solid and dissolved aspirin (Figure 5.11).

- The equilibrium between a solid and its dissolved solute can be calculated from K_s, the solubility constant (Figure 5.12).

- For a reaction of the form aspirin.solid $\leftrightarrows$ aspirin.solute, K_s = [aspirin.solute]$_{eq}$ (or 1/[aspirin.solute]$_{eq}$ if the reaction is assumed to proceed in the opposite direction) (Figure 5.12a).

- Solubility is usually expressed in terms of mass of solute per volume of *solvent* (e.g., 0.25 g of solute / 100 mL of solvent). Concentration is expressed as mass of solute to volume of *solution*. The two are equivalent only if the volume of solvent remains essentially unchanged by solute addition. *To the degree that solvent volume remains unchanged by solute addition, the equilibrium distribution of a solute between two immiscible solvents can be calculated from relative solubilities instead of relative equilibrium concentrations* (Figure 5.12b, Equation 5.42):

$$K = \frac{[\text{aspirin.water}]_{eq}}{[\text{aspirin.ether}]_{eq}} \sim \frac{S_w}{S_o} = D \qquad (5.42)$$

where K is the equilibrium constant for the reaction aspirin.ether $\leftrightarrows$ aspirin.water, [aspirin.ether]$_{eq}$ and [aspirin.water]$_{eq}$ are the equilibrium concentrations of aspirin in contact with the solid in

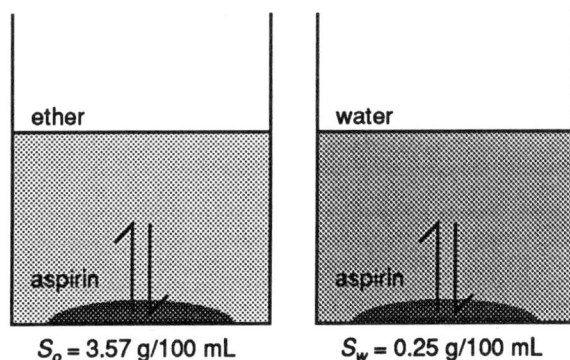

Figure 5.10

Aspirin is added to a mixture of water and diethyl ether with vigorous stirring. Water and diethyl ether are not miscible and will separate into two phases with the dissolved aspirin distributed between them in a dynamic equilibrium.

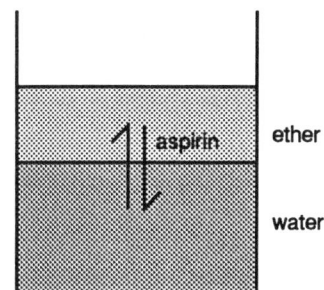

S_o = 3.57 g/100 mL S_w = 0.25 g/100 mL

Figure 5.11

Aspirin is soluble in both diethyl ether and water, and, in the presence of excess solid, a dynamic equilibrium exists between solid aspirin and its dissolved solute. At room temperature, a maximum of 3.57 g of aspirin can dissolve in ether before excess solid remains. In 100 mL of water, a maximum of 0.25 g of aspirin will dissolve. The solubility of a solid in an organic solvent is often designated S_o, whereas the solubility of a solid in water is designated S_w. For aspirin therefore, S_o = 3.57 g/100 mL and S_w = 0.25 g/100 mL.

Figure 5.12

(a) The concentrations of dissolved aspirin in equilibrium with its solid can be calculated from a knowledge of its solubility product constant K_{sp}.

(b) The equilibrium concentrations of solute distributed between two solvents can be calculated from a knowledge of the relative solubility of solute in the two solvents if it is assumed that the volume of solvent increases negligibly with added solute.

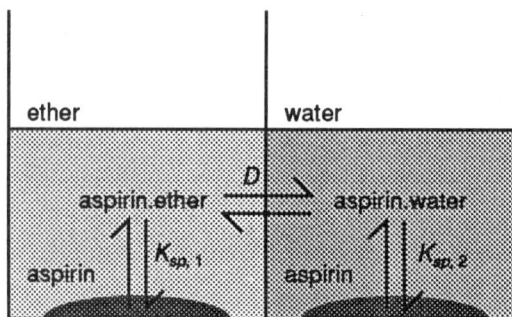

(a)
$$K_{sp,1} = \frac{1}{[\text{aspirin.ether}]_{eq}} \qquad K_{sp,2} = [\text{aspirin.water}]_{eq}$$

$$\text{ether} \rightleftharpoons \text{solid} \rightleftharpoons \text{water}$$

(b)
$$K = \frac{[\text{aspirin.water}]_{eq}}{[\text{aspirin.ether}]_{eq}} \sim \frac{S_w}{S_o} = D$$

$$\text{ether} \rightleftharpoons \text{water}$$

ether and water, S_w and S_o are the solubilities of aspirin in water and ether (o is for organic), and D is the so-called *distribution coefficient* that is used to replace "equilibrium constant."

The distribution of solute between two immiscible solvents is often of considerable practical value. For example, relatively nonpolar organic compounds often can be extracted from aqueous solutions and thus separated from more polar impurities. The natural compounds used in perfumes, drugs, and other products are often isolated, at least in part, by extraction.

LESSONS

1. Build and verify Model 5.10. Model 5.10 calculates the equilibrium amounts and concentrations of solute partitioned between two immiscible solvents. Examine the structure of this model and verify the following statements concerning its structure.

   a. Use cells B2–C4 to enter the solvent volumes, solubilities, and initial solute amounts of the partitioned system. Although no dimensions have been included in the model, the initial values are appropriate for a system in which 100 mL of water (containing 0.15 g of aspirin), and 50 mL of ethyl ether (containing 0.10 g of aspirin) have mixed together and allowed to reach equilibrium.

b. Cells D2 and D4 calculate the total volume of solvent (water plus ether) and total amount of solute.

c. Cell D6 calculates the partition coefficient of solute (the ratio of *concentrations* in the two phases at equilibrium) from the solubilities in cells B3 and C3. In cell D7, the solubilities are multiplied by the volumes of the corresponding solvents to give the ratio of *amounts* of solute in the two phases.

d. Most of the rest of the model should be self-explanatory, but perhaps some explanation for the formula in cell B9 is in order. This cell first calculates the fraction of the solute that is in the organic phase, $x/(x+y)$, where x is the amount in the aqueous phase and y is the amount in the organic phase. Dividing both the numerator and denominator of that fraction by y, you obtain $(x/y)/[(x/y)+1]$. This fraction is then multiplied by the total amount of solute (cell D4) to obtain the amount of solute in the aqueous phase.

Build the on-screen graphic described in Figure 5.13 to help in your studies.

2. When aspirin is partitioned between water and diethyl ether in a dynamic equilibrium, some features of this equilibrium can be surprising.

a. Suppose that 0.15 g of aspirin is dissolved in 100 mL of water and then mixed with 100 mL of ether. What are the equilibrium concentrations of aspirin in each phase?

It probably makes sense that some of the aspirin moves from the solvent in which it is less soluble (water) into the solvent in which it is most soluble (ether), but read on.

b. Suppose that 0.15 g of aspirin is dissolved in 100 mL of ether and then mixed with 100 mL of water. What are the equilibrium concentrations of aspirin in each phase? Explain why some of the aspirin moves from the solvent in which it is more soluble (ether) into the solvent in which it is less soluble (water).

c. Suppose that 0.20 g of aspirin is distributed at equilibrium between 100 mL of water and 100 mL of ether. Verify that the equilibrium amounts of aspirin in water and ether are 0.0131 g and 0.1869 g, respectively.

Figure 5.13

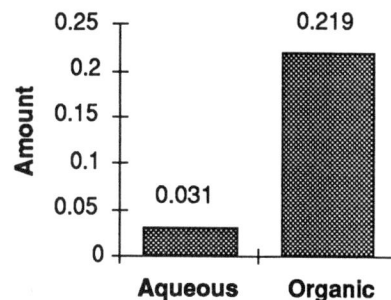

BAR CHART

Data
Horizontal (x) axis: cells B1–C1 (labels)
Vertical (y) axis: cells B9–C9 (equilibrium amounts)

Notice the numeric readout above the bar. Can you get your program to do this?

Figure 5.14

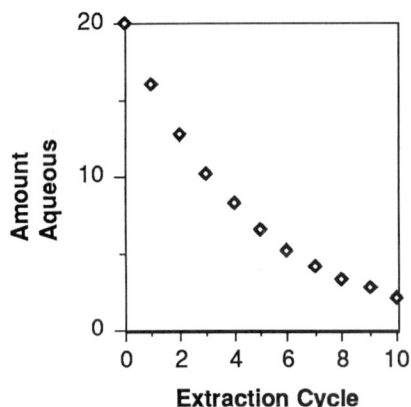

XY–PLOT

Data
Horizontal (x) axis: cells A10–A20
(extraction cycle)
Vertical (y) axis: cells D10–D20
(solute remaining in aqueous phase)

Figure 5.15

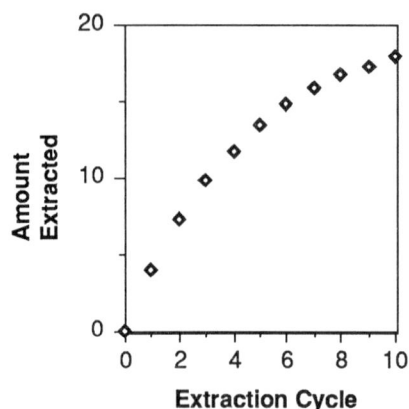

XY–PLOT

Data
Horizontal (x) axis: cells A10–A20
(extraction cycle)
Vertical (y) axis: cells F10–F20 (total
solute extracted)

Now enter these amounts as initial concentrations in your model (i.e., 0.0131 in cell B4 and 0.1867 in cell C4). Your initial and final amounts of aspirin in each solvent should, of course, be the same. Now add 0.01 g of aspirin to the aqueous phase (change cell B4 to 0.231). Notice that most of the added aspirin moves from the aqueous to the ether phase *against a concentration gradient*. Explain how this can happen.

d. Suppose that you have synthesized a valuable product that is mixed in an aqueous solution with several highly polar by-products that are essentially insoluble in organic solvents. Your product has a water solubility S_w of 0.79 g/100 mL and a solubility in ethyl ether S_o of 2.72 g/100 mL. What volume of ether is required to extract 99% of this product from 75 mL of aqueous solution?

e. Explore your system with different solvents and solutes (real or imagined) and vary their amounts to obtain a feel for partitioning in such simple systems.

3. You may have been disappointed by your results in Lesson 2d and the amount of organic solvent necessary to extract most of your synthesized product. There is an easier way.

a. Build and verify Model 5.11. Model 5.11 calculates the effects of multiple extractions in which fresh organic solvent is repeatedly applied to the same batch of aqueous solution. Explore the structure of Model 5.11 until its logic is clear to you.

b. Build two on-screen graphics (Figures 5.14 and 5.15) to help you in your explorations.

4. Suppose that you have synthesized a valuable product that is mixed in an aqueous solution with several highly polar by-products that are essentially insoluble in organic solvents. Your product has a water solubility S_w of 0.79 g/100 mL and a solubility in ethyl ether S_o of 2.72 g/100 mL.

a. What total volume of ether is required to extract 99% of this product from 75 mL of aqueous solution in five extractions? Compare your result with the volume of ether required in Lesson 2d.

b. What total volume of ether is required to extract 99% of this product from 75 mL of aqueous solution in ten extractions? Notice that the additional savings is rather small.

In a real extraction, you must balance extraction efficiency against the time and inconvenience of carrying out many extractions.

5. Repeated extractions make it possible to thoroughly remove solute from a solution, even when it is less soluble in the extracting solvent. Consider a solute that has a solubility of 4 g/L in water and 2 g/L in toluene.

Suppose you have 100 mg of solute in 50 mL of water that is being repeatedly extracted with 50 mL of fresh toluene. How many extractions are required to remove 95% of the solute? 99%? (You may need to add a few additional rows to your model.)

6. The balance between extraction efficiency vs inconvenience is worth one final exercise. In a system containing the solute and solvents described in Lesson 5, how much solute can you withdraw from the aqueous solution in a single extraction with 50 mL of toluene? How much solute can you withdraw in two extractions of 25 mL of toluene each? Five extractions, each with 10 mL of toluene? Ten extractions, each with 5 mL of toluene?

How many additional extractions would you be willing to perform for such small increases in yield?

PROBLEMS

1. Caffeine has a distribution coefficient of 0.032 at room temperature between water and ether. An aqueous solution containing 14 mg of caffeine in a volume of 50 mL is shaken with 50 mL of diethyl ether. How much caffeine is extracted into the organic phase? If the extraction is repeated two more times, how much caffeine will remain in the aqueous layer?

If you were asked to extract 90% of the caffeine present in 50 mL of water, what protocol would you use (i.e., what volumes of ether, and how many extraction cycles)?

2. Trichlorfon, an active ingredient in many commercial insecticides, has the following solubilities at 25°C: 15 g/100 mL in water, 17 g/100 mL in diethyl ether. A large vat contains an aqueous solution of trichlorfon, but due to evaporation its concentration is no longer

known. To determine the amount of trichlorfon present, 100 mL of the aqueous solution is extracted four times with 25 mL of ether. No other components of the aqueous solution are ether-soluble.

When the ether is pooled and evaporated, a dry residue weighing 1.16 g is obtained. If this residue is pure trichlorofon, what is the concentration (in g/100 mL) of trichlorfon in the original solution?

3. Phenacetin is an analgesic with the following solubilities at 25°C: 0.10 g/100 mL in water, 1.2 g/100 mL in diethyl ether. A solution is prepared by dissolving 0.50 g of phenacetin in 500 mL of water.

   a. What volume of ether is required to extract 90% of the phenacetin at one time?

   b. How many extractions are required to remove 99% of the phenacetin if 50 mL of fresh ether is used in each extraction?

MODELS

Model 5.10

BUILD and verify Model 5.10.

ENTER data using cells B2 and C2 (volumes of aqueous and organic solvents), B3 and C3 (solubilities of solute in aqueous and organic solvents), and B4 and C4 (amounts of solute initially dissolved in aqueous and organic phases).

READ output from the model at cells cells B9–C9 (amounts of solute in each phase at equilibrium), B10–C10 (concentrations of solute in each phase at equilibrium), and B11–C11 (mole fractions of solute in each phase at equilibrium).

FORMULAS

	A	B	C	D
1		Aqueous	Organic	Total
2	Volumes =	100	50	=B2+C2
3	Solubilities =	0.25	3.57	
4	Initial Amounts =	0.15	0.1	=B4+C4
5				
6	=(D7/(D7+1))*D4		D = [aq]/[org] =	=B3/C3
7			Amt(aq)/Amt(org) =	=(B3*B2)/(C3*C2)
8				
9	Equil. Amount =		=D4-B9	=B9+C9
10	[Solute]eq =	=B9/B2	=C9/C2	
11	X(solute) =	=B9/D4	=C9/D4	=B11+C11

VALUES

	A	B	C	D
1		Aqueous	Organic	Total
2	Volumes =	100	50	150
3	Solubilities =	0.25	3.57	
4	Initial Amounts =	0.15	0.1	0.25
5				
6			D = [aq]/[arg] =	0.07003
7			Amt(aq)/Amt(org) =	0.14006
8				
9	Equil. Amount =	0.03071	0.21929	0.25
10	[Solute]eq =	0.00031	0.00439	
11	X(solute) =	0.12285	0.87715	1

FORMULAS

	A	B	C	D	E	F	G
1		Aqueous	Organic				
2	Volumes =	10	5				
3	Solubilities =	1	0.5				
4	Initial Amount =	20					
5							
6	Amt(aq)/Amt(org) =	=(B3*B2)/(C3*C2)					
7							
8		Starting	Starting	Equil	Equil	Total	Volume
9	Extraction Cycle	Aqueous	Organic	Aqueous	Organic	Extracted	Extracted
10	0			=B4	0	0	0
11	1	=D10	0				
12	2	=D11	0				
13	3	=D12	0	Formula	Formula	Formula	Formula
14	4	=D13	0	Set	Set	Set	Set
15	5	=D14	0	I	II	III	IV
16	6	=D15	0				
17	7	=D16	0				
18	8	=D17	0				
19	9	=D18	0				
20	10	=D19	0				

VALUES

	A	B	C	D	E	F	G
1		Aqueous	Organic				
2	Volumes =	10	5				
3	Solubilities =	1	0.5				
4	Initial Amount =	20					
5							
6	Amt(aq)/Amt(org) =	4					
7							
8		Starting	Starting	Equil	Equil	Total	Volume
9	Extraction Cycle	Aqueous	Organic	Aqueous	Organic	Extracted	Extracted
10	0			20	0	0	0
11	1	20	0	16	4	4	5
12	2	16	0	12.8	3.2	7.2	10
13	3	12.8	0	10.24	2.56	9.76	15
14	4	10.24	0	8.192	2.048	11.808	20
15	5	8.192	0	6.5536	1.6384	13.4464	25
16	6	6.5536	0	5.24288	1.31072	14.75712	30
17	7	5.24288	0	4.194304	1.048576	15.8057	35
18	8	4.194304	0	3.355443	0.838861	16.64456	40
19	9	3.355443	0	2.684355	0.671089	17.31565	45
20	10	2.684355	0	2.147484	0.536871	17.85252	50

Formula Set I Prototype Cell is D11	Formula Set II Prototype Cell is E11
=(**B6**/(**B6**+1))*(D10+E10)	=D10+E10-D11

Formula Set III Prototype Cell is F11	Formula Set IV Prototype Cell is G11
=F10+E11	=A11***C2**

Model 5.11

BUILD and verify Model 5.11.

ENTER data using cells B2 and C2 (volumes of aqueous and organic solvents), B3 and C3 (solubilities of solute in aqueous and organic solvents), and B4 (amount of solute initially dissolved in aqueous phase).

READ output from the model at cells A10–A20 (extraction cycle), D10–D20 (equilibrium solute concentration in aqueous solvent at end of extraction), E10–E20 (equilibrium solute concentration in organic solvent at end of extraction), F10–F20 (total solute extracted), and G10–G20 (total volume of organic solvent used).

NOTICE that each extraction is assumed to be with fresh organic solvent and that the volume of organic solvent used in each extraction is assumed to be constant.

CHAPTER **6**

ACIDS, BASES, AND IONS

We live in a world whose oceans, lakes, and ponds are filled with water, not benzene; and a critically important vapor in our atmosphere is H_2O, not NH_3. In this chapter, you round out your understanding of dynamic equilibrium by studying the behavior of acids, bases, and ions in aqueous solution. Such behavior forms the basis of most complex chemical reactions that underlie living systems (including you); the chemical reactions occurring in your brain at this moment are responsible for many of your experiences as you read this book.

The fundamental chemical reaction that supports the behavior of acids and bases in aqueous solution is a reversible reaction in which a proton is transferred from one water molecule to another (Equation 6.1):

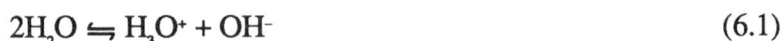

$$2H_2O \rightleftharpoons H_3O^+ + OH^- \tag{6.1}$$

The equilibrium constant for this reaction can be represented as

$$K = \frac{[H_3O^+][OH^-]}{[H_2O]^2} \tag{6.2}$$

Although free protons do not exist in appreciable numbers in aqueous solution, we will pretend that they do and rewrite Equations 6.1 and 6.2 as Equations 6.3 and 6.4:

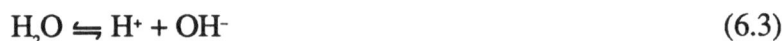

$$H_2O \rightleftharpoons H^+ + OH^- \tag{6.3}$$

$$K = \frac{[H^+][OH^-]}{[H_2O]} \tag{6.4}$$

Because the concentration of water remains essentially unchanged at $55M$ in this reaction, both sides of Equation 6.4 can be multiplied by $[H_2O]$ to yield

$$K[H_2O] = K_w = 10^{-14} = [H^+][OH^-] \tag{6.5}$$

where K_w is called the *dissociation constant* for water and has a value of 10^{-14} at $25°C$.

In aqueous solution, strong acids such as HCl are completely dissociated into H^+ and Cl^- ions. If HCl is added to pure water to form a $10^{-3}M$ solution, you can use Equation 5.5 to calculate that $[OH^-] = 10^{-11}M$ in the solution (i.e., $[H^+][OH^-] = [10^{-3}][10^{-11}] = 10^{-14}$). This kind of calculation occurs so often in aqueous chemistry that a separate notation has evolved in which K, $[H^+]$, and $[OH^-]$ are represented as their negative logarithms (to base 10) and in which $-\log K_w = pK_w$, $-\log [H^+] = pH$, and $-\log [OH^-] = pOH$.

In this notation (and at 25°C) Equation 6.5 is written

$$pK_w = 14 = pH + pOH \qquad (6.6)$$

and a $10^{-3}M$ aqueous solution of HCl is said to have a pH of 3.[1]

- A solution with a pH of 7 has equal numbers of H^+ and OH^- ions and is *neutral*.

- A solution with a pH below 7 has more H^+ than OH^- ions and is *acidic*.

- A solution with a pH above 7 has fewer H^+ than OH^- ions and is *basic*.

This notation becomes especially useful when studying weak acids. Acetic acid, for example, is a weak acid — it only partially dissociates in aqueous solution (Equation 6.7):

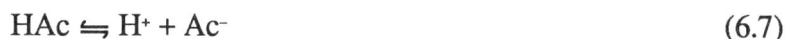

$$HAc \rightleftharpoons H^+ + Ac^- \qquad (6.7)$$

where HAc is the undissociated form of acetic acid and Ac^- is its anion.

This equilibrium can be written

$$K = \frac{[H^+][Ac^-]}{[HAc]} \qquad (6.8a)$$

or

$$[H^+] = K\left(\frac{[HAc]}{[Ac^-]}\right) \qquad (6.8b)$$

Equation 6.8b can also be written in pH notation as

$$pH = pK + \log\left(\frac{[Ac^-]}{[HAc]}\right) \qquad (6.9)$$

Chemical handbooks usually describe the dissociative behavior of weak acids and bases in terms of the pKs rather than their equilibrium constants.

Equation 6.9 is the Henderson–Hasselbalch equation, and you will be living with it quite intimately in the exercises that follow.

ACIDIC DISSOCIATION: THE HENDERSON–HASSELBALCH EQUATION

Weak acids are only partially dissociated in aqueous solution (Equation 6.10):

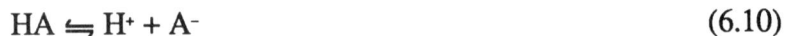

$$HA \rightleftharpoons H^+ + A^- \tag{6.10}$$

where HA and A^- are the undissociated and dissociated forms of a weak acid and H^+ is a proton. H^+, of course, does not really exist as a free proton but rather is transferred to water to form H_3O^+ (or larger) complexes (Equation 6.11):

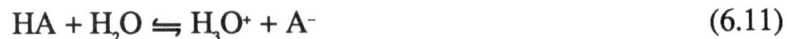

$$HA + H_2O \rightleftharpoons H_3O^+ + A^- \tag{6.11}$$

A weak acid that donates a proton becomes a *conjugate base* (A^-) that is capable of accepting a proton. A base that accepts a proton becomes a *conjugate acid* (HA) that is capable of donating a proton.

The equilibrium constant K for the reaction described in Equation 6.10 is defined by Equation 6.12, but, as noted in the introduction, a more convenient measure of dissociation can be obtained by taking the negative logarithms (to base 10) of each factor and rearranging the terms (Equation 6.13):

$$K = \frac{[H^+][A^-]}{[HA]} \qquad (6.12)$$

$$pH = pK + \log\left(\frac{[A^-]}{[HA]}\right) \qquad (6.13)$$

Equation 6.13 is called the Henderson–Hasselbalch equation, and it is the cornerstone of all work involving weak acids and bases in aqueous solutions. An alternate form of Equation 6.13 that is also very useful is

$$\frac{[A^-]}{[HA]} = 10^{(pH-pK)} \qquad (6.14)$$

Equation 6.14 is especially useful when the pH is known and you wish to know the concentrations of conjugate acid and base. The ratio of $[A^-]$ to $[HA]$ obtained in Equation 6.14 can be converted to mole fractions of A^- and HA by using Equations 6.15 and 6.16).

$$X_{HA} = \frac{[HA^-]}{[A^-] + [HA]} = \frac{1}{\frac{[A^-]}{[HA]} + 1} \qquad (6.15)$$

$$X_A = 1 - X_{HA} \qquad (6.16)$$

where X_{HA} and X_A are the mole fractions of HA and A^- in a mixture consisting of only these two forms.

Equations similar to Equation 6.15 can be obtained even when the weak acid contains more than a single dissociable group (as you will see).

LESSONS

1. Suppose that you prepare an aqueous solution of $0.01M$ acetic acid (HAc). Taking the solution to your laboratory bench, you monitor pH with a pH meter while stirring and adding known amounts of sodium hydroxide (NaOH).[1] At the end of your experiment, you prepare a *titration curve* of pH vs added base.

   NaOH is a strong base, and HAc is a weak acid. As NaOH is added to the solution, virtually all added OH^- combines with H^+ (from partially dissociated HAc) to form H_2O. As $[H^+]$ is reduced by added OH^-, HAc further dissociates to produce more H^+. The reaction is shown in Equation 6.17:

$$Na^+ + OH^- + HAc \rightarrow Na^+ + Ac^- + H_2O \qquad (6.17)$$

Figure 6.1

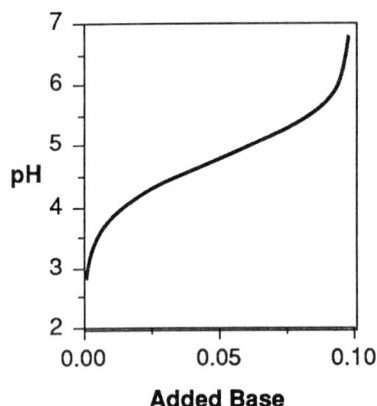

Added Base

XY–PLOT

Data
Horizontal (*x*) axis: cells A9–A107
(base consumed)
Vertical (*y*) axis: cells E9–E107 (pH)

Notice that the endpoints of the range
of calculated values are not plotted.
The endpoints contain error flags
when the model is initialized with pure
conjugate acid or conjugate base, a
common situation.

Build and verify Model 6.1. Model 6.1 represents the titration of a weak acid with a strong base.

a. Examine the structure of Model 6.1. The model creates a table of values for added base (column A) and calculates a range of values for [A⁻] and [HA] in an aqueous solution from these values [added strong base stoichiometrically increases conjugate base (column B) and reduces conjugate acid (column C) as predicted by Equation 6.17]. The model then calculates the pH of the resulting solution using the Henderson–Hasselbalch equation (Equation 6.13).

Explore the model until you understand its structure and notice especially the following features: The pK of the simulated weak acid or base is entered into cell B1, and the starting concentrations of conjugate base and conjugate acid are entered into cells B2 and B3.[2] Cell B4 defines the increment in strong acid used to set up column A. Columns B and C calculate the changes in concentration of conjugate acid and conjugate base. Columns D and E represent the Henderson–Hasselbalch equation (Equation 6.13).

Notice that this kind of structure can yield error flags at the endpoints of the model (cells E8 and E108) when the initial conditions assume pure conjugate acid or base. Why?

b. Build an on-screen graphic that plots pH vs added base (Figure 6.1). Notice that there is a region of this plot where pH changes only slightly as NaOH is added. Weak acids are often called *buffers* because of their ability to reduce the pH change that results from added strong base. Different weak acids serve as buffers over different pH ranges. Enter other values of pK (cell B1) into your model in order to see other buffered pH ranges.

What is the relationship between the pK of a weak acid and the range of pH values that it can buffer?

c. Model 6.1 represents the titration of a weak acid by a strong base. How would you change the model so that it simulates the titration of a weak base by a strong acid?

2. Model 6.1 calculates pH from a knowledge of [A⁻] and [HA]. The amount of strong base consumed in the reaction is calculated from the change on [A⁻]. Look at these relationships from a different

perspective by calculating [A⁻] and [HA] from a knowledge of pH (Equations 6.14–6.16).

Build and verify Model 6.2. Model 6.2 examines the titration of a weak acid by a strong base by calculating the mole fraction of conjugate acid X_{HA} and the amount of strong base consumed as a function of pH.

a. Examine the structure of Model 6.2. The model generates a range of pH values (column A) and then computes [A⁻]/[HA] via Equation 6.14 (column B). Equation 6.15 then computes X_{HA} (column C), and the amount of strong base consumed is found from the change in X_{HA} (column D).

Explore the model until you understand its structure then build the two on-screen graphics described in Figures 6.2 and 6.3.

b. Using the initial values in the model, notice that for pH ranges far from the pK of a weak acid, large changes in pH can occur with almost no change in added strong acid or base. *Weak acids are often added to solutions as buffers to stabilize the pH of an ongoing chemical reaction.* Examine the titration curves of the weak acids in the table below. Which weak acid would you use to control the pH of a chemical reaction requiring an environment of pH 9.4?

Acid	pK
Acetic acid	4.75
Acetoacetic cid	3.58
Glycine	9.87
Iodic acid	0.77
Periodic acid	1.64
Saccharine	11.68

c. What are the concentrations of the conjugate acid and conjugate base of acetic acid in a solution of pH = 2? pH = 9? Over what range of pH values do the greatest changes in composition occur?

3. Many weak acids contain more than a single donatable proton. Phosphoric acid (H_3PO_4), for example, has three different protons ($pK_1 = 2.12$, $pK_2 = 7.28$, $pK_3 = 12.32$). Computer models designed to track the behavior of such acids rely only on the Henderson–Hasselbalch equation (Equation 6.14) and partition equations similar to those in Equation 6.15 and 6.16.

Figure 6.2

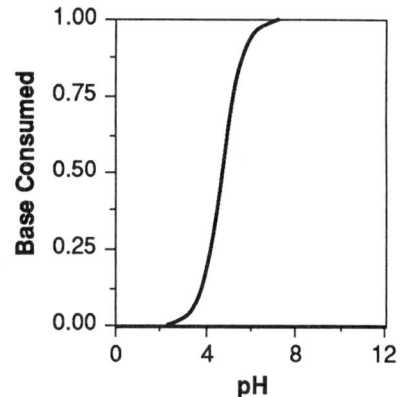

XY–PLOT

Data
Horizontal (x) axis: cells A6–A61 (pH)
Vertical (y) axis: cells D6–D61 (base consumed)

Figure 6.3

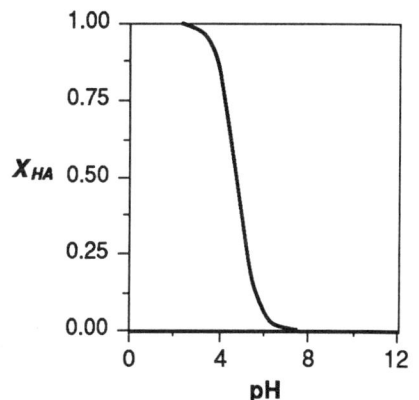

XY–PLOT

Data
Horizontal (x) axis: cells A6–A61 (pH)
Vertical (y) axis: cells C6–C61 (mole fraction of HA)

Figure 6.4

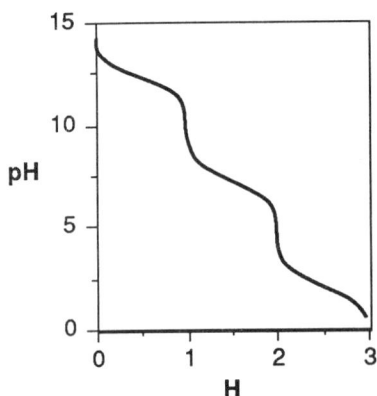

XY PLOT

Data
Horizontal (x) axis: cells I7–I63 (total H)
Vertical (y) axis: cells A7–A63 (pH)

The organization of Model 6.3 is somewhat inconvenient for Excel users because Excel generates *xy*-plots with the *x*-axis data in the left-most row. To get around this feature, either add a duplicate pH row in column J and plot normally or an *xy*-plot of any convenient data and then edit the edit line with new addresses.

Figure 6.5

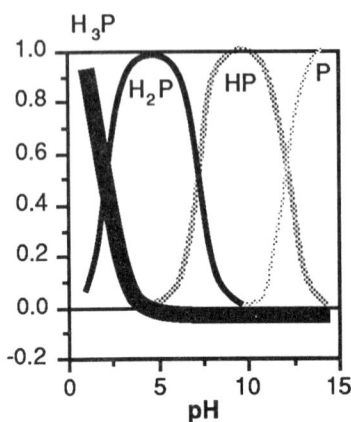

XY PLOT

Data
Horizontal (x) axis: cells A7–A63 (pH)
Vertical (y) axis: cells E7–E63, F7–F63, G7–G63, and H7–H63 (mole fractions of ionic forms)

Consider a weak acid such as H_3PO_4 with three dissociable groups. In aqueous solution, the mole fraction of H_3A, with respect to all forms of the weak acid, can be calculated using Equation 6.18. Equation 6.18 can, with a bit of work, rearrange to Equation 6.19:

$$X_{H_3A} = \frac{[H_3A]}{[H_3A] + [H_2A^-] + [HA^{2-}] + [A^{3-}]} \quad (6.18)$$

$$X_{H_3A} = \frac{1}{1 + \left(\frac{[H_2A^-]}{[H_3A]}\right) + \left(\frac{[HA^{2-}]}{[H_2A^-]}\right)\left(\frac{[H_2A^-]}{[H_3A]}\right) + \left(\frac{[A^{3-}]}{[HA^{2-}]}\right)\left(\frac{[HA^{2-}]}{[H_2A^-]}\right)\left(\frac{[H_2A^-]}{[H_3A]}\right)} \quad (6.19)$$

From the value of $[H_3A]$ in Equation 6.19, the values of the appropriate concentration ratios (Equation 6.15), and the relationships below (Equations 6.20–6.22), the concentrations or mole fractions of each of the weak acid forms can be computed.

$$[H_2A^-] = \frac{[H_2A^-]}{[H_3A]}[H_3A] \quad (6.20)$$

$$[HA^{2-}] = \frac{[HA^{2-}]}{[H_2A^-]}[H_2A^-] \quad (6.21)$$

$$[A^{3-}] = \frac{[A^{3-}]}{[HA^{2-}]}[HA^{2-}] \quad (6.22)$$

These equations work together to provide an understandable and accurate picture of the behavior of a buffer with three dissociable groups.

a. Build and verify Model 6.3. The model calculates the acid consumed (column J) and the mole fraction of each ionic form of a weak acid with three dissociable groups (columns F, G, H) at a variety of pH values (column A).

    Explore the model and its structure until you understand how it works.

b. Build and verify two on-screen graphics capable of monitoring the behavior of the model (Figures 6.4 and 6.5).

c. There are three pH values at which H_3PO_4 exhibits maximum buffering capability (Figure 6.4). What are they?

d. Use your on-screen graphic (Figure 6.5) to identify the pH regions where the mole fractions of two or more forms of the weak acid are undergoing rapid change. How do these regions correlate with the regions of maximum buffering and the pK values of the weak acid?

e. Phosphate is an important buffer system in human cells and is used as a buffer in many biochemical experiments. If the pH of your cells is 7.2, which ionization of the phosphate system is involved in buffering? Which ionic forms of phosphate occur in the highest concentrations in your body?

f. Many organic molecules that are important in your body contain two or more ionizable groups. The amino acid arginine, for example, has three donatable protons with pK values of 2.3, 9.6, and 12.5. Modify Model 6.3 to represent the dissociation behavior of arginine. Does arginine contribute significantly to buffering in living cells? Why or why not?

MODELS

Model 6.1

BUILD and verify Model 6.1.

ENTER data using cells B1–B4 (pK, starting [A⁻], starting [HA], and added strong base).

READ output from the model at cells A8–A108 (added base), B8–B108, C8–C108 (concentrations of conjugate base and acid), and E8–E108 (pH).

FORMULAS

	A	B	C	D	E
1	pK =	4.75			
2	Start [A] =	0			
3	Start [HA] =	0.1			
4	Δ Base =	0.001			
5					
6	Added Base				
7	Consumed	[A]	[HA]	[A]/[HA]	pH
8	0	=B2	=B3		
9					
10					
11	Formula Set I	Formula Set II	Formula Set III	Formula Set IV	Formula Set V
106					
107					
108					

VALUES

	A	B	C	D	E
1	pK =	4.75			
2	Start [A] =	0			
3	Start [HA] =	0.1			
4	Δ Base =	0.001			
5					
6	Added Base				
7	Consumed	[A]	[HA]	[A]/[HA]	pH
8	0	0	0.1	0	#NUM!
9	0.001	0.001	0.099	0.0101	2.75436
10	0.002	0.002	0.098	0.02041	3.0598
11	0.003	0.003	0.097	0.03093	3.24035

	A	B	C	D	E
106	0.098	0.098	0.002	49	6.4402
107	0.099	0.099	0.001	99	6.74564
108	0.1	0.1	-7E-17	-1E+15	#NUM!

Formula Set I
Prototype Cell is A9

=A8+**B4**

Formula Set III
Prototype Cell is C9

=**B3**-A9

Formula Set II
Prototype Cell is B9

=**B2**+A9

Formula Set IV
Prototype Cell is D8

=B8/C8

Formula Set V
Prototype Cell is E8

=**B1**+LOG10(D8)

FORMULAS

	A	B	C	D
1	pK =	7		
2	Start pH =	1		
3	Incr pH =	0.2		
4				Base
5	pH	[A]/[HA]	X(HA)	consumed
6	=B2			
7				
8	Formula	Formula	Formula	Formula
9	Set I	Set II	Set III	Set IV
59				
60				
61				

Model 6.2

BUILD and verify Model 6.2.

ENTER data using cells B1 (pK) and B2 and B3 (starting pH, increment in pH).

READ output from the model at cells A6–A61 (pH), and C6–C61 (mole fraction of HA), and D6–D61 (base consumed).

VALUES

	A	B	C	D
1	pK =	7		
2	Start pH =	1		
3	Incr pH =	0.2		
4				Base
5	pH	[A]/[HA]	X(HA)	consumed
6	1	0.000001	0.999999	
7	1.2	1.58489E-06	0.999998415	5.84892E-07
8	1.4	2.51189E-06	0.999997488	1.51188E-06
9	1.6	3.98107E-06	0.999996019	2.98106E-06

	A	B	C	D
59	11.6	39810.71706	2.51182E-05	0.999973882
60	11.8	63095.73445	1.58487E-05	0.999983151
61	12	100000	9.9999E-06	0.999989

Formula Set I Prototype Cell is A7
=A6+**B3**

Formula Set II Prototype Cell is B6
=10^(A6-**B1**)

Formula Set III Prototype Cell is C6
=1/(B6+1)

Formula Set IV Prototype Cell is D7
=**C6** -C7

Model 6.3

BUILD and verify Model 6.3.

ENTER data using cells B1–B3 (three pK values for weak acid) and D1 and D2 (starting pH, increment in pH)

READ output from the model at cells A7–A60 (pH), E7–E63, F7–F63, G7–G63, H7–H63 (mole fractions of ionic forms of weak acid), and I7–I63 (total H reacted).

FORMULAS

	A	B	C	D	E	F	G	H	I
1	pK1 =	2.12	Start pH =	0					
2	pK2 =	7.28	Incr pH =	0.25					
3	pK3 =	12.32							
4									
5			——————[A]/[HA] Ratios——————			——————-mol fraction——————		
6	pH	[H2A]/[H3A]	[HA]/[H2A]	[A]/[HA]	H3A	H2A	HA	A	Tot H
7	=D1								
8									
9	Form Set I	Formula Set II	Formula Set III	Formula Set IV	Formula Set V	Formula Set VI	Formula Set VII	Formula Set VIII	Formula Set IX
61									
62									
63									

VALUES

	A	B	C	D	E	F	G	H	I
1	pK1 =	2.12	Start pH =	0					
2	pK2 =	7.28	Incr pH =	0.25					
3	pK3 =	12.32							
4									
5			——————[A]/[HA] Ratios——————			——————-mol fraction——————		
6	pH	[H2A]/[H3A]	[HA]/[H2A]	[A]/[HA]	H3A	H2A	HA	A	Tot H
7	0	0.00758578	5.2481E-08	4.7863E-13	0.99247	0.00753	4E-10	1.9E-22	2.99247
8	0.25	0.01348963	9.3325E-08	8.5114E-13	0.98669	0.01331	1.2E-9	1.1E-21	2.98669
9	0.5	0.02398833	1.6596E-07	1.5136E-12	0.97657	0.02343	3.9E-9	5.9E-21	2.97657

	A	B	C	D	E	F	G	H	I
61	13.5	2.3988E+11	1659586.91	15.1356125	1.6E-19	3.7E-08	0.06197	0.93803	0.06197
62	13.75	4.2658E+11	2951209.23	26.915348	2.8E-20	1.2E-08	0.03582	0.96418	0.03582
63	14	7.5858E+11	5248074.6	47.8630092	5.1E-21	3.9E-09	0.02047	0.97953	0.02047

Formula Set I Prototype Cell is A8	Formula Set IV Prototype Cell is D7	Formula Set VII Prototype Cell is G7
=A7+**D2**	=10^(A7-**B3**)	=C7*F7

Formula Set II Prototype Cell is B7	Formula Set V Prototype Cell is E7	Formula Set VIII Prototype Cell is H7
=10^(A7-**B1**)	=1/(1+B7+C7*B7+D7*C7*B7)	=D7*G7

Formula Set III Prototype Cell is C7	Formula Set VI Prototype Cell is F7	Formula Set IX Prototype Cell is I7
=10^(A7-**B2**)	=B7*E7	=G7+2*F7+3*E7

ACIDIC DISSOCIATION: ISOELECTRIC POINTS

When a weak acid in aqueous solution dissociates and donates a proton to water, it also loses a positive charge. The conjugate base therefore becomes either less positively charged, neutral, or more negatively charged. While all weak acids examined in Exercise 6.1 have this property (go back and look!), this exercise focuses upon a particularly interesting and important class of molecules — amino acids. Amino acids are organic molecules containing at least two ionizable groups: a carboxyl group and an amino group. α-amino acids are essential components of all living organisms on Earth and are the components out of which all proteins are made.[1]

Glycine is the simplest member of the amino acid family (Figure 6.6). At pH values very much below 2, glycine has a net positive charge, and both of its ionizable groups are conjugate acids. In a broad range of pH values between 2.3 and 9.8, a particularly interesting species occurs that is highly polar despite the fact that it has no net charge. Above pH values of 10, most glycine molecules in aqueous solution carry a net negative charge.

Exercise 6.1 contains all the tools you will need to explore these interesting molecules.

TO GET THE MOST OUT OF THIS EXERCISE, YOU SHOULD ALREADY KNOW

The definition of the terms:
dynamic equilibrium
logarithm
acid
base

Figure 6.6

Glycine is a member of a family of amino acids called α-amino acids. It has an amino group that can serve as a conjugate base (–NH$_2$) or a conjugate acid (–NH$_3^+$) and a carboxyl group that can do likewise (–COO$^-$ and –COOH).

The pK of the carboxyl group has a value of 2.3. The pK of the amino group has a value of 9.8. The spacing between these two pK values makes glycine a *dipolar ion* at neutral pH (i.e., it contains both a positive and a negative charge but has a net charge of zero). The concentration of neutral glycine molecules, NH$_2$CH$_2$COOH, is extremely low (but, recalling the principles of dynamic equilibrium, not zero).

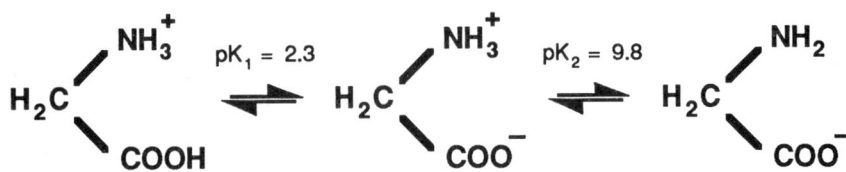

LESSONS

1. Model 6.4 simulates the behavior of the amino acid glycine in an aqueous environment. It is quite similar to Models 6.2 and 6.3. A new column that calculates the net charge of the molecule for a given pH has been added.

   Build and verify Model 6.4. Also build the on-screen graphics described in Figures 6.7 and 6.8.

   a. What two columns of cells contain the Henderson–Hasselbalch equation?

   b. What is the role of the cells in columns D, E, and F? Derive these equations.

   c. Explain the function of the formulas in columns G and H.

2. Examine your plots (and the data tables from which they are derived, when appropriate) to answer the following questions:

   a. What forms of glycine exist at what values of pH? What are the net charges on each of these ionic forms?

   b. The isoelectric point of a molecule is the pH value at which the molecule exhibits no net charge. What is the isoelectric point of glycine? Can you write an algebaic expression that calculates this pH from pK$_1$ and pK$_2$?

   Notice that, when the pK values are far apart (as in amino acids), there is a broad range around the isoelectric point in which nearly all molecules have no net charge.

   c. What is the predominant ionic form of glycine is at its isoelectric point? From your knowledge of the principles of dynamic equilibrium, would you say that glycine's other forms are totally absent?

Figure 6.7

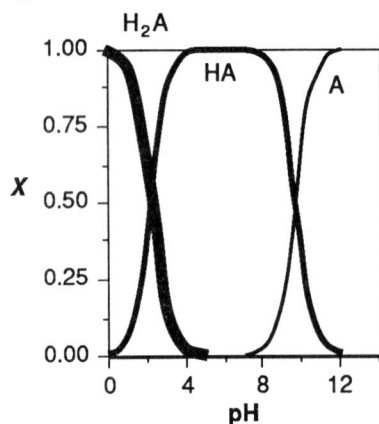

XY–PLOT

Data
Horizontal (x) axis: cells A11–A67 (pH)
Vertical (y) axis: cells D11–D67, E11–E67, and F11–F67 (mole fractions of ionic forms)

3. Below is a table of several amino acids and their pK values. Plot the net charge of each amino acid as a function of pH. Find the isoelectric point of each amino acid and the predominant ionic form at pH = 7.2. (In each case, the net charge of the fully protonated form is +1.)

Amino Acid	pK_1	pK_2
Leucine	2.4	9.6
Methionine	2.3	9.2
Phenylalanine	1.8	9.1
Proline	2.0	10.6

4. Arginine, also an amino acid, is somewhat more complex than glycine. Arginine has an additional ionizable group, called a guanidino group, that increases the complexity of its titration curve (Figure 6.9). You can obtain a titration curve of arginine by simply replacing the pK values in Model 6.3 with those of arginine (pK_1 = 2.3, pK_2 = 9.1, pK_3 = 12.5). To examine the net charge of the molecule as a function of pH, Model 6.3 must be slightly modified.

a. Build and verify Model 6.5. If you are facile with the editing capabilities of your spreadsheet program, modify Model 6.3 by inserting additional rows (for the table of charges for each ionic form) and adding new formulas in column J.

Explore Model 6.5 thoroughly until you understand its structure. How is Model 6.5 different from Model 6.3?

Build on-screen graphics to support your interrogation of this model (Figures 6.10 and 6.11).

b. What forms of arginine exist at what values of pH? What are the net charges on each of these ionic forms?

Figure 6.8

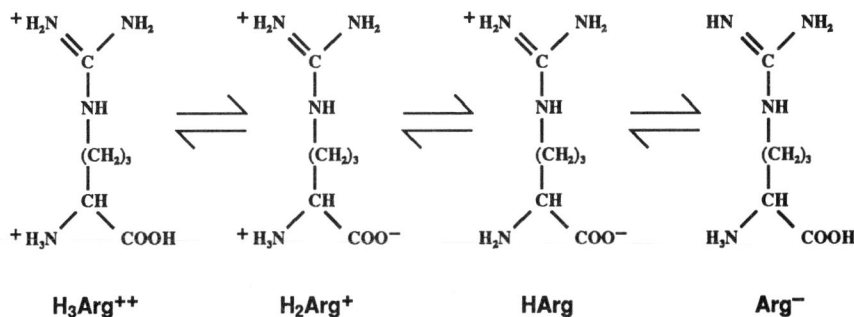

XY–PLOT

Data
Horizontal (x) axis: cells A11–A67 (pH)
Vertical (y) axis: cells H11–H67 (net charge of molecule)

Figure 6.9

Arginine is an amino acid with three ionizable groups. Because the pKs of these groups are widely spaced, only four of the possible ionic forms ever exist in appreciable concentrations.

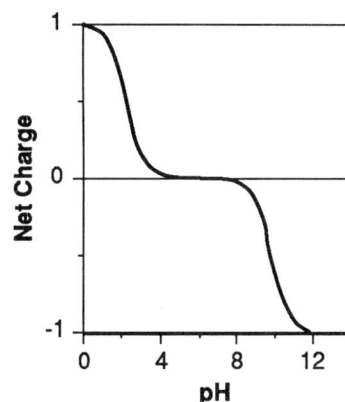

Figure 6.10

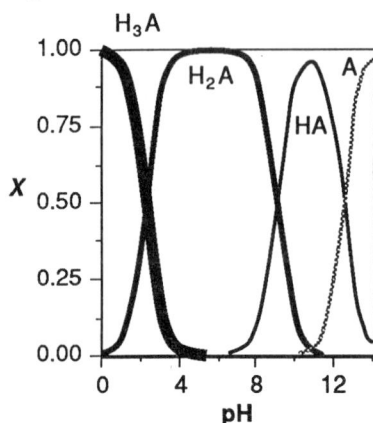

XY–PLOT

Data

Horizontal (x) axis: cells A13–A69 (pH)

Vertical (y) axis: cells E13–E69, F13–F69, G13–G69, and H13–H69 (mole fractions of ionic forms)

Figure 6.11

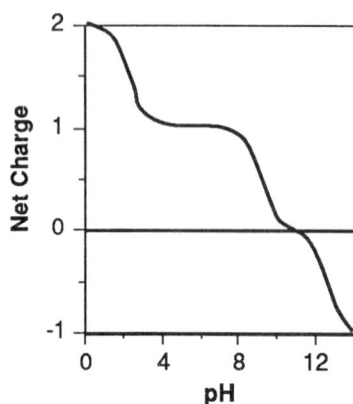

XY–PLOT

Data

Horizontal (x) axis: cells A13–A69 (pH)

Vertical (y) axis: cells J13–J69 (net charge of molecule)

c. What is the isoelectric point of arginine, and what is the predominant ionic form of arginine at its isoelectric point?

d. Below is a table of several amino acids and their pK values. The net charge of the fully protonated form is found in parentheses after the name of the amino acid.

Plot the net charge of each amino acid as a function of pH. Find the isoelectric point of each amino acid and the predominant ionic form at pH = 7.2.

Amino Acid	pK_1	pK_2	pK_3
Lysine (+2)	2.2	9.2	10.8
Histidine (+2)	1.8	6.0	9.2
Tyrosine (+1)	2.2	9.1	10.9

5. Amino acids can be assembled into large, complex molecules called proteins whose three-dimensional structures mediate many of the vital functions required of a living organism (Figure 6.12). Amino acids polymerize into proteins (under the direction of the genetic code) by forming long chains in which the amino end of one amino acid reacts with the carboxyl end of another amino acid. The ionization behavior of proteins are mediated by the ionizing groups on the side chains of seven of the twenty amino acids and by the single amino and carboxyl groups located at the ends of each polymerized chain. Titration curves of proteins (and their smaller sisters, polypeptides) can be complex because of these ionizing groups incorporated into their structures. Although interactions between neighboring ionizing groups can affect ionization behavior, a close approximation to actual titration curves can be created by assuming that each group behaves independently.

Consider an amino acid containing a side chain with a charge of –1 when in its conjugate *base* form. You should be able to verify that the charge z contributed by n such groups in a polypeptide chain equals

$$z = -n \left[\frac{10^{(pH - pK)}}{1 + 10^{(pH - pK)}} \right] \tag{6.23}$$

On the other hand, an amino acid containing a side chain with a charge of +1 when in its conjugate *acid* form, contributes a charge z that equals

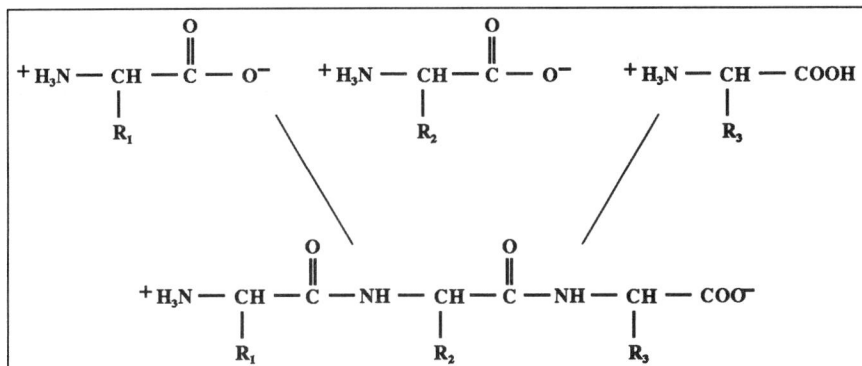

$$z = +n \left[1 - \frac{10^{(pH - pK)}}{1 + 10^{\,(pH - pK)}} \right] \qquad (6.24)$$

a. Build and verify Model 6.6. Model 6.6 calculates the net charge on a protein or polypeptide over a range of pH values. Cells B3–J5 contain the ionization properties of seven ionizing side groups of the amino acids aspartic acid, glutamic acid, histidine, lysine, arginine, tyrosine, and cysteine, as well as the ionization properties of the terminal amino and carboxyl groups of a polypeptide chain. Cells B2–J2 contain cells into which you can put the number of residues of each amino acid present in your simulated protein or polypeptide.

Explore your model and verify that you understand its structure; then build the on-screen graphic represented by Figure 6.13.

b. Lysozyme, a protein found in egg white and whose structure has been extensively studied, is composed of 129 amino acids, 32 of which contain ionizable side chains. There are 2 glutamate, 8 aspartate, 1 histidine, 6 lysine, 11 arginine, and 4 tyrosine residues in the molecule.

Generate a plot of net charge vs pH for lysozyme and find its isoelectric point. (Because the pK values of each ionizing group can be affected by neighboring chemical groups, your plot will be only approximately correct.)

c. Lysozyme contains eight cysteine residues, all of which are cross-linked via disulfide bonds in the native protein. What happens to the isoelectric point of lysozyme if these disulfide bridges are broken and eight –SH groups become available? (The –SH group of cysteine has a pK value of 8.3.)

Figure 6.12

Amino acids can be assembled by polymerization into much larger molecules called proteins. Proteins are essential to all known forms of life on Earth and serve as the structural elements, catalysts (i.e., enzymes), and motors (i.e., contractile proteins) that make life possible. Amino acids are assembled into proteins under the direction of the genetic code by joining the carboxyl end of one amino acid to the amino group of the next. In this way, long chains of amino acids assemble to form molecules with remarkable properties. Twenty different amino acids participate in the assembly of proteins. They are distinguished by the nature of the side chain (R_1, R_2, ...) attached to the carbon adjacent to the carboxyl group of the amino acid. Once assembled, only the ionizing groups of the side chains and the amino and carboxyl groups at the end of each peptide participate in the titration behavior of the protein.

Figure 6.13

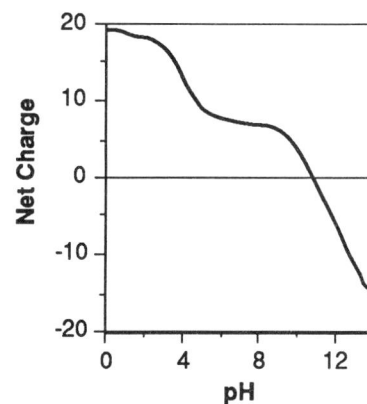

XY–PLOT

Data
Horizontal (*x*) axis: cells A9–A37 (pH)
Vertical (*y*) axis: cells K9–K37 (net charge of molecule)

Suppose that, by chemical reaction, the ionizing groups of tyrosine and cysteine in lysozyme are blocked. Generate a new plot of net charge vs pH and find the isoelectric point of chemically treated lysozyme.

d. Suppose that, by chemical reaction, three of lysozyme's aspartate residues are converted to neutral asparagine residues. What is the isoelectric point of the resulting molecule?

6. The amino acid contents and isoelectric points of many important proteins are known and available in chemical handbooks. Use Model 6.6 to predict the isoelectric points of several such proteins and compare your predicted value with its actual value. What are the causes of the deviations between simulation and reality?

PROBLEMS

1. Lesson 4 assumes that pK values of each of the ionizable groups of arginine are so widely spaced that only four ionic species are found in significant amounts. Extend the model to include all imaginable ionic forms and plot the mole fraction of each form as a function of pH. At pH = 2, 7, and 10, what percentage of the total arginine is present in these minor forms? Compare these results with the situation where two of the ionization constants are relatively close together (e.g., cysteine: $pK_1 = 1.8$, $pK_2 = 8.3$, $pK_3 = 10.8$).

2. Predict the titration curve of dipeptide glycyl–aspartate. What is the isoelectric point of glycylaspartate?

3. Predict the titration curve of the tripeptide aspartyl–glycyl–histidine. What is the isoelectric point of this molecule?

MODELS

FORMULAS

	A	B	C	D	E	F	G	H
1	pK1 =	2.3	Start pH =	0				
2	pK2 =	9.8	Incr pH =	0.25				
3								
4	Form	Charge						
5	H2A	1						
6	HA	0						
7	A	-1						
8								
9			------[A]/[HA] Ratios------			-----------mol fraction---------			
10	pH	[HA]/[H2A]	[A]/[HA]	H2A	HA	A	Tot H	Charge
11	=D1							
12								
13	Form Set I	Formula Set I	Formula Set I	Formula Set I	Formula Set I	Formula Set I	Formula Set I	Formula Set I
65								
66								
67								

VALUES

	A	B	C	D	E	F	G	H
1	pK1 =	2.3	Start pH =	0				
2	pK2 =	9.8	Incr pH =	0.25				
3								
4	Form	Charge						
5	H2A	1						
6	HA	0						
7	A	-1						
8								
9			------[A]/[HA] Ratios------			-----------mol fraction---------			
10	pH	[HA]/[H2A]	[A]/[HA]	H2A	HA	A	Tot H	Charge
11	0	0.00501187	1.585E-10	0.995	0.005	7.9E-13	1.99501	0.995
12	0.25	0.00891251	2.818E-10	0.9912	0.0088	2.5E-12	1.99117	0.9912
13	0.5	0.01584893	5.012E-10	0.9844	0.0156	7.8E-12	1.9844	0.9844

	A	B	C	D	E	F	G	H
65	13.5	1.5849E+11	5011.8723	1E-15	0.0002	0.9998	0.0002	-0.9998
66	13.75	2.8184E+11	8912.5094	4E-16	0.0001	0.99989	0.00011	-0.9999
67	14	5.0119E+11	15848.932	1E-16	6E-05	0.99994	6.3E-05	-0.9999

Formula Set I Prototype Cell is A12	**Formula Set II** Prototype Cell is B11	**Formula Set III** Prototype Cell is C11
=A11+**D2**	=10^(A11-**B1**)	=10^(A11-**B2**)

Formula Set IV Prototype Cell is D11	**Formula Set V** Prototype Cell is E11	**Formula Set VI** Prototype Cell is F11
=1/(1+B11+C11*B11)	=B11*D11	=C11*E11

Formula Set VII Prototype Cell is G11	**Formula Set VIII** Prototype Cell is H11
=2*D11+E11	=**B5***D11+**B6***E11+**B7***F11

Model 6.4

BUILD and verify Model 6.4.

ENTER data using cells B1 and B2 (pK values), B5–B7 (charges of ionic forms), and D1 and D2 (starting pH, pH increment).

READ output from the model at cells A11–A67 (pH); D11–D67, E11–E67, F11–F67 (mole fractions of ionic forms), G11–G67 (total H consumed or donated), and H11–H67 (net charge of molecule).

Model 6.5

BUILD and verify Model 6.5.

ENTER data using cells B1 and B3 (pK values), B6–B9 (charges of ionic forms), and D1 and D2 (starting pH, pH increment).

READ output from the model at cells A13–A69 (pH); E13–E69, F13–F69, G13–G69, H13–H69 (mole fractions of ionic forms), I13–I69 (total H consumed or donated), and J13–J69 (net charge of molecule).

FORMULAS

	A	B	C	D	E	F	G	H	I	J
1	pK1 =	2.3	Start pH =	0						
2	pK2 =	9.1	Incr pH =	0.25						
3	pK3 =	12.5								
4										
5	Form	Charge								
6	H3A	2								
7	H2A	1								
8	HA	0								
9	A	-1								
10										
11			[A]/[HA] Ratios			mol fraction				Total
12	pH	[H2A]/[H3A]	[HA]/[H2A]	[A]/[HA]	H3A	H2A	HA	A	Tot H	Charge
13	=D1									
14										
15	Form Set I	Formula Set II	Formula Set III	Formula Set IV	Formula Set V	Formula Set VI	Formula Set VII	Formula Set VIII	Formula Set IX	Formula Set X
67										
68										
69										

VALUES

	A	B	C	D	E	F	G	H	I	J
1	pK1 =	2.3	Start pH =	0						
2	pK2 =	9.1	Incr pH =	0.25						
3	pK3 =	12.5								
4										
5	Form	Charge								
6	H3A	2								
7	H2A	1								
8	HA	0								
9	A	-1								
10										
11			[A]/[HA] Ratios			mol fraction				Total
12	pH	[H2A]/[H3A]	[HA]/[H2A]	[A]/[HA]	H3A	H2A	HA	A	Tot H	Charge
13	0	0.00501187	7.9433E-10	3.1623E-13	0.99501	0.00499	4E-12	1.3E-24	2.99501	1.99501
14	0.25	0.00891251	1.4125E-09	5.6234E-13	0.99117	0.00883	1.2E-11	7E-24	2.99117	1.99117
15	0.5	0.01584893	2.5119E-09	1E-12	0.9844	0.0156	3.9E-11	3.9E-23	2.9844	1.9844

	A	B	C	D	E	F	G	H	I	J
67	13.5	1.5849E+11	25118.8643	10	2.3E-17	3.6E-06	0.09091	0.90909	0.09092	-0.9091
68	13.75	2.8184E+11	44668.3592	17.7827941	4.2E-18	1.2E-06	0.05324	0.94676	0.05324	-0.9468
69	14	5.0119E+11	79432.8235	31.6227766	7.7E-19	3.9E-07	0.03065	0.96935	0.03065	-0.9693

Formula Set I
Prototype Cell is A14

=A13+ **D2**

Formula Set II
Prototype Cell is B13

=10^(A13-**B1**)

Formula Set VI
Prototype Cell is F13

=B13*E13

Formula Set VII
Prototype Cell is G13

=C13*F13

Formula Set III
Prototype Cell is C13

=10^(A13-**B2**)

Formula Set IV
Prototype Cell is D13

=10^(A13-**B3**)

Formula Set VIII
Prototype Cell is H13

=D13*G13

Formula Set IX
Prototype Cell is I13

=G13+2*F13+3*E13

Formula Set V
Prototype Cell is E13

=1/(1+B13+C13*B13+D13*C13*B13)

Formula Set X
Prototype Cell is J13

=E13* **B6** +F13* **B7** +G13* **B8** +H13* **B9**

FORMULAS

	A	B	C	D	E	F	G	H	I	J	K
1		(-COOH)	(-NH2)	Asp	Glu	His	Lys	Arg	Tyr	Cys	
2	# Resid. =	1	1	8	2	1	6	11	4	0	
3	pK =	2.1	9.7	3.9	4.3	6	10.8	12.5	10.1	8.3	
4	Charge (A) =	-1	0	-1	-1	0	0	0	-1	-1	
5	Charge (HA) =	0	1	0	0	1	1	1	0	0	
6											
7			—Mole Fraction of Charged Group—>								Total
8	pH	(-COOH)	(-NH2)	Asp	Glu	His	Lys	Arg	Tyr	Cys	Charge
9	0										
10	0.5										
11	1										
12	1.5										

	Form Set I	Form Set II	Form Set III	Form Set IV	Form Set V	Form Set VI	Form Set VII	Form Set VIII	Form Set IX	Form Set X

	A
35	13
36	13.5
37	14

VALUES

	A	B	C	D	E	F	G	H	I	J	K
1		(-COOH)	(-NH2)	Asp	Glu	His	Lys	Arg	Tyr	Cys	
2	# Resid. =	1	1	8	2	1	6	11	4	0	
3	pK =	2.1	9.7	3.9	4.3	6	10.8	12.5	10.1	8.3	
4	Charge (A) =	-1	0	-1	-1	0	0	0	-1	-1	
5	Charge (HA) =	0	1	0	0	1	1	1	0	0	
6											
7			—Mole Fraction of Charged Group—>								Total
8	pH	(-COOH)	(-NH2)	Asp	Glu	His	Lys	Arg	Tyr	Cys	Charge
9	0	0.00788	1	0.00013	5E-05	1	1	1	7.9E-11	5E-09	18.991
10	0.5	0.0245	1	0.0004	0.00016	1	1	1	2.5E-10	1.6E-08	18.972
11	1	0.07359	1	0.00126	0.0005	0.99999	1	1	7.9E-10	5E-08	18.9153
12	1.5	0.20076	1	0.00397	0.00158	0.99997	1	1	2.5E-09	1.6E-07	18.7643

	A	B	C	D	E	F	G	H	I	J	K
35	13	1	0.0005	1	1	1E-07	0.00627	0.24025	0.99874	0.99998	-12.3141
36	13.5	1	0.00016	1	1	3.2E-08	0.00199	0.09091	0.9996	0.99999	-13.9863
37	14	1	5E-05	1	1	1E-08	0.00063	0.03065	0.99987	1	-14.6585

Formula Set I
Prototype Cell is B9

=10^(A9-**B3**)/(1+10^(A9-**B3**))

Formula Set IV
Prototype Cell is E9

=10^(A9-**E3**)/(1+10^(A9-**E3**))

Formula Set VII
Prototype Cell is H9

=1-10^(A9-**H3**)/(1+10^(A9-**H3**))

Formula Set II
Prototype Cell is C9

=1-10^(A9-**C3**)/(1+10^(A9-**C3**))

Formula Set V
Prototype Cell is F9

=1-10^(A9-**F3**)/(1+10^(A9-**F3**))

Formula Set VIII
Prototype Cell is I9

=10^(A9-**I3**)/(1+10^(A9-**I3**))

Formula Set III
Prototype Cell is D9

=10^(A9-**D3**)/(1+10^(A9-**D3**))

Formula Set VI
Prototype Cell is G9

=1-10^(A9-**G3**)/(1+10^(A9-**G3**))

Formula Set IX
Prototype Cell is J9

=10^(A9-**J3**)/(1+10^(A9-**J3**))

Formula Set X
Prototype Cell is K9

=B2·**B4**·B9+C2·**C5**·C9+D2·**D4**·D9+E2·**E4**·E9+F2·**F5**·F9+G2·**G5**·G9+H2·**H5**·H9+I2·**I4**·I9+J2·**J4**·J9

Model 6.6

BUILD and verify Model 6.6.

ENTER data using cells B2 and J2 (number of residues of each ionizable amino acid). The remaining values in the data table (cells B3–J5) are fixed by Nature.

READ output from the model at cells A9–A37 (pH) and K9–K37 (net charge of molecule).

BUFFERS

Take a moment to review what you have learned about acids, bases, and the reactions they undergo.

Strong acids and strong bases are ionic materials that are completely dissociated in aqueous solution. In aqueous solution, strong acids such as HCl and H_2SO_4 transfer their protons quantitatively to water to form H_3O^+ (or larger) complexes (Figure 6.14a, Reaction A). Strong bases such as KOH and NaOH are ionic materials that ordinarily do not exist as molecules at all and exist in aqueous solution as OH^- ions and K^+ and Na^+ ions (Figure 6.14a, Reaction B).

The titration of a strong base by a strong acid, or vice versa, is merely the reaction of H^+ (as H_3O^+) with OH^- to form H_2O (Figure 6.14a, Reaction C). H_3O^+ is often represented simply as H^+ (Figure 6.14b). The equilibrium constant for the reaction $H^+ + OH^- \rightleftharpoons H_2O$ is about 10^{14} ($1/K_w$), and virtually all reactants (H^+ and OH^-) are consumed to form product (H_2O).

By contrast, weak acids are only partially dissociated in aqueous solution, and matters are more complex. A solution of acetic acid contains both of its potential forms (Equation 6.25):

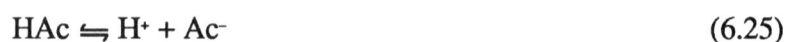

$$HAc \rightleftharpoons H^+ + Ac^- \qquad (6.25)$$

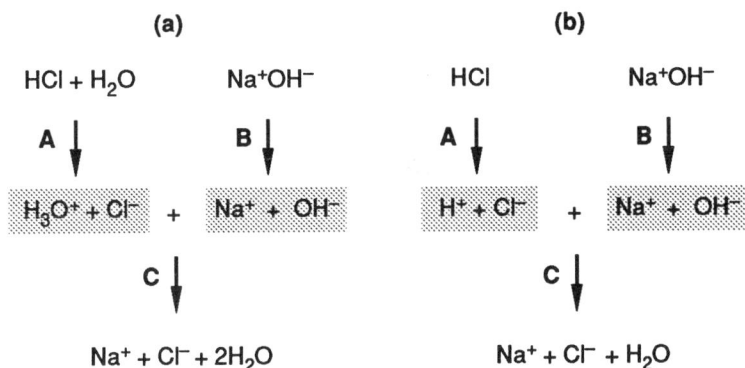

(a)   **(b)**

HCl + H₂O   Na⁺OH⁻   HCl   Na⁺OH⁻

A ↓   B ↓   A ↓   B ↓

$H_3O^+ + Cl^-$ + $Na^+ + OH^-$ $H^+ + Cl^-$ + $Na^+ + OH^-$

         C ↓   C ↓

   $Na^+ + Cl^- + 2H_2O$ $Na^+ + Cl^- + H_2O$

Figure 6.14

Strong acids and strong bases are essentially completely dissociated in aqueous solution. HCl, for example, dissociates in water to form H_3O^+ complexes and Cl⁻ anion [Reaction A of (a)]. Often, however, the products of dissociation are represented simply as H⁺ and Cl⁻ [Reaction A of (b)]. The titration of HCl by NaOH yields Na⁺, Cl⁻, and H_2O. The reaction is essentially complete (K ~ 10¹⁴) (Reaction C).

If NaOH, a strong base, is added to this solution, OH⁻ reacts with H⁺ to form H_2O, and the added base is neutralized. If HCl, a strong acid, is added to this solution, however, the concentration of H⁺ simply increases because very little Ac⁻ is present to react with it. This is because the equilibrium constant K for the dissociation of HAc is quite low (i.e., $[H^+][Ac^-]/[HAc] = 10^{-4.75}$) and there would be little conjugate base (Ac⁻) present to react with the added acid.

Contrast this situation with the behavior of a solution of acetic acid (HAc) and sodium acetate (Na⁺ and Ac⁻) (Figure 6.15).

- If NaOH is added to this aqueous solution, OH⁻ (from the strong base) reacts with H⁺ (from the weak acid) to form H_2O. The added base is neutralized (although the reduction in [H⁺] forces further dissociation of HAc to yield increased conjugate base Ac⁻). *The net effect is a partial neutralization of the added base.*

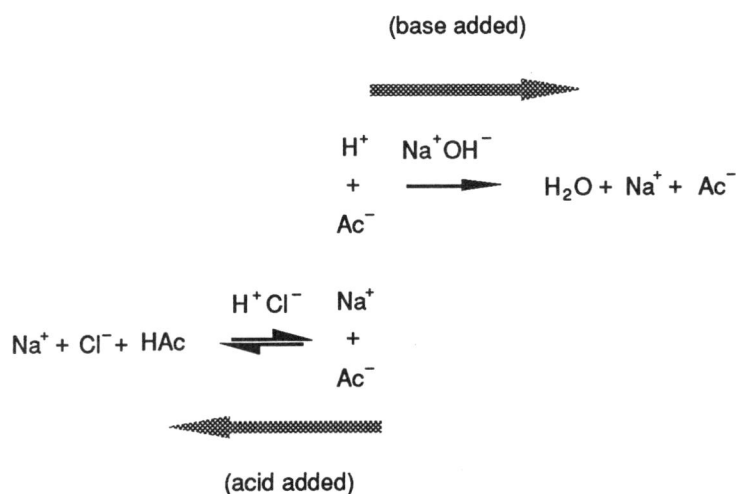

(base added)

⟹⟹⟹⟹⟹⟹⟹⟹⟹⟹⟹

H⁺   Na⁺OH⁻
+   ⟶ $H_2O + Na^+ + Ac^-$
Ac⁻

              H⁺Cl⁻   Na⁺
$Na^+ + Cl^- + HAc$ ⇌   +
                      Ac⁻

⟸⟸⟸⟸⟸⟸⟸⟸⟸⟸⟸

(acid added)

Figure 6.15

Weak acids in aqueous solution with their salts are called buffers. A solution of acetic acid (HAc) and sodium acetate (Na⁺ and Ac⁻), for example, can neutralize both added acid and added base.

- If base is added, it combines with H⁺ (dissociated from the weak acid) to form H_2O.

- If acid is added, it combines with Ac⁻ (present as a salt) to form HAc.

Figure 6.16

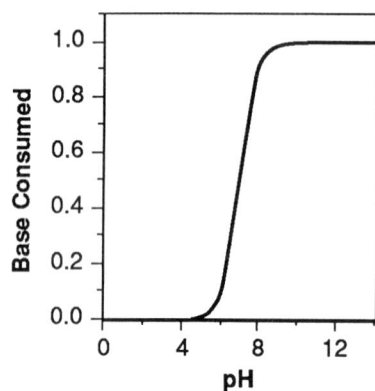

XY-PLOT

Data
Horizontal (x) axis: cells A6–A61 (pH)
Vertical (y) axis: cells D6–D61 (base
consumed)

- If HCl is added to this solution, H^+ (from the strong acid) reacts with Ac^- (from NaAc) to form HAc. *The added acid is partially neutralized.*

Solutions of weak acids and their salts therefore have the capacity to partially neutralize the effects of added acid and added base. Such solutions are *buffers*.

The important properties of buffers are readily obtained from the Henderson–Hasselbalch equation. If the concentration of a buffer (conjugate acid plus conjugate base) has a value 1, the Henderson–Hasselbalch equation can be written

$$pH = pK + \log\left(\frac{A^-}{1 - A^-}\right) \tag{6.26}$$

It is an easily verifiable fact that functions of the form $y = x/(1-x)$ change least when $x = 1 - x = 0.5$.[1] Therefore,

1. Weak acids are good buffers when the dissociated and undissociated forms of the weak acid exist in roughly equal concentrations (i.e., $[A^-] \sim [HA]$ or $[A^-]/[HA] \sim 1$).

2. The pK of a weak acid equals the pH of maximum buffering (if $[A^-]/[HA] = 1$, $\log([A^-]/[HA]) = 0$ and $pH = pK + 0$).

LESSONS

1. Modify Model 6.1 by adding a new column to plot buffer strength (Model 6.7). An excellent measure of the effectiveness of the buffer (buffer strength) is to plot the amount of consumed base required to change pH by some constant increment.

   a. Explore Model 6.7 and verify that it is identical to Model 6.1 except for the addition of column E. Examine column E and see how it implements a method for measuring buffer strength.

   b. Build two on-screen graphics to plot pH vs base consumed (Figure 6.16) and pH vs buffer strength (Figure 6.17).

   c. In the laboratory, you obtain a titration curve (Figure 6.18) for a solution buffered with an unknown weak acid selected from the following table. What is the identity of the weak acid?

Figure 6.17

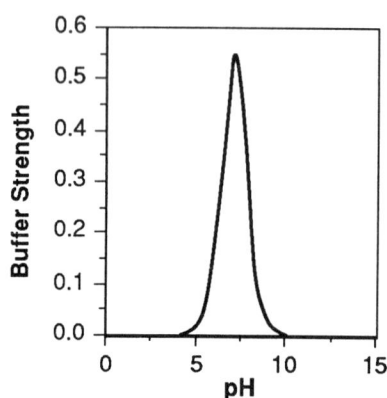

XY-PLOT

Data
Horizontal (x) axis: cells A6–A61 (pH)
Vertical (y) axis: cells E6–E61 (buffer
strength)

Acid	pK
Acetic acid	4.75
Acetoacetic cid	3.58
Glycine	9.87
Iodic acid	0.77
Periodic acid	1.64
Saccharine	11.68

d. Suppose that you buffer a reaction (at pH 6.5) that liberates H^+. Should you choose a buffer with a pK above, below, or equal to 6.5?

Use a chemical handbook to identify several weak acids that are suitable for buffering this reaction and plot their titration curves (Figure 6.16).

e. How would you stabilize the pH of a reaction mixture at pH 6.5 if you know that the reaction will consume protons? Should you choose a buffer with a pK above, below, or equal to 6.5?

Use a chemical handbook to identify several weak acids that are suitable for buffering this reaction and plot their titration curves (Figure 6.16).

2. Buffer solutions often contain more than one group that can donate or accept protons. This situation may occur because each molecule of the buffer compound contains two or more such groups or because two or more buffer compounds are present. You have already built several models capable of tracking such buffers (e.g., Model 6.3 and 6.4).

a. Modify Model 6.3 to include a measure of buffer strength identical to that used in Model 6.7. Also build on-screen graphics capable of plotting pH vs acid consumed, mole fraction of each ionic form vs pH, and buffer strength vs pH.

b. Use this model and these plots in the lessons that follow.

3. Recall from Exercise 6.1 that a common inorganic buffer molecule containing three ionizable groups is phosphate. The four forms are H_3PO_4, $H_2PO_4^-$, HPO_4^{2-}, and PO_4^{3-}. The pK values for the three ionizations are 2.12, 7.28, and 12.32. Enter these pK values into the model you constructed in Lesson 2.

Figure 6.18

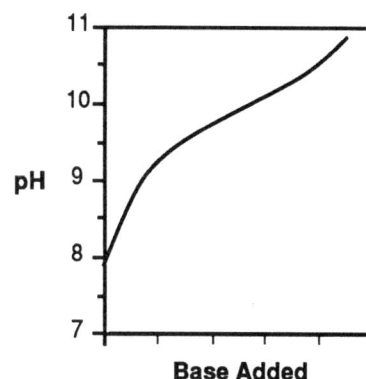

A solution buffered with an unknown weak acid gives the titration curve shown. What is the identity of the weak acid?

a. What are three pH values at which maximum buffering occurs?

b. Identify the pH regions where the mole fractions of two or more of the ionic forms of PO_4 are undergoing rapid change. How do these regions correlate with the regions of maximum buffering and the pK values of the weak acid?

c. Phosphate is an important buffer system in the cells of your body and is used as a buffer in many biochemical experiments. If the pH of your cells is 7.2, which ionization of the phosphate system is involved in buffering? Which ionic forms will occur in cells?

d. Is phosphate a good buffer throughout the pH range of 2 to 12? Why or why not?

4. Many organic molecules that are important in your body contain two or three ionizable groups.

a. You have already explored some of the properties of the amino acid arginine. Recall that this amino acid has pK values of 2.3, 9.6, and 12.5. Does arginine contribute significantly to buffering in living cells? Why or why not? At what pH ranges is arginine an effective buffer?

b. Another common amino acid, glutamic acid, has pK values of 2.3, 4.0, and 9.7. Does glutamic acid aid in biological buffering? At what pH ranges is glutamic acid an effective buffer?

c. Use your model to examine the buffering strengh of citric acid ($pK_1 = 3.08$, $pK_2 = 4.74$, $pK_3 = 5.40$) and arsenic acid ($pK_1 = 2.25$, $pK_2 = 6.77$, $pK_3 = 11.60$). Which of these two weak acids provides the widest range of continuous buffering action?

PROBLEMS

1. It is, of course, impossible to change pK values of a compound, but you can mix buffer systems with different pK values. This is often done to buffer a solution over a wider range than would be possible with a single buffer system.

   Assign pK values of 6, 7, and 8 to the model you built in Lesson 2. What is the resulting buffer pattern?

   Now try 5, 7, and 9. Notice that buffering is still fairly strong but over a wider range.

Explore different relative values of pK. If you want to observe the behavior of a compound containing two ionizable groups or a solution containing a mixture of two buffers, you can set pK_3 to 0 or a negative value.

In situations where several different molecules cooperate to buffer a solution, titration curves and buffer-strength curves obtained from these models are valid, but curves describing mole fractions of ionic species vs pH are not. Why?

2. Assume that you plan to carry out a reaction for which the pH must be held between 5.2 and 5.8. You calculate that the reaction, in a volume of 1.0 L, will yield 50 mmol of H^+. Also assume that two buffers are available with pK values of 5.3 (Buffer A) and 6.3 (Buffer B). Calculate for each buffer what concentration of total buffer will be required to hold the pH within the desired range.

    Suppose that you wish to run this reaction on a moderately large scale. The molecular weights of buffers A and B are equal, but their prices are $5.00 and $1.45 per kg, respectively. Which buffer should you use?

Model 6.7

BUILD and verify Model 6.7.

ENTER data using cells B1 (pK) and
B2 and B3 (starting pH, increment in
pH).

READ output from the model at cells
A6–A61 (pH), C6–C61 (mole fraction
of HA), D7–D61 (base consumed),
and D8–D61 (buffer strength).

NOTICE that "Base consumed" in
column D can be represented simply
as the current value of [A⁻] less the
starting value of [A⁻].

MODELS

FORMULAS

	A	B	C	D	E
1	pK =	7			
2	Start pH =	1			
3	Incr pH =	0.25			
4				Base	Buffer
5	pH	[A]/[HA]	X(HA)	consumed	Strength
6	=B2				
7	Formula	Formula	Formula	Formula	
8	Set	Set	Set	Set	Formula
9	I	II	III	IV	Set
10					V

VALUES

	A	B	C	D	E
1	pK =	7			
2	Start pH =	1			
3	Incr pH =	0.25			
4				Base	Buffer
5	pH	[A]/[HA]	X(HA)	consumed	Strength
6	1	0.000001	0.999999		
7	1.25	1.77828E-06	0.999998222	7.78277E-07	
8	1.5	3.16228E-06	0.999996838	2.16227E-06	5.53597E-06
9	1.75	5.62341E-06	0.999994377	4.62338E-06	9.84446E-06
10	2	0.00001	0.99999	8.9999E-06	1.75061E-05

Formula Set I Prototype Cell is A7	Formula Set III Prototype Cell is C6
=A6+**B3**	=1/(B6+1)

Formula Set II Prototype Cell is B6	Formula Set IV Prototype Cell is D7
=10^(A6-**B1**)	=**C6** -C7

Formula Set V Prototype Cell is E8
=(D8-D7)/(A8-A7)

pH, EXTRACTIONS, AND SOLUBILITY

As your knowledge of chemistry increases, you will discover increasing opportunities to pull together information from different sources and to unify them into new understanding. In this exercise, you bring together your knowledge of solubility (Exercise 5.4) and your knowledge of the behavior of acids and bases (Exercises 6.1–6.3).

We've chosen just two arenas in which to help you exercise your knowledge of chemistry — the effect of pH on the distribution of weak acid between aqueous and organic solvents and (in Problems) the effect of pH on the solubility of metal ions. There are many others. After you finish this exercise, you may want to put this book aside for awhile and examine your primary chemistry textbook for model building opportunities.

First, recall that when two immiscible liquids are in contact, solute moves between them and establishes a dynamic equilibrium. The equilibrium concentrations of solute dissolved in each solvent depends on the relative solubilities of solute in each solvent (Equation 6.27):

$$D = \frac{S_o}{S_w} \tag{6.27}$$

TO GET THE MOST OUT OF THIS EXERCISE, YOU SHOULD ALREADY KNOW

The definition of the terms:
conjugate acid
conjugate base

How to simulate acid–base titrations using electronic spreadsheets.

How to use solubilities and solubility product constants to determine the concentrations of substances in different solvents.

Figure 6.19

The equilibrium distribution of a weak acid between an organic and aqueous phase varies with pH. If only the conjugate acid is soluble in the organic phase, the equilibrium relationships used to calculate this distribution are as shown.

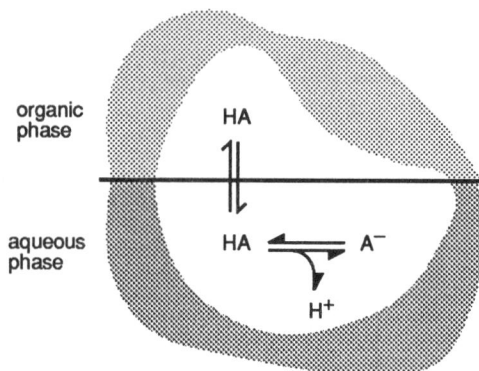

where D is the distribution coefficient between the organic and aqueous phases and S_o and S_w are the solubilities of the solute in the organic and aqueous phases.

Futhermore, the equilibrium amounts of solute dissolved in each solvent also depends on the volume of each solvent (Equation 6.28):

$$\frac{\text{solute amount in organic phase}}{\text{solute amount in aqueous phase}} = D\left(\frac{V_o}{V_w}\right) \tag{6.28}$$

where V_o and V_w are the volumes of organic solvent and water.

Now suppose that the solute is a weak acid with a single donatable proton. If the conjugate acid is uncharged (e.g., HA) and the conjugate base has a single negative charge (e.g., A⁻), it is probable that the conjugate base will be much less soluble in the organic solvent than in water. For simplicity's sake, presume that HA has a measurable solubility in both phases but that A⁻ is only measurably soluble in water. You should be able to picture the equilibrium relationships that exist between HA in the aqueous and organic environments and between HA and A⁻ in the aqueous environment (Figure 6.19). You should also be able to derive a partition equation that calculates the mole fraction X_{HA_w} that HA in the aqueous phase constitutes of all forms of weak acid in either phase (Equations 6.29 and 6.30):

$$X_{HA_w} = \frac{HA_w}{HA_w + A_w + HA_o} \tag{6.29}$$

or

$$X_{HA_w} = \frac{1}{1 + \dfrac{A_w}{HA_w} + \dfrac{HA_o}{HA_w}} \tag{6.30}$$

where HA_w and A_w are the amounts of conjugate acid and base in aqueous phase and HA_o is the amount of conjugate acid in the organic phase.

Because A_w/HA_w refers to amounts in the same volume of solvent, this fraction can substitute concentrations for amounts without changing its value: $[A^-]_w/[HA]_w$. Thus, the second term in the denominator can be calculated using the Henderson–Hasselbalch equation (Equation 6.31):

$$\frac{[A^-]_w}{[HA]_w} = 10^{(pH-pK)} \tag{6.31}$$

The second term in the denominator of Equation 6.30 can be calculated using Equation 6.28 (Equation 6.32):

$$\frac{HA_o}{HA_w} = D\left(\frac{V_o}{V_w}\right) \tag{6.32}$$

The mole fractions of of the other forms of weak acid can be calculated in a straightforward manner (or can be "reverse engineered" from an examination of Model 6.8).

You will shortly have the opportunity to combine these relationships into a simulation of the behavior of a weak acid distributed between two solvents. The issues surrounding metal solubility, your second example, are postponed to the problems in which they play a part.

LESSONS

1. Build and verify Model 6.8. Also build the on-screen graphic described in Figure 6.20. Examine the structure of Model 6.8 and verify the following features:

    a. Cells D9 and D12 represent Equation 6.31. Cell D10 represents Equation 6.27.

    b. Cell D13 represents Equation 6.32.

    c. Cell B16 represents Equation 6.30, and the remaining cells in this region of the spreadsheet correctly represent the indicated concentrations and amounts.

2. The pK of acetic acid, vinegar, is about 4.7. Suppose that 10 mL of a $0.7M$ solution of acetic acid and sodium acetate exhibits a pH of 2 and is in equilibrium with 10 mL of an organic solvent in which the conjugate acid of acetic acid is five times more soluble than in

Figure 6.20

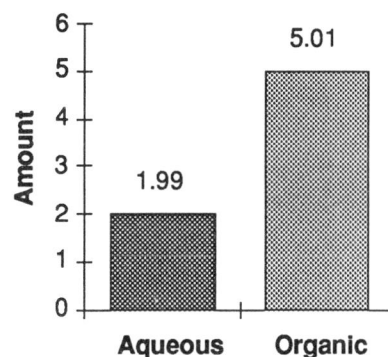

BAR CHART

Data
Horizontal (x) axis: cells B4–C4 (labels)
Vertical (y) axis: cells B22–C22 (solute amounts in aqueous and organic solvents)

NOTE: Different spreadsheets may set up this bar chart in different ways.

water. (Acetate ion is presumed to be insoluble in the organic solvent.)

a. What are the equilibrium amounts of acetic acid in the two solvents?

b. How does the distribution change if the pH is raised to 7? Why?

c. When the pH of the aqueous solution equals the pK of acetic acid, the concentrations of conjugate acid and conjugate base in the aqueous phase should be equal. Does the solute equilibrate equally between the phases? Why or why not?

d. Under the conditions of Lesson 2c, what volume of organic solvent causes equal amounts of acetic acid to distribute between the two phases?

e. Perform a series of experiments with your computer in which 50 mL of organic solvent are used to extract acetic acid from aqueous solutions of different pH values. In one set of experiments, perform single extractions using 50 mL of solvent. In another set of experiments, perform two extractions of 25 mL each. In a final set of experiments, perform five extractions of 10 mL each.

   Write your results (with explanations) in the form of a laboratory report. (Be sure to include in your report a series of titration curves in which the equilibrium distribution of acetic acid amounts is plotted against pH. The DATA TABLE command of your electronic spreadsheet will be useful here.)

3. Explore the distribution behavior of several different weak acids in several different water–organic systems. Use your imagination or take real data from chemical handbooks, at your discretion.

PROBLEMS

1. Imagine that you must analyze samples of galvanized-metal buckets that are known to contain (mostly) iron and zinc. You dissolve samples of each bucket in acid and have a series of solutions of Fe^{3+} and Zn^{2+} that you must analyze for their metal content. Before measuring metal amounts, however, you must separate each ion into their own containers by finding a pH value in which one metal is soluble and the other is not.

The equilibrium relationships for these two metal ions are

$$Fe(OH)_3 \rightleftharpoons Fe^{3+} + 3OH^- \qquad (6.33a)$$

$$K_{sp[Fe(OH)_3]} = [Fe^{3+}][OH^-]^3 = 1.1 \times 10^{-36} \qquad (6.33b)$$

and

$$Zn(OH)_2 \rightleftharpoons Zn^{2+} + 2OH^- \qquad (6.34a)$$

$$K_{sp[Zn(OH)_2]} = [Zn^{2+}][OH^-]^2 = 1.8 \times 10^{-14} \qquad (6.33b)$$

A quick consideration of these equations points out that changes in pH (and therefore [OH$^-$]) can affect the solubility of these ions in a predictable manner (Equations 6.35 and 6.36).

$$[Fe^{3+}]\left(\frac{K_w}{[H^+]}\right)^3 = 1.1 \times 10^{-36} \qquad (6.35)$$

$$[Zn^{3+}]\left(\frac{K_w}{[H^+]}\right)^2 = 1.8 \times 10^{-14} \qquad (6.36)$$

Solving Equations 6.35 and 6.36 for metal ion concentrations and then using your electronic spreadsheet to plot each metal ion concentration vs pH will quickly show what pH ranges support the precipitation of one ion but not the other.

a. What pH should you use to separate iron from zinc?

b. Which ion is in the precipitate and which remains soluble?

[NOTE: Your plots will be hard to read if you allow the range of *y*-axis values to be specified automatically. Command your spread-sheet program to lock the *y*-axis (metal ion concentration) into the range of 0 to 4*M*.]

2. Another way to separate zinc from iron involves the somewhat more complex equilibrium of H$_2$S (Equation 6.37):

$$H_2S \rightleftharpoons HS^- + H^+ \rightleftharpoons S^{2-} + 2H^+ \qquad (6.37)$$

where K_1 (the first proton's dissociation constant) $= 9.5 \times 10^{-8}$ and K_2 (the second proton's dissociation constant) $= 1.0 \times 10^{-19}$.

a. Use Model 6.4 to verify that at very low pH values S^{2-} is a very small number but is not zero.

You can generate an algebraic expression for $[S^{2-}]$ by using the trick that $[S^{2-}]$ is so low at low pH that *virtually* all of this weak acid is present as H_2S. Verify that this assumption can be used to combine the equilibrium expressions for K_1 and K_2 implicit in Equation 6.37 into the single expression

$$[S^{2-}] = \frac{K_1 K_2 [H_2S]}{[H^+]^2} \tag{6.38}$$

b. The fact that S^{2-} does exist at low pH values (even if at extraordinarily low concentrations) can be exploited to separate Zn from Fe by using the equilibria

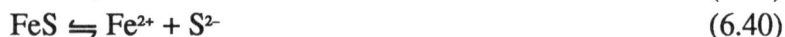

$$ZnS \rightleftharpoons Zn^{2+} + S^{2-} \tag{6.39}$$
$$FeS \rightleftharpoons Fe^{2+} + S^{2-} \tag{6.40}$$

Show that Equation 6.38 can be combined with the equilibrium expressions implicit in Equations 6.39 and 6.40 to yield

$$[Zn^{2+}] = \frac{K_{sp(ZnS)} [H^+]}{K_1 K_2 [H_2S]} \tag{6.41}$$

$$[Fe^{2+}] = \frac{K_{sp(FeS)} [H^+]}{K_1 K_2 [H_2S]} \tag{6.42}$$

c. Build a model that plots $[Zn^{2+}]$ and $[Fe^{2+}]$ vs pH in the presence of H_2S gas. (NOTE: An aqueous solution into which H_2S gas is continuously bubbled has a concentration of H_2S of about $0.1M$.)

d. What range of pH values allows effective separation of these two metal ions in the presence of H_2S?

e. Which ion is in the precipitate, and which remains soluble? Is this the same order found in Lesson 1?

3. For weak bases, the conjugate acid usually bears a positive charge. Thus, for bases, it is usually the conjugate base, rather than the conjugate acid, that is less polar and more soluble in organic solvents. Modify Model 6.8 to calculate the equilibrium distributions of weak bases between water and nonpolar organic solvents.

4. Use your electronic spreadsheet program to assess the separation of Fe^{3+} from Al^{3+} (in the absence of H_2S) in aqueous solutions of different pH. Obtain the values of K_{sp} from any chemical handbook. Watch out for the reaction $Al(OH)_3 + OH^- \rightleftharpoons Al(OH)_4^-$. ($K_f = 3.7 \times 10^{18}$ at 20°C for $Al(OH)_4^-$.)

5. How would you separate Cr^{3+}, Cu^{2+}, Cu^+, Pb^{2+}, Ag^+, and Zn^{2+} in aqueous solution in both the presence and absence of H_2S gas? Is one approach better than the other?

MODEL

FORMULAS

	A	B	C	D
1	pK =	3.5		
2	pH =	6		
3				
4		Aqueous	Organic	Total
5	Volume =	5	20	
6	Sol (HA) =	0.02	4	
7	Initial Amounts =			7
8				
9			[A]w/[HA]w =	=10^(B2-B1)
10			D = [HA]o/[HA]w =	=C6/B6
11				
12			(Amt.A)w/(Amt.HA)w =	=D9
13			(Amt.HA)o/(Amt.HA)w =	=D10*C5/B5
14	=1/(1+D12+D13)			
15		Aqueous	Organic	
16	(X.HA)eq =		=B16*D13	
17	(X.A)eq =	=B16*D12		
18				
19	Amt.HA at Equil =	=B16*D7	=C16*D7	
20	Amt.A at Equil =	=B17*D7		
21				
22	Total	=B19+B20	=C19	

VALUES

	A	B	C	D
1	pK =	3.5		
2	pH =	6		
3				
4		Aqueous	Organic	Total
5	Volume =	5	20	
6	Sol (HA) =	0.02	4	
7	Initial Amounts =			7
8				
9			[A]w/[HA]w =	316.2278
10			D = [HA]o/[HA]w =	200
11				
12			(Amt.A)w/(Amt.HA)w =	316.2278
13			(Amt.HA)o/(Amt.HA)w =	800
14				
15		Aqueous	Organic	
16	(X.HA)eq =	0.000895	0.716058	
17	(X.A)eq =	0.283047		
18				
19	Amt.HA at Equil =	0.006266	5.012407	
20	Amt.A at Equil =	1.981328		
21				
22	Total	1.987593	5.012407	

Model 6.8

BUILD and verify Model 6.8.

ENTER data using cells B1 and B2 (pK and pH), B5 and C5 (volumes of water and organic solvent), B5 and C6 (solubilities of conjugate acid in water and organic solvent), and D7 (total solute present).

READ output from the model at cells B22 and C22 (amounts of solute in aqueous and organic phases at equilibrium).

CHAPTER 7

KINETICS

The rates at which chemical reactions proceed are important for two principal reasons:

- Both in and out of the laboratory, it is often important to know how long a reaction will take to go to (apparent) completion or to reach equilibrium.

- The influence of reactant concentrations on reaction rate can provide valuable clues regarding the mechanism of the reaction.

Chemists use *rate laws* to describe how reaction rates vary as a function of reactant concentration. Bodenstein, for example, studied the gas-phase association of H_2 and I_2 to form HI (Equation 7.1):

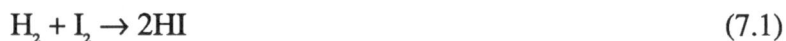

$$H_2 + I_2 \rightarrow 2HI \tag{7.1}$$

Equation 7.2 is the rate law for this reaction:

$$\frac{d[\text{HI}]}{dt} = k[H_2][I_2] \tag{7.2}$$

where [HI] is the concentration of HI, $d[\text{HI}]/dt$ is the rate of change in concentration of HI; and k is the reaction-rate constant.

Rate laws are determined by experimental observation and cannot be derived from the stoichiometry of the reaction. The stoichiometry of the reaction of hydrogen with gaseous bromine, for example, is simple (Equation 7.3), but the rate law is complex (Equation 7.4):

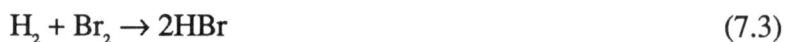

$$H_2 + Br_2 \rightarrow 2HBr \tag{7.3}$$

$$\frac{d[\text{HBr}]}{dt} = \frac{k[H_2][Br_2]^{1/2}}{m + [\text{HBr}]/[Br_2]} \tag{7.4}$$

Expressions for the reaction rates of elementary steps, however, *can* be determined from the stoichiometry of such steps. According to the law of mass action, the reaction rate of an elementary reaction of the form

$$a\text{A} + b\text{B} + ... \rightarrow d\text{D} + e\text{E} + ... \tag{7.5}$$

can be expressed as

$$\frac{d[\text{A}]}{dt} = -k[\text{A}]^a[\text{B}]^b ... \tag{7.6}^1$$

Equation 7.2 therefore is consistent with the view that two molecules of HI collide to form single molecules of H_2 and I_2. Clearly, however, the reaction of H_2 with Br_2 does not consist of so elementary a process.

Rate laws depict the *rate of change* in the concentration of a reactant or product as a function of the concentration of the reactants. Because laboratory methods typically measure reactant concentrations rather than their rates of change, you must be able to convert these rates of change to actual reactant concentrations. There are two principal methods:

- Analytic methods in which equations such as Equation 7.6 are integrated exactly using calculus to yield an expression where reactant concentration is a function of time

- Numeric methods in which computers are used to calculate reactant concentrations over time by using the rate expression directly (and avoiding calculus)

Both methods play an important role in chemical kinetics, but analytic methods are limited to those rate expressions that can be integrated symbolically. In the exercises that follow, you will have ample opportunity to explore a wide range of chemical reactions using both analytic and numeric techniques. Much of this material, although central to an understanding of chemical kinetics, was unavailable to students prior to the appearance of personal computers.

RATE LAWS: ANALYTIC METHODS

Rate laws often assume the form

$$\frac{d[A]}{dt} = -k[A]^a[B]^b \dots \tag{7.7}$$

where [A] and [B] are concentrations of reactants A and B; k, a and b are constants; and $d[A]/dt$ is the rate of change of [A] with respect to time. k is said to be the rate constant of the reaction, whereas a and b are constants associated with the order of the reaction.

If exponent a has a value of 1, then the reaction is first-order with respect to A. If exponent b has a value of 2, then the reaction is second-order with respect to B. If a and b both have values of 1, then the reaction is first-order with respect to both A and B, but is second-order overall.

Consider first the rate law for a simple, first-order reaction (Equation 7.8):

$$\frac{d[A]}{dt} = -k[A] \tag{7.8}$$

To follow [A] as a function of time, Equation 7.8 can be integrated to give

$$\int_{[A]_0}^{[A]} \frac{d[A]}{[A]} = - \int_0^t k \, dt \tag{7.9}$$

or

$$\ln \frac{[A]}{[A]_0} = -kt \tag{7.10}$$

where $[A]_0$ is the starting concentration of A.

Solving Equation 7.10 for [A] yields the expression required (Equation 7.11):

$$[A] = [A]_0 e^{-kt} \tag{7.11}$$

Consider next the second-order reaction $2A \rightarrow B$ having the rate law

$$\frac{d[A]}{dt} = - k[A]^2 \tag{7.12}$$

Equation 7.12 can be integrated to yield

$$\int_{[A]_0}^{[A]} \frac{d[A]}{[A]^2} = -\int_0^t k \, dt \tag{7.13}$$

or

$$\frac{1}{[A]} - \frac{1}{[A]_0} = kt \tag{7.14}$$

Solving Equation 7.14 for [A] yields

$$[A] = \frac{[A]_0}{1 + kt[A]_0} \tag{7.15}$$

In this exercise, you explore the kinetic behavior of reactions amenable to analytic methods. You examine rate laws of different order and explore their properties using the graphics capability of your electronic spreadsheet. In following exercises, you will have the opportunity to explore a much broader range of reactions by using numeric methods. Numeric methods usually depend on computers for the many calculations required and must be used when the rate law has no analytic solution.

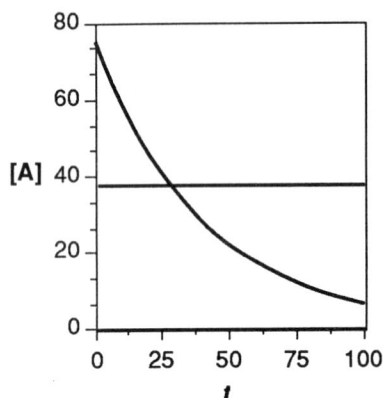

Figure 7.1

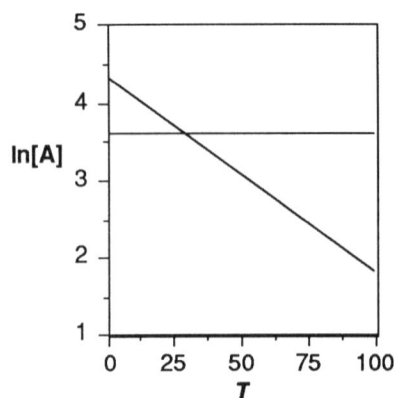

XY-PLOT

Data
Horizontal (*x*) axis: cells A6–A106
(time)
Vertical (*y*) axis: cells B6–B106 ([A])
and C6–C106 ([A]$_o$/2)

Figure 7.2

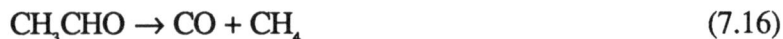

XY-PLOT

Data
Horizontal (*x*) axis: cells A6–A106
(time)
Vertical (*y*) axis: cells D6–D106
(ln[A]), and E6–E106 (ln{[A]$_o$/2})

LESSONS

Consider the thermal decomposition of acetaldehyde to carbon monoxide and methane (Equation 7.16):

$$CH_3CHO \rightarrow CO + CH_4 \tag{7.16}$$

If this reaction follows first-order kinetics, you would expect the plot of ln [CH$_3$CHO] vs time to be a straight line (Equation 7.17):

$$\ln[CH_3CHO] = \ln[CH_3CHO]_0 - kt \tag{7.17}$$

If, on the other hand, the decomposition of acetaldehyde is a second-order reaction, you would obtain a straight line by plotting 1/[CH$_3$CHO] vs time (Equation 7.18):

$$\frac{1}{[CH_3CHO]} - \frac{1}{[CH_3CHO]_0} = kt \tag{7.18}$$

Actual data obtained by Hinshelwood in 1928 showed the following:[1]

Time (min)	[CH$_3$CHO] (mm Hg)
0	225
10	145
12	135
15	123
19.5	107
23	97.5
26	91.5
31	82
36	75
41	70

In the lessons that follow, you first explore the general behavior of first- and second-order reactions and then determine whether the decomposition of acetaldehyde is a first- or second-order reaction.

1. Build and verify Model 7.1. Model 7.1 calculates [A] (Equation 7.11) and ln[A] as a function of time for a first-order reaction. Also build two on-screen graphics to plot [A] vs *t* (Figure 7.1) and ln[A] vs *t* (Figure 7.2). These graphics also draw a horizontal line across the plot at the point where [A] is reduced by one-half of its starting concentration.

   a. Examine your plots using a number of different values of *k* and [A]$_0$ to verify that [A] always decreases exponentially with time, whereas ln[A] always decreases linearly with time. You may need to adjust the value of Δt in order to get a good view of the resulting curve. (Keep in mind, however, that as Δt increases the accuracy of the numeric approximation decreases.)

b. Exponential decay is characterized by a half-life (i.e., the time required to reduce the concentration of [A] by one-half). The half-life of a property undergoing exponential decay is independent of the initial magnitude of the property. If you substitute [A] = x and [A]$_0$ = $2x$ into Equation 7.16, you obtain an expression for the half-life of a first-order reaction:

$$t_{1/2} = \frac{\ln 2}{k} \tag{7.19}$$

Cell D1 of Model 7.1 uses Equation 7.19 to calculate the half-life of [A]. Examine your plots with several values of [A]$_0$ and k and verify that [A] equals [A]$_0$/2 when $t = t_{1/2}$.

2. Build and verify Model 7.2. Model 7.2 calculates [A] (Equation 7.15) as a function of time for a second-order reaction. Also build on-screen graphics to plot [A] vs t (Figure 7.3) and 1/[A] vs t (Figure 7.4).

a. Examine your plots using a number of different values of [A]$_0$ and k. You may need to adjust the value of Δt in order to get a good view of the resulting curve.

b. Do you understand why all plots of 1/[A] vs t are straight lines?

3. Enter Hinshelwood's 1928 data for the decomposition of acetaldehyde into a new spreadsheet model and add columns to calculate ln[CH$_3$CHO] and 1/[CH$_3$CHO] (our answer is Model 7.3).

a. Plot ln[CH$_3$CHO] vs t and 1/[CH$_3$CHO] vs t on your computer screen.

b. Is the decomposition of acetaldehyde a first- or second-order reaction?

c. What is the rate constant for this reaction?

d. Do your data rule out any possible mechanisms for the reaction?

Figure 7.3

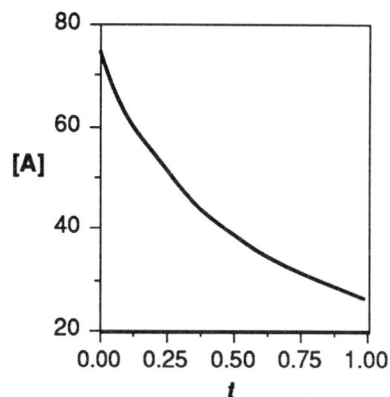

XY-PLOT

Data
Horizontal (x) axis: cells A6–A106 (time)
Vertical (y) axis: cells B6–B106 ([A])

Figure 7.4

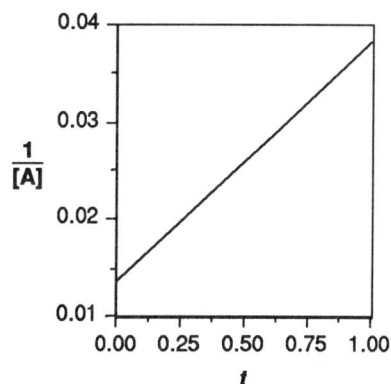

XY-PLOT

Data
Horizontal (x) axis: cells A6–A106 (time)
Vertical (y) axis: cells C6–C106 (1/[A])

PROBLEMS

1. A reaction $A + B \rightarrow C$ having the rate law

    $$\frac{d[A]}{dt} = -k[A][B] \qquad (7.20)$$

    yields the integrated expression

    $$kt = \frac{1}{[A]_0 - [B]_0} \ln\left(\frac{[A][B]_0}{[A]_0[B]}\right) \qquad (7.21)$$

    a. Solve Equation 7.21 for [A].

    b. Build and verify a model that plots [A] and [B] as a function of of t.

    c. What happens to the plot of ln[A] vs t when [B] is in large excess? Why?

2. At the turn of the century, Bodenstein studied the gas-phase formation of HI from H_2 and I_2 as described in Equation 7.22. The formation of HI is favored at high temperatures. At 556°C, for example, $k_1 = 2.04 \times 10^{-5}\ M^{-1}\ s^{-1}$ and $k_{-1} = 3.10 \times 10^{-7}\ M^{-1}\ s^{-1}$.

    $$H_2 + I_2 \underset{k_{-1}}{\overset{k_1}{\rightleftharpoons}} 2HI \qquad (7.22)$$

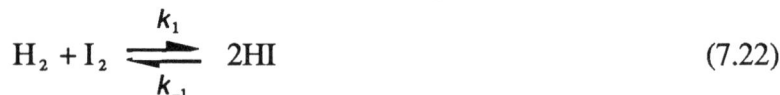

    The rate law for this reaction is

    $$\frac{d[HI]}{dt} = k_1\left([H_2]_0 - \frac{[HI]}{2}\right)\left([I_2]_0 - \frac{[HI]}{2}\right) - k_{-1}[HI]^2 \qquad (7.23)$$

    where $d[HI]/dt =$ the rate of change in concentration of HI.

    While Equation 7.23 does have an analytic solution, you can simplify your task enormously by exploring only the initial rates of reaction when [HI] = 0.

    a. What is the rate law for the formation of HI when [HI] is assumed to be zero? Does the law of mass action predict this expression? Is the rate law consistent with the view that one molecule of H_2 collides with one molecule of I_2 to form two molecules of HI?

    b. Build and verify a model that follows [HI] with time during the initial time periods of the reaction.

c. Assuming that Equation 7.22 accurately reflects the actual steps in this reaction (i.e., one molecule of H_2 collides with one molecule of I_2 to form two molecules of HI), what is the equilibrium constant K for this reaction?

d. Use your spreadsheet as a scratchpad to calculate equilibrium concentrations of HI for a variety of initial pressures of $[H_2]$ and $[I_2]$.

e. Re-examine Equation 7.23 and discuss the contribution of [HI] to the reaction rate throughout the reaction. Take into account in your discussion the equilibrium concentration of HI and the value of k_{-1}.

Numeric methods make it possible to easily integrate expressions such as Equation 7.23 without using simplifying assumptions. You will discover this for yourself in Exercise 7.4, "Approach to Equilibrium."

MODELS

FORMULAS

	A	B	C	D	E
1	k =	0.025	t(1/2) =	=LN(2)/B1	
2	[A]o =	75			
3	Δt =	1			
4					
5	Time	[A]	[A]o/2	ln[A]	ln([A]o/2)
6	0	Formula	=B2/2	=LN(B6)	=LN(C6)
7	=B3+A6	Set	=C6	=LN(B7)	=E6
8	=B3+A7	I	=C6	=LN(B8)	=E7

VALUES

	A	B	C	D	E
1	k =	0.025	t(1/2) =	27.72589	
2	[A]o =	75			
3	Δt =	1			
4					
5	Time	[A]	[A]o/2	ln[A]	ln([A]o/2)
6	0	75	37.5	4.317488	3.624341
7	1	73.14824	37.5	4.292488	3.624341
8	2	71.34221	37.5	4.267488	3.624341

```
Formula Set I
Prototype Cell is B6

=B2*EXP(-B1*A6)
```

Model 7.1

BUILD and verify Model 7.1.

ENTER data using cells B1–B3 (equilibrium constant, initial concentration of A, and Δt).

READ output from the model at cells A6–A106 (time t), B6–B106 ([A]), C6–C106 (one-half the initial concentration [A]$_o$), D6–D106 (ln[A]), and E6–E106 (the natural logarithm of one-half the initial concentration [A]$_o$).

NOTICE that cell D1 contains a formula that calculates the half-life of a compound disappearing according to first-order kinetics.

Model 7.2

BUILD and verify Model 7.2.

ENTER data using cells B1–B3 (equilibrium constant, initial concentration of A, and Δt).

READ output from the model at cells A6–A106 (time t), B6–B106 ([A]), and C6–C106 (1/[A]).

FORMULAS

	A	B	C
1	k =	0.025	
2	[A]o =	75	
3	Δt =	0.01	
4			
5	Time	[A]	1/[A]
6	0	Formula	=1/B6
7	=A6+B3	Set	=1/B7
8	=A7+B3	I	=1/B8

VALUES

	A	B	C
1	k =	0.025	
2	[A]o =	75	
3	Δt =	0.01	
4			
5	Time	[A]	1/[A]
6	0	75	0.013333
7	0.01	73.61963	0.013583
8	0.02	72.28916	0.013833

Formula Set I Prototype Cell is B6
$=B2/(1+(B2*B1*A6))$

FORMULAS

	A	B	C	D
1	Time, min	[CH3CHO]	1/[CH3CHO]	ln[CH3CHO]
2	(min)	(mm, Hg)		
3	0	225	=1/B3	=LN(B3)
4	10	145	=1/B4	=LN(B4)
5	12	135	=1/B5	=LN(B5)
6	15	123	=1/B6	=LN(B6)
7	19.5	107	=1/B7	=LN(B7)
8	23	97.5	=1/B8	=LN(B8)
9	26	91.5	=1/B9	=LN(B9)
10	31	82	=1/B10	=LN(B10)
11	36	75	=1/B11	=LN(B11)
12	41	70	=1/B12	=LN(B12)

VALUES

	A	B	C	D
1	Time, min	[CH3CHO]	1/[CH3CHO]	ln[CH3CHO]
2	(min)	(mm, Hg)		
3	0	225	0.00444444	5.4161004
4	10	145	0.00689655	4.97673374
5	12	135	0.00740741	4.90527478
6	15	123	0.00813008	4.81218436
7	19.5	107	0.00934579	4.67282883
8	23	97.5	0.01025641	4.57985238
9	26	91.5	0.01092896	4.51633897
10	31	82	0.01219512	4.40671925
11	36	75	0.01333333	4.31748811
12	41	70	0.01428571	4.24849524

Model 7.3

BUILD and verify Model 7.3.

ENTER experimental data using columns A (time t) and B (concentration of remaining acetaldehyde).

PLOT the output of columns C and D as an xy-plot. If the plot of ln[A] vs t is linear, the reaction *may* be a first-order reaction. If the plot of 1/[A] vs t is linear, the reaction *may* be a second-order reaction.

RADIOACTIVE DECAY:
AN EXAMPLE OF FIRST-ORDER KINETICS

TO GET THE MOST OUT OF THIS
EXERCISE, YOU SHOULD
ALREADY KNOW

The definition of the terms:
radioactive decay
first-order kinetics

Radioactive decay is an excellent example of first-order kinetics. Consider substance A, at least some of whose molecules are radioactive. If A^*_0 is the radioactivity of A (in disintegrations per second) of a sample at time 0 and if A^* is the radioactivity of the sample at time t, then (as noted earlier in Equation 7.11)

$$A^* = A^*_0 e^{-kt} \qquad (7.24)$$

The rate of decay of a radioactive sample is usually measured in terms of its half-life $t_{1/2}$ rather than a rate constant. Equation 7.19 can be modified to express k in terms of $t_{1/2}$ (Equation 7.25). This approach is useful for many problems involving radioactive decay:

$$k = \frac{\ln 2}{t_{1/2}} \qquad (7.25)$$

Radioactive isotopes are often used to trace atoms in chemical or biological processes. Because of radioactive decay, however, the radioactivity recovered at the end of an experiment may not accurately measure the actual number of atoms recovered. The radioactive isotope ^{14}C has a half-life of 5730 years (so that its decrease in specific activity is more likely to be of interest to geologists than chemists), but other frequently used radioisotopes have somewhat more ephemeral existences (^{32}P has a half-life of 14.3 days, and ^{13}N has a half-life of only 10 minutes).

Thus, chemists must often rely on Equation 7.24 to correct their measurements for the effects of radioactive decay.

LESSON

Build and verify the model described in Figure 7.5 using only the information contained in the figure. Check your answer by comparing it to Model 7.4.

Cell B2 contains the value of the half-life of the radioisotope of A.

Cell B3 contains the time t during which A decays.

Cell B4 contains the total radioactivity A^*_0 at the beginning of the experiment.

Cell B5 contains the radioactivity recovered at the end of the experiment (not A^*, but A^* times the fraction of A recovered).

Cell B7 calculates the rate constant k from the half-life $t_{1/2}$ (Equation 7.25).

Cell B9 contains a formula to calculate the total radioactivity A^* remaining at the end of the experiment (at time t) (Equation 7.24).

Cell B10 contains a formula to calculate the fraction of total radioactivity remaining at the end of the experiment (at time t).

	A	B
1	Decay Calculations:	
2	t(1/2) =	20
3	t =	55
4	A*0 =	245000
5	Recovered A* at t =	4320
6		
7	k =	0.03465736
8		
9	A* =	36419.4679
10	A*/A*0 =	0.14865089
11		
12	Corrections:	
13	Recovered A* if no decay =	29061.3801
14	Fraction [A] recovered =	0.11861788

Figure 7.5

This might be the display of a model designed to perform radioisotope half-life calculations on an electronic spreadsheet. Build this model using only the labels in the model as your guide. Our answer is contained in Model 7.4.

Figure 7.6

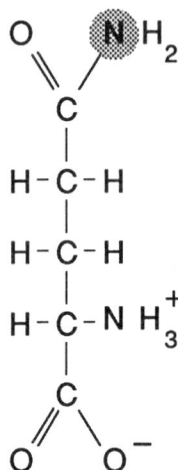

^{13}N is incorporated into the amide position of glutamine and then used in a study of nitrogen metabolism. Because ^{13}N has a half-life of only 10 minutes, the first-order decay of the specific activity of ^{13}N must be part of recovery calculations.

Cell B13 contains a formula to calculate the radioactivity that would have been recovered if no radioactive decay had occurred.

Cell B14 contains the fraction of A recovered in product after corrections for radioactive decay have been made.

Figure 7.5 might represent the calculations for the following experiment: A reactant containing radioactive-labeled atoms measuring 245,000 counts per minute (cpm) enters into a chemical reaction, and the radioactivity of the product is measured 55 minutes later. At that time, the radioactivity in the product is found to be 4320 cpm. At 55 minutes, the total radioactivity remaining (after radioactive decay) is determined by calculation to be 36,400 cpm or 14.9% of the initial activity. Because 4320 cpm are found in the product, 11.9% of the labeled material was incorporated into the product.

PROBLEM

The only radioisotope of nitrogen that is available for use as a chemical tracer is ^{13}N, which has a half-life of 10 minutes.

In a study of nitrogen metabolism, the amino acid glutamine was labeled with ^{13}N in the amide position (Figure 7.6). A 0.01-mL sample, counted in a scintillation counter at zero time (zero time in this experiment is the midpoint of the 1-minute counting period), contains 57,600 cpm of radioactivity. At 4 minutes, a 0.1-mL sample of labeled glutamine was added to a cell suspension, and the cells were killed by addition of acid at 9 minutes.

By quick chromatography on a short column, residual glutamine, ammonium ion, and the amino acids alanine and aspartic acid were separated and counted separately. The radioactivity and the time of the count is documented in the table below:

Time (min)	Compound	cpm
21	Alanine	38,800
23	Aspartic acid	20,500
25	NH_4^+	53,600
27	Glutamine	89,800

What percentage of the amide nitrogen atoms that were in glutamine when the sample was added to the cells were still in glutamine at the moment the cells were killed?

What percentage of the amine nitrogen atoms were in each of the products at the moment the cells were killed?

Assuming a constant rate of degradation, what fraction of glutamine was degraded each minute of the incubation?

MODEL

FORMULAS

	A	B
1	Decay Calculations:	
2	t(1/2) =	20
3	t =	55
4	A*0 =	245000
5	Recovered A* at t =	4320
6		
7	k =	=LN(2)/B2
8		
9	A* =	=B4*EXP(-B7*B3)
10	A*/A*0 =	=B9/B4
11		
12	Corrections:	
13	Recovered A* if no decay =	=B5/B10
14	Fraction [A] recovered =	=B13/B4

VALUES

	A	B
1	Decay Calculations:	
2	t(1/2) =	20
3	t =	55
4	A*0 =	245000
5	Recovered A* at t =	4320
6		
7	k =	0.03465736
8		
9	A* =	36419.4679
10	A*/A*0 =	0.14865089
11		
12	Corrections:	
13	Recovered A* if no decay =	29061.3801
14	Fraction [A] recovered =	0.11861788

Model 7.4

BUILD and verify Model 7.4.

ENTER data using cells B2–B5. Cell B2 contains the half-life of the radioisotope; cell B3 contains the time t during which radioactive decay occurs; cell B4 contains the total radioactivity of the radioisotope at time 0; and cell B5 contains the radioactivity recovered at time t.

The rate constant k is calculated in cell B7 from the half-life $t_{1/2}$.

READ output from the model at cells B9 and B10 (the total radioactivity remaining at time t and the fraction of total radioactivity remaining at time t) and B13 and B14 (the recovered radioactivity that had no first-order decay and the fraction of labeled material that was incorporated into the product).

RATE LAWS: NUMERIC METHODS

TO GET THE MOST OUT OF THIS
EXERCISE, YOU SHOULD
ALREADY KNOW

The definition of the terms:
reaction rate
rate law

How to calculate the change in
concentrations of reactants and
products with time, using analytic
techniques.

Rate laws of the form

$$\frac{d[A]}{dt} = - k[A]^a[B]^b \dots \tag{7.26}$$

can be approximated by equations of the form

$$\frac{\Delta[A]}{\Delta t} = - k[A]^a[B]^b \dots \tag{7.27}$$

where $\Delta[A]$ is the change in concentration of A over the time interval Δt. The accuracy of $\Delta[A]/\Delta t$ as an estimate for reaction rate $d[A]/dt$ increases as Δt decreases. As you will see, the only cost associated with increasing the accuracy of $\Delta[A]/\Delta t$ is an increase in computing time.[1]

Consider the reaction

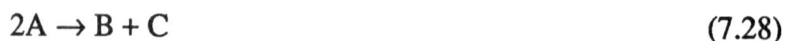

$$2A \rightarrow B + C \tag{7.28}$$

having the rate law

$$\frac{d[A]}{dt} = - k[A]^2 \tag{7.29}$$

Where no reverse reaction exists, the change in concentration $\Delta[A]$ over the the interval Δt is

$$\frac{\Delta[A]}{\Delta t} = -k[A]^2 \quad \text{or} \quad \Delta[A] = -k[A]^2\Delta t \qquad (7.30)$$

If the concentration of A at $t = 0$ is $[A]_0$, then the concentration of A at $t = \Delta t$ is

$$[A]_{\Delta t} = [A]_0 + \Delta[A] = [A]_0 - k[A]^2\Delta t \qquad (7.31)$$

Use Equation 7.31 to determine the concentration of $[A]$ over as many intervals Δt as you wish by replacing $[A]_0$ with the value of $[A]$ calculated at the end of the previous time period (Figure 7.7).

Numeric techniques are usually not feasible in the absence of digital computers, but they are remarkably accurate and can be applied to any rate law without exception — they are not limited to those reactions that have analytic solutions. You can use your personal computer to explore a wide range of chemical processes that would normally be beyond your reach.

In this exercise, you learn to use use numeric methods to solve rate laws for reactant concentration. In the next exercise, you will use the technique to observe reactions as they approach equilibrium and will compare your results with those previously obtained in Chapter 5.

A word of caution is in order. Numeric methods involve approximations, and, under some circumstances, the errors in these approximations can suddenly increase or decrease without limit or exhibit other unrealistic behavior. Such behaviors are not hard to identify and can usually be solved by choosing a smaller value for Δt.

LESSONS

1. The gas-phase decomposition of azomethane ($CH_3N_2CH_3 \rightarrow CH_3CH_3 + N_2$) is a first-order reaction with a rate constant k of $3.6 \times 10^{-4}\,s^{-1}$ at 325°C. The rate law for this reaction can be approximated by

$$\frac{\Delta[CH_3N_2CH_3]}{\Delta t} = -k[CH_3N_2CH_3] \qquad (7.32)$$

or

$$\Delta[CH_3N_2CH_3] = -k[CH_3N_2CH_3]\Delta t \qquad (7.33)$$

Figure 7.7

$$[A]_0$$

$$[A]_{\Delta t} = [A]_0 - k[A]_0^2\Delta t$$

$$[A]_{2\Delta t} = [A]_{\Delta t} - k[A]_{\Delta t}^2\Delta t$$

$$[A]_{3\Delta t} = [A]_{2\Delta t} - k[A]_{2\Delta t}^2\Delta t$$

Equation 7.31 can be used repeatedly to approximate the value of [A] at any time point. $[A]_0$ is inserted into the equation to find the value at Δt. The value at Δt is inserted into the equation to find the value at $2\Delta t$, etc.

Figure 7.8

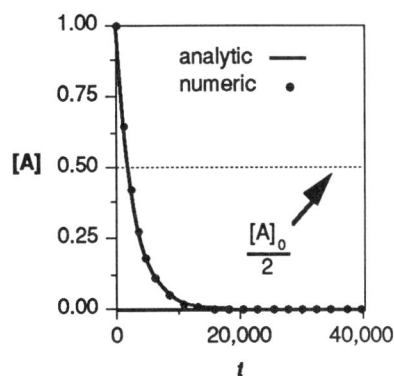

XY-PLOT

Data
Horizontal (x) axis: cells A7–A107 (time)
Vertical (y) axis: cells C7–C107 (analytic [A]), F7–F107 (numeric [A]), and I7–I107 ($[A]_0/2$)

The numeric calculations yield lower values of [A] at all time points.

Figure 7.9

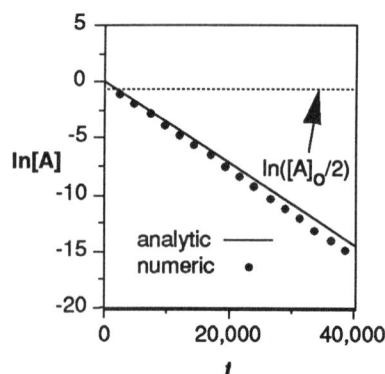

XY-PLOT

Data
Horizontal (x) axis: cells A7–A107
(time)
Vertical (y) axis: cells D7–D107
(analytic ln[A]), G7–G107 (numeric
ln[A]), and J7–J107 (ln[A]$_o$/2)

The numeric data have lower values
of ln [A] at all time points. In this
instance, the difference between
analytic and numeric results becomes
significant only when the reaction is
extended far beyond what would be
practical in the laboratory.

a. Build and verify Model 7.5. Model 7.5 uses numeric techniques to calculate [A] and ln[A] as a function of time where A is any reactant that disappears according to a first-order rate law. Model 7.5 includes a calculation for the analytic expression (Exercise 7.1) so that you can compare the numeric and analytic results. Examine the model carefully until you understand how Equation 7.33 is implemented. You may find it helpful to refer to Figure 7.7.

b. Build two on-screen graphics to plot [A] vs t (Figure 7.8) and ln[A] vs t (Figure 7.9), using both numeric and analytic techniques. ([A]$_o$/2 is also shown.)

Vary Δt and observe the correspondence between the numeric and analytic methods. You will find that the plot of ln[A] vs t more readily displays the slight differences between the two methods.

Can you find conditions in which the numeric approximation "misbehaves" (i.e., enters into the chaotic behavior mentioned earlier)?

c. If you wish to increase the accuracy of your estimates of [A], would you increase or decrease the value of Δt?

d. By visual inspection of your graph, what is the half-life of azomethane under the conditions described?

2. Consider the reaction $2NOBr \rightarrow 2NO + Br_2$. The rate law for this reaction can be approximated by

$$\frac{\Delta[NOBr]}{\Delta t} = -k[NOBr]^2 \tag{7.34}$$

or

$$\Delta[NOBr] = -k[NOBr]^2\Delta t \tag{7.35}$$

where $k = 0.80$ s^{-1}.

a. Build and verify Model 7.6. Model 7.6 uses numeric techniques to calculate [A] as a function of time where A is any reactant that disappears according to a second-order rate law. Also build two on-screen graphics to plot [A] vs t (Figure 7.10) and 1/[A] vs t (Figure 7.11) using both numeric and analytic techniques.

b. Find a value for Δt such that the analytic and numeric methods give equivalent results to within 1%.

PROBLEMS

1. Reaction A + B → C has the rate law

$$\frac{d[A]}{dt} = -k[A][B] \qquad (7.36)$$

Design and build a model that calculates [A] and [B] as a function of t, using a numeric technique. Such a model can be used to represent, for example, $H_2 + I_2 \rightarrow 2HI$. (The rate constant for the formation of HI is $2.04 \times 10^{-5}\ M^{-1}\ s^{-1}$ at 556 K.

2. In Exercise 5.1, you explored the spontaneous dimerization of NO_2, an airborne pollutant generated by automobiles and some industries.

Design, build, verify, and explore models of the forward and reverse reactions for this process (Equation 7.37).

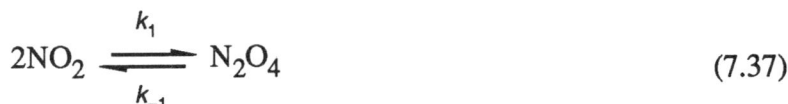

$$2NO_2 \underset{k_{-1}}{\overset{k_1}{\rightleftharpoons}} N_2O_4 \qquad (7.37)$$

Although the values of the rate constants are unknown, it is known that $k_1/2k_{-1} = 9$. Choose $k_1 = 0.9$ and $k_{-1} = 0.05$ in arbitrary units.

The rate laws for the forward and reverse reactions are

$$-\frac{d[NO_2]}{dt} = k_1[NO_2]^2 \qquad (7.38)$$

$$-\frac{d[N_2O_4]}{dt} = k_{-1}[N_2O_4] \qquad (7.39)$$

If you build both models on a single spreadsheet, you can add a column that calculates the net result of both reactions over each time interval. For example, Equation 7.38 defines the rate of conversion of NO_2 to N_2O_4, while Equation 7.39 describes the rate of conversion of N_2O_4 to NO_2. If the change in $[NO_2]$ from both reactions is added together for each time period, you can watch $[NO_2]$ as it approaches equilibrium. (Keep in mind, however, that NO_2 is consumed at twice the rate that N_2O_4 is produced. Such relationships become critical when building models of this sort.)

Figure 7.10

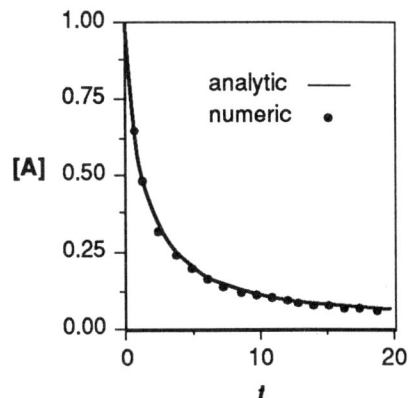

XY-PLOT

Data
Horizontal (x) axis: cells A7–A107 (time)
Vertical (y) axis: cells C7–C107 (analytic [A]) and F7–F107 (numeric [A])

The numeric data have lower values of [A] at all time points.

Figure 7.11

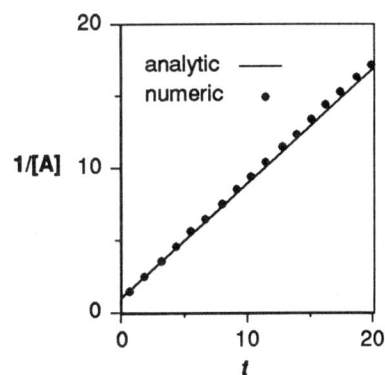

XY-PLOT

Data
Horizontal (x) axis: cells A7–A107 (time)
Vertical (y) axis: cells D7–D107 (analytic 1/[A]) and G7–G107 (numeric 1/[A])

The numeric data have higher values of 1/[A] at all time points.

A full exploration of this kind of model (and how to build it) is postponed until Exercise 7.4, but you might wish to try to build and explore such a model now.

3. Consider the reaction

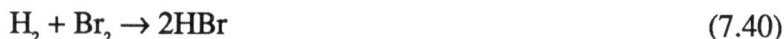

$$H_2 + Br_2 \rightarrow 2HBr \qquad (7.40)$$

having the rate law

$$\frac{d[HBr]}{dt} = \frac{k[H_2][Br_2]^{1/2}}{m + [HBr]/[Br_2]} \qquad (7.41)$$

where $k = 1 \times 10^{-10}$ and $m = 0.12$.

Build and explore a model of this reaction.

4. Numerous reactions given in your chemistry textbook can be studied using numeric techniques. Build models for some of them now.

MODELS

Model 7.5

BUILD and verify Model 7.5.

ENTER data using cells B1–B3 (rate constant, reactant initial concentration, time interval).

READ output from the model from columns A, C, D, F, G, I, and J (time, concentration of A and ln[A] obtained analytically, concentration of A and ln[A] obtained numerically, one-half of initial reactant concentration, and its natural logarithm).

FORMULAS

	A	B	C	D	E	F	G	H	I	J
1	k =	3.6E-04								
2	[A]o =	1								
3	Δt =	400								
4										
5			Analytic			Numeric				
6	Time		[A]	ln [A]		[A]	ln [A]		[A]o/2	ln ([A]o/2)
7	0					=B2			=B2/2	=LN(I7)
8	Form Set I		Form Set II	Form Set III		Form Set IV	Form Set V		=I7	=LN(I8)
9									=I7	=LN(I9)
10									=I7	=LN(I10)

VALUES

	A	B	C	D	E	F	G	H	I	J
1	k =	3.6E-04								
2	[A]o =	1								
3	Δt =	400								
4										
5			Analytic			Numeric				
6	Time		[A]	ln [A]		[A]	ln [A]		[A]o/2	ln ([A]o/2)
7	0		1	0		1	0		0.5	-0.693147
8	400		0.866	-0.14		0.856	-0.16		0.5	-0.693147
9	800		0.75	-0.29		0.733	-0.31		0.5	-0.693147
10	1200		0.649	-0.43		0.627	-0.47		0.5	-0.693147

Formula Set I Prototype Cell is A8	Formula Set III Prototype Cell is D7
=A7+B3	=LN(C7)

Formula Set II Prototype Cell is C7	Formula Set IV Prototype Cell is F8
=B2*EXP(-B1*A7)	=F7-(B3*B1*F7)

Formula Set V Prototype Cell is G7
=LN(F7)

FORMULAS

	A	B	C	D	E	F	G
1	k =	0.8					
2	[A]o =	1					
3	Δt =	0.2					
4							
5				Analytic		Numeric	
6	Time		[A]	1/[A]		[A]	1/[A]
7	0			=1/C7			=1/F7
8	Form		Form	=1/C8		Form	=1/F8
9	Set		Set	=1/C9		Set	=1/F9
10	I		II	=1/C10		III	=1/F10

VALUES

	A	B	C	D	E	F	G
1	k =	0.8					
2	[A]o =	1					
3	Δt =	0.2					
4							
5				Analytic		Numeric	
6	Time		[A]	1/[A]		[A]	1/[A]
7	0		1	1		1	1
8	0.2		0.862	1.16		0.84	1.19
9	0.4		0.758	1.32		0.727	1.375
10	0.6		0.676	1.48		0.643	1.556

Formula Set I **Prototype Cell is A8**	**Formula Set II** **Prototype Cell is C7**
=A7+**B3**	=**B2**/(1+(**B2*****B1***A7))

Formula Set III **Prototype Cell is F8**
=F7-(**B3*****B1***F7*F7)

Model 7.6

BUILD and verify Model 7.6.

ENTER data using cells B1–B3 (rate constant, reactant initial concentration, time interval).

READ output from the model at columns A, C, D, F, and G (time, concentration of A and 1/[A] obtained analytically, concentration of A and 1/[A] obtained numerically).

APPROACH TO EQUILIBRIUM

When a chemical reaction is in dynamic equilibrium, the forward and reverse reactions rates are equal. This simple fact indicates that the rate laws governing a reaction and the equilibrium state of a reaction are closely related. The numeric techniques described in Exercise 7.3 can be used to study chemical reactions as they approach equilibrium.

Consider a reaction that is first-order in both the forward and reverse directions (Equation 7.42). Considering the forward and reverse reactions separately, the rate laws are given in Equations 7.43–7.44:

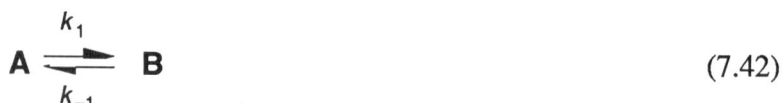

$$A \underset{k_{-1}}{\overset{k_1}{\rightleftharpoons}} B \tag{7.42}$$

$$-\frac{d[A]_{fwd}}{dt} = \frac{d[B]_{fwd}}{dt} = k_1[A] \tag{7.43}$$

$$-\frac{d[B]_{rev}}{dt} = \frac{d[A]_{rev}}{dt} = k_{-1}[B] \tag{7.44}$$

Notice that [A] decreases as it is consumed in the forward reaction and increases as it is formed in the reverse reaction. The net change in [A] can be expressed as the sum of these contributions (Equation 7.45). Similarly the net change in [B] is expressed in Equation 7.46:

$$\frac{d[A]_{net}}{dt} = \frac{d[A]_{fwd}}{dt} + \frac{d[A]_{rev}}{dt} = -k_1[A] + k_{-1}[B] \tag{7.45}$$

$$\frac{d[B]_{net}}{dt} = \frac{d[B]_{fwd}}{dt} + \frac{d[B]_{rev}}{dt} = -k_{-1}[B] + k_1[A] \tag{7.46}$$

Using numeric techniques, the equations to be iterated on each row of your spreadsheet are

$$\Delta[A]_{net} = -\Delta t k_1[A] + \Delta t k_{-1}[B] \tag{7.47}$$

$$\Delta[B]_{net} = -\Delta t\, k_{-1}[B] + \Delta t\, k_1[A] \tag{7.48}$$

Although this approach can be used to solve any problem involving an approach to equilibrium, recall our earlier caution. Numeric methods involve approximations, and, under some circumstances, the errors in these approximations can suddenly increase or decrease without limit or exhibit other unrealistic behavior. Such behaviors are not hard to identify and can always be solved by choosing a smaller value for Δt.

LESSON

Build and verify Model 7.7. Model 7.7 simulates the behavior of a reversible, first-order chemical reaction (Equation 7.42) by calculating the forward and reverse reaction rates and the concentrations of reactant and product as a function of time. Cells D1, D2, and D4 are used to calculate the equilibrium concentrations of A and B from the principles described in Exercise 5.2 (Equations 7.49 and 7.50), and these values are repeated in columns F and G for graphing purposes.

$$K = \frac{[B]_0 + x}{[A]_0 - x} \tag{7.49}$$

or

$$x = \frac{-[B]_0 + K[A]_0}{1 + K} \tag{7.50}$$

where $[A]_0$ and $[B]_0$ are the initial concentrations of A and B, K is the equilibrium constant, and x is the difference in concentration between initial and equilibrium concentrations of A and B. The equilibrium concentrations of A and B, $[A]_{eq}$ and $[B]_{eq}$, are simply $[A]_0 - x$ and $[B]_0 + x$.

Initialize your model such that $[A]_0 = 1$, $[B]_0 = 0$, $k_1 = 0.3$, and $k_{-1} = 0.1$ and use $\Delta t = 0.2$. (K is calculated from k_1 and k_2 using Equation 7.52.) Build on-screen graphics to plot forward and reverse reaction rates vs t (Figure 7.12) and [A] and [B] vs t (Figure 7.13).

Figure 7.12

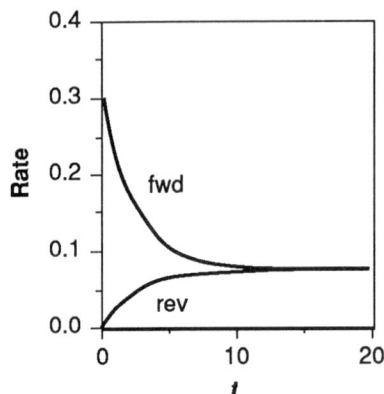

XY-PLOT

Data
Horizontal (*x*) axis: cells A12–A112 (*t*)
Vertical (*y*) axis: cells B12–B112 and
C12–C112 (forward and reverse
reaction rates)

a. At $t = 0$, $[A] = 1$, $[B] = 0$, and only the forward reaction $A \rightarrow B$ occurs. What happens to $[A]$ and $[B]$ as time proceeds? What happens to the forward and reverse reaction rates? What is the relationship between the forward and reverse reaction rates when equilibrium is established and reactant and product concentrations no longer change? Do you see why chemical equilibrium is often called *dynamic* equilibrium?

Examine your model using a wide range of initial values for $[A]_0$, $[B]_0$, k_1, and k_{-1}.

b. Figure 7.12 demonstrates that, at equilbrium, $[A]$ and $[B]$ do not change with time because the forward and reverse reaction rates are equal. $[A]$, for example, does not change at equilibrium because

$$\frac{d[A]_{fwd}}{dt} = \frac{d[A]_{rev}}{dt} \quad \text{or} \quad -k_1[A] = -k_{-1}[B] \tag{7.51}$$

Rearranging,

$$\frac{k_1}{k_{-1}} = \frac{[B]}{[A]} \tag{7.52}$$

The right side of Equation 7.52 is the expression of the equilibrium constant of the reaction, and so, in this instance, k_1/k_{-1} is equivalent to the equilibrium constant K. *This relationship holds only when forward and reverse reactions are elementary reactions whose rate laws can be derived from the law of mass action.*

Examine your model using a wide range of initial values for $[A]_0$, $[B]_0$, k_1, and k_{-1} and verify that the equilibrium predicted using Equation 7.50 (columns F and G) is indeed the equilibrium that is approached by using numeric methods to approximate reaction rates (columns B–E).

Figure 7.13

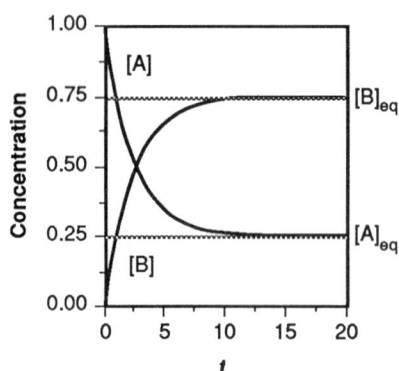

XY-PLOT

Data
Horizontal (*x*) axis: cells A12–A112
(time)
Vertical (*y*) axis: cells D12–D112 ([A]),
E12–E112 ([B]), F12–F112 ([A]$_{eq}$), and
G12–G112 ([B]$_{eq}$)

c. Explore the behavior of your model for starting concentrations of A and B listed below. In each case, note the initial concentrations and initial reaction rates and describe in words how these values change as the reaction approaches equilibrium.

$[A]_0$	$[B]_0$
1.0	0.0
1.0	0.5
1.0	1.0
1.0	2.0
0.5	2.0
0.0	2.0

d. Explore the behavior of your model for other values of the rate constants when $[A]_0 = 1$ and $[B]_0 = 0$.[1]

k_1	k_{-1}
0.3	0.1
0.3	0.3
0.3	0.9
0.3	0.1
0.6	0.2
0.9	0.3

e. In Exercise 7.1 you learned that $\ln[A]$ vs t is linear for the first-order reaction $A \rightarrow B$. In Equation 7.42 two first-order reactions, $A \rightarrow B$ and $B \rightarrow A$, are linked as forward and reverse components of the same overall reaction. In this case, would you expect $\ln[A]$ vs t and $\ln[B]$ vs t to be linear?

To check your answer, add columns to Model 7.7 that calculate $\ln[A]$ and $\ln[B]$. Also design and build on-screen graphics that plot $\ln[A]$ vs t and $\ln[B]$ vs t. Examine these plots for the values of k_1 and k_{-1} such as those below. Modify $[A]_0$ and $[B]_0$ as appropriate.

k_1	k_{-1}
0.01	0.01
0.01	0.001
0.01	0.0001
0.01	0.00001

Apparently $\ln[A]$ vs t and $\ln[B]$ vs t are linear only under certain circumstances. What are those circumstances, and why do they have this effect?

f. $\ln[A]$ vs t is linear only when $k_1 \gg k_{-1}$. Can you propose a function that would be linear with t even when $k_1 = k_{-1}$?

Think it through. $[A]$ asymptotically approaches zero for a first-order reaction of the form $A \rightarrow B$, when there is no reverse reaction. For a reaction of the form $A \rightleftharpoons B$, $[A]$ asymptotically approaches its equilibrium concentration, $[A]_{eq}$. By analogy you might expect that $[A] - [A]_{eq}$ vs t might be a linear function. Modify your model and build on-screen graphics to test this hypothesis.

g. Does $\ln[A]$ vs. t approximate the shape of $\ln([A] - [A]_{eq})$ when $k_1 \gg k_{-1}$? Why or why not?

Figure 7.14

(a) $A \underset{k_{-1}}{\overset{k_1}{\rightleftharpoons}} B$

(b) $A + A \underset{k_{-1}}{\overset{k_1}{\rightleftharpoons}} B + A$

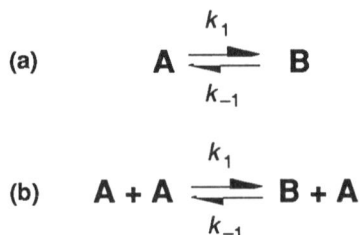

Both reactions (a) and (b) have the same net outcome, $A \to B$. However, (a) occurs through the conversion of a single molecule of A to B, whereas in (b) two molecules of A must collide in order for one of them to convert to B.

Reaction (a) exhibits a first-order rate law. Reaction (b) exhibits a second-order rate law.

PROBLEMS

1. Reconsider the reaction described by Equation 7.42 in a situation where the forward reaction ($A \to B$) follows a second-order rate law and the reverse reaction ($B \to A$) follows a first-order rate law. This might happen, for example, if the forward reaction consisted of several smaller steps.

   a. What is the rate law for the forward reaction?

   b. What is the rate law for the reverse reaction?

   c. Derive an expression for the equilibrium concentrations of [A] and [B] (similar to Equations 7.49 and 7.50).

   d. What expression for K belongs in cell B4?

   Modify Model 7.7 to reflect this new kinetic situation and explore it.

2. Design, build, verify, and explore a model of the reaction

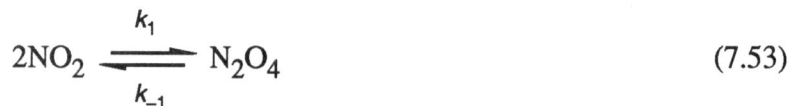

$$2NO_2 \underset{k_{-1}}{\overset{k_1}{\rightleftharpoons}} N_2O_4 \qquad (7.53)$$

   The equilibrium constant K for this reaction is 9 (when the standard state is chosen to be a partial pressure of 1 atm). The values of the rate constants are unknown, but $K = k_1/2k_{-1}$ and so (in arbitrary time units) you can use $k_1 = 9$ and $k_{-1} = 0.5$.

   The rate laws for the forward and reverse reactions are

$$-\frac{d[NO_2]}{dt} = k_1[NO_2]^2 \qquad (7.54)$$

$$-\frac{d[N_2O_4]}{dt} = k_{-1}[N_2O_4] \qquad (7.55)$$

   Recall that, because of the stoichiometry of the reaction, NO_2 disappears at twice the rate that N_2O_4 appears and you must take such matters into account when building models.

   In Chapter 3, you discovered that the ratio of NO_2 to N_2O_4 at equilibrium is dependent on the total pressure in the reaction vessel. Use your kinetic model of this reaction to verify again this important fact.

3. At the turn of the century, Bodenstein studied a gas-phase reaction involving the collision of H_2 and I_2 to form HI (Equation 7.56). The formation of HI is favored at high temperatures and at 556°C, for example, $k_1 = 2.04 \times 10^{-5}\ M^{-1}\ s^{-1}$ and $k_{-1} = 3.10 \times 10^{-7}\ M^{-1}\ s^{-1}$.

$$H_2 + I_2 \underset{k_{-1}}{\overset{k_1}{\rightleftharpoons}} 2HI \tag{7.56}$$

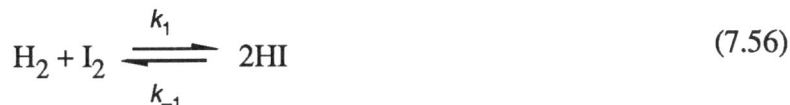

The rate laws for the forward and reverse reactions can be written

$$-\frac{d[H_2]}{dt} = -\frac{d[I_2]}{dt} = k_1[H_2][I_2] \tag{7.57}$$

$$-\frac{d[HI]}{dt} = k_{-1}[HI]^2 \tag{7.58}$$

a. What is the relationship between K, k_1, and k_{-1} for this reaction?

b. What happens to the equilibrium pressure of HI if the initial pressures of both H_2 and I_2 are doubled?

c. When the initial concentration of I_2 is very much larger than the initial concentration of H_2, the forward reaction ($H_2 + I_2 \rightarrow 2HI$) appears to be first-order with respect to $[H_2]$. Use your model to verify this fact and then explain why.

d. What happens when the initial concentration of H_2 is very much larger than the initial concentration of I_2?

4. There are numerous other reactions that can be modeled using the techniques described in this chapter. If you can't find many in your primary textbook, you may find them in the kinetics chapter of any textbook on physical chemistry. Search out such reactions and build models that predict their behavior.

Model 7.7

BUILD and verify Model 7.7.

ENTER data using cells B1 and B2 (starting concentrations of A and B], B5 and B6 (forward and reverse rate constants), and B8 (Δt).

READ output from the model in the appropriately labeled data columns. Cells D1 and D2 calculate the equilibrium concentrations of A and B according to the principles established in Chapter 3. Columns F and G reproduce these values in a form suitable for graphing.

NOTICE that the equilibrium constant K (in cell B4) is calculated from the values of the rate constants. This is a valid calculation only when both the forward and reverse reactions constitute elementary chemical processes that obey the law of mass action.

MODEL

FORMULAS

	A	B	C	D	E	F	G
1	[A]o =	1	[A]eq =	=B1-D4			
2	[B]o =	0	[B]eq =	=B2+D4			
3							
4	Keq =	=B5/B6	x =	=(-B2+(B4*B1))/(1+B4)			
5	k =	0.3					
6	k-1 =	0.1					
7							
8	Δt =	0.2					
9							
10		-d[A]/dt	d[A]/dt				
11	t	Forward	Reverse	[A]	[B]	[A]eq	[B]eq
12	0.0			=B1	=B2	=D1	=D2
13	Formula	Formula	Formula	Formula	Formula	=D1	=D2
14	Set	Set	Set	Set	Set	=D1	=D2
15	I	II	III	IV	V	=D1	=D2

VALUES

	A	B	C	D	E	F	G
1	[A]o =	1	[A]eq =	0.25			
2	[B]o =	0	[B]eq =	0.75			
3							
4	Keq =	3	x =	0.75			
5	k =	0.3					
6	k-1 =	0.1					
7							
8	Δt =	0.2					
9							
10		-d[A]/dt	d[A]/dt				
11	t	Forward	Reverse	[A]	[B]	[A]eq	[B]eq
12	0.0			1	0	0.25	0.75
13	0.2	0.3	0	0.94	0.06	0.25	0.75
14	0.4	0.282	0.006	0.8848	0.1152	0.25	0.75
15	0.6	0.26544	0.01152	0.83402	0.16598	0.25	0.75

Formula Set I
Prototype Cell is A13

=A12+**B8**

Formula Set III
Prototype Cell is C13

=**B6***E12

Formula Set II
Prototype Cell is B13

=**B5***D12

Formula Set IV
Prototype Cell is D13

=D12-(**B8*****B5***D12)+(**B8*****B6***E12)

Formula Set V
Prototype Cell is E13

=E12+(**B8*****B5***D12)-(**B8*****B6***E12)

EQUILIBRIUM VERSUS KINETICS: THE CONTROL OF A CHEMICAL REACTION

Consider a reaction flask containing both reactants and products in specific concentrations. The direction in which the reaction can proceed and the extent of the reaction at equilibrium can be determined from a knowledge of the equilibrium constant K. The reaction rate at all times during the reaction can be determined from a knowledge of the rate constants (k_1, k_2, etc.) and their associated rate laws.

From any starting condition in the real world, there are usually several or many possible reactions. It is the relative rates of reaction that usually determine which reaction predominates. In Figure 7.15, for example, K_1 may be very much larger than K_2, and therefore at equilibrium [B] is very much larger than [C]. However, if k_1 is very small compared to k_3, the reaction A $\rightarrow$ C may be the only significant (or even detectable) reaction.

If you expose a solution of sucrose (common sugar) to a weak acid in the presence of O_2, it will slowly hydrolyze to yield its two component sugars, glucose and fructose. At equilibrium, however, sucrose (in the presence of O_2) should be converted quantitatively to CO_2 and H_2O. This latter reaction is never seen because it proceeds so slowly.

The ratio of products in the reaction mixture depends on the time at which the reaction is stopped. In many cases, the concentrations of

TO GET THE MOST OUT OF THIS EXERCISE, YOU SHOULD ALREADY KNOW

The definition of the terms:
equilibrium
equilibrium constant
reaction rate
rate constant

How to find the equilibrium concentrations of reversible reactions.

How to use rate laws to calculate the concentrations of reactants and products as they change with time.

Figure 7.15

(a)

(b)

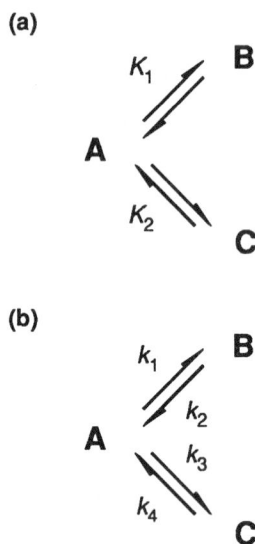

Suppose that the reactant A can engage in two alternative reactions: A → B and A → C. The reverse reactions A ← B and A ← C are also possible.

(a) The relative concentrations of A, B, and C at equilibrium are determined by the equilibrium constants K_1 and K_2. It follows that for any given set of initial concentrations — $[A]_i$, $[B]_i$, $[C]_i$ — the direction of each reaction is determined by K_1 and K_2. These conclusions follow directly from the definition of an equilibrium constant (e.g., $K_1 = [B]_{eq}/[A]_{eq}$ where $[B]_{eq}$ and $[A]_{eq}$ are the concentrations of B and A at equilibrium).

(b) The relative rates of the reactions A → B, A → C, B → A, and C → A are determined by the rate laws and the rate constants k_1, k_3, k_2, and k_4, respectively.

It is quite possible for one of the reactions (e.g., A → C) to proceed so slowly that for all practical purposes it can be ignored.

reactants and products are determined by reaction rates at short reaction times and by equilibrium constants at long reaction times.[1]

In this exercise, you explore the interconversion of three organic acids: *cis*-aconitic acid, citric acid, and isocitric acid (Figure 7.16a). This reaction is catalyzed by the enzyme aconitase and is occurring in every cell of your body at this moment. The equations governing the rates of these reactions (Equations 7.59–7.61) are assembled from the equations governing the individual forward and reverse reaction rates. The rate of change of [C] with time ($d[C]/dt$), for example, is simply the rate of the reaction A → C minus the rate of the reaction C → A.

$$\frac{d[A]}{dt} = k_{-c}[C] + k_{-i}[I] - k_c[A] - k_i[A] \qquad (7.59)$$

$$\frac{d[C]}{dt} = k_c[A] - k_{-c}[C] \qquad (7.60)$$

$$\frac{d[I]}{dt} = k_i[A] - k_{-i}[I] \qquad (7.61)$$

where the reactants, products, and rate constants are defined in Figure 7.16b and $d[A]/dt$, $d[C]/dt$, and $d[I]/dt$ are the rates of change with time of the concentrations of A, C, and I, respectively.

Both reactions A → C and A → I occur at the same rate ($k_c = k_i = 0.1$).[2] The back reaction C → A, however, is much slower than I → A ($k_{-c} = 0.0044$, $k_{-i} = 0.067$). For these reactions $K_1 = k_c/k_{-c}$ and $K_2 = k_i/k_{-i}$ and so the equilibrium constant for A ⇋ C, K_1, must be greater than the equilibrium constant for A ⇋ I, K_2.

A few moments with pencil and paper will tell you that an equilibrium mixture of A, C, and I contains about 90% C, 6% I, and 4% A. Nevertheless, a reaction vessel that begins with only A will contain 30% I if the reaction is stopped at the appropriate time. You will explore this apparently impossible result in this exercise.

LESSONS

1. Build and verify Model 7.8. Initialize your model with the values $k_c = k_i = 0.1$, $k_{-c} = 0.0044$, and $k_{-i} = 0.067$. Set [A] to 100 and all other concentrations to 0.

    a. Trace the concentrations of A, C, and I with reaction time by constructing an on-screen graphic (Figure 7.17).

    Notice that while [A] declines and [C] increases steadily throughout the reaction, [I] first increases until it constitutes

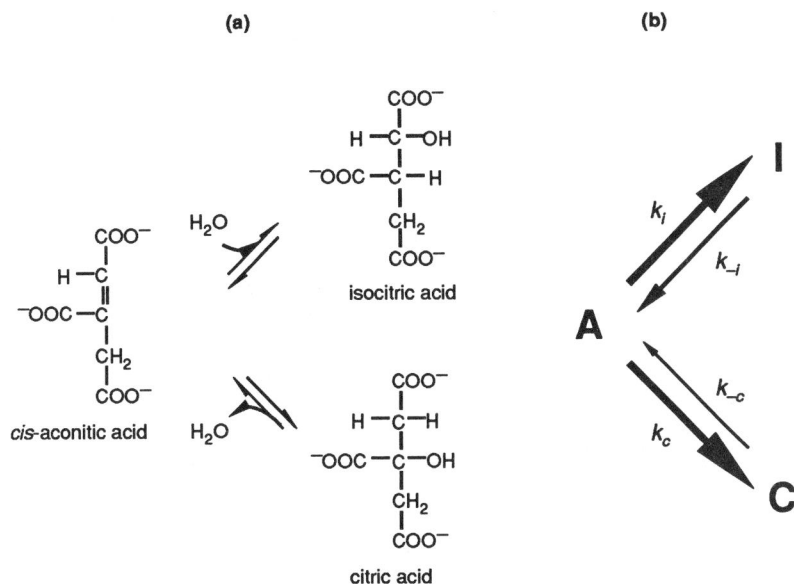

Figure 7.16

(a) *Cis*-aconitic acid (A) can be converted to either citric acid (C) or isocitric acid (I).

(b) The forward reaction rate constants are identical ($k_c = k_i = 0.1$, in arbitrary units), but the rate constant for the reverse reaction I $\rightarrow$ A is greater than the rate constant for the reverse reaction C $\rightarrow$ A ($k_{-i} = 0.067$, $k_{-c} = 0.0044$). These values lead to intriguing kinetic behavior.

about 30% of the contents of the reaction vessel and only then slowly declines to its much lower equilibrium value.

At what time during the reaction is the maximum yield of I achieved?

What is the equilibrium value of [I]?

[NOTE: To support the on-screen graphics capabilities of most electronic spreadsheet programs, the model was limited to $t = 0$ to 100 in increments of 1. You must extend your model (e.g., $t = 0$ to 300 in increments of 1) to obtain a close approximation of equilibrium concentrations.]

b. Why is [I] approximately equal to [C] early in the reaction? What happens to this relationship later in the reaction?

Your data indicate that it is possible for [I] to rise above its equilibrium value early in the reaction. To understand how this is possible, turn to an examination of the various reaction rates in this system.

2. Use Model 7.8 to construct an on-screen graphic that displays the rates of the reactions A $\rightarrow$ I, A $\rightarrow$ C, I $\rightarrow$ A, C $\rightarrow$ A (Figure 7.18).

a. Examine the reaction rates for A $\rightarrow$ I and A $\rightarrow$ C at all times in the reaction and notice that they are identical.

Figure 7.17

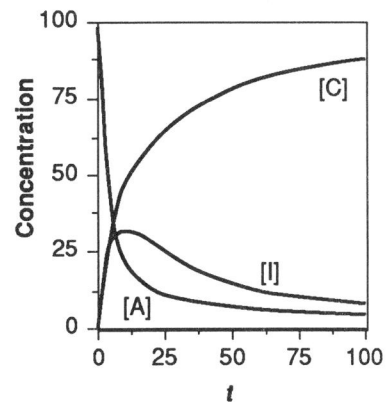

XY-PLOT

Data
Horizontal (*x*) axis: cells A7–A107 (time)
Vertical (*y*) axis: cells B7–B107, C7–C107, and C7–C107 (concentrations of A, C, and I, respectively)

Figure 7.18

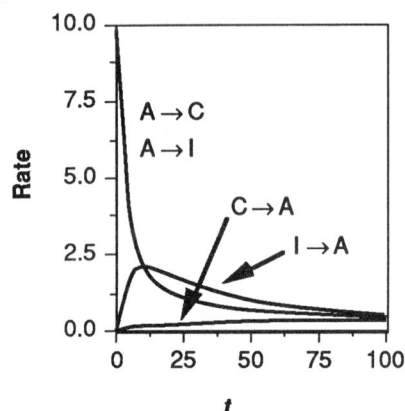

XY-PLOT

Data
Horizontal (*x*) axis: cells A7–A107
(time)
Vertical (*y*) axis: cells E7–E107,
F7–F107, G7–G107, and H7–H107
(reaction rates)

Examine the rate laws for A → I and A → C (Equations 7.62 and 7.63, respectively) and explain why the rates for these two forward reactions must be equal. Explore your model for various (purely hypothetical) situations where k_i is not equal to k_c until you're certain that you understand what controls the relative rates of these two reactions.

$$\frac{d[A]}{dt} = k_i[A] \tag{7.62}$$

$$\frac{d[A]}{dt} = k_c[A] \tag{7.63}$$

What effect does [A] have on these two reactions?

b. Restore the initial input values to your model ($k_c = k_i = 0.1$, $k_{-c} = 0.0044$, $k_{-i} = 0.067$, [A] = 100, [C] = [I] = 0) and examine the reaction rates for I → A and C → A early in the reaction.

At the beginning of the reaction [C] = [I] = 0, and so the reverse reactions I → A and C → A are also zero. Once [C] and [I] are not equal to zero, which reaction is faster? Explain why this must be so in terms of the rate laws for I → A and C → A (Equations 7.64 and 7.65, respectively).

$$\frac{d[C]}{dt} = k_{-c}[C] \tag{7.64}$$

$$\frac{d[I]}{dt} = k_{-i}[I] \tag{7.65}$$

c. Explore the profiles of hypothetical reactions in which $k_{-i} < k_{-c}$ and $k_{-i} = k_{-c}$.

What substance (C or I) accumulates to concentrations above equilibrium concentrations when $k_{-i} < k_{-c}$?

Does either [C] or [I] accumulate to above equilibrium concentrations when $k_{-i} = k_{-c}$?

d. Under what circumstances can either [C] or [I] initially accumulate to concentrations greater than the equilibrium concentration?

3. Your model shows that it is possible for a chemical reaction to produce more product than that present at equilibrium. Respond to a skeptic who argues that anything can be programmed into a computer but that in the real world yields cannot exceed those

specified by conditions at equilibrium. Consider both the forward and reverse reaction rates in your answer.

4. In the example explored above, the maximal obtainable yield of I is about five times greater than the equilibrium yield. If the equilibrium ratio of C to I were much larger (say, 50:1 instead of 15:1), would the ratio of maximal obtainable yield to equilibrium be greater or smaller than in your example? Change the rate constants of the model to check your intuitive answer.

5. Think about how the kinetic and equilibrium features of the aconitase reaction interact. The peculiar kinetic behavior depends entirely on the properties of the catalyst used (i.e., the enzyme aconitase). Isocitrate would not necessarily accumulate to a concentration higher than its equilibrium concentration if some other catalyst were used, because the ratio of k_c to k_i could have any value. However, the final equilibrium ratios are independent of catalyst and will be precisely the same for all possible cases.

Prove this fact for yourself by changing the values of k_c and k_i in your model. (Be sure to change k_{-c} and k_{-i} proportionately so that the equilibrium constants don't change.)

PROBLEM

In many situations, equilibrium is never attained (Figure 7.19). For example, laboratory situations arise in which products are swept away by a stream of solvent or where the reverse reaction is negligible at all attainable concentrations of reactant. Here again, the interaction between thermodynamic and kinetic factors can lead to surprising results.

a. Build and verify Model 7.9. The model simulates the behavior of the reactions described in Figure 7.19. Explore the formulas in the model and verify that they make the appropriate calculations.

b. Construct two on-screen graphics that plot the concentrations of A, C, I, and A + C + I vs t (Figure 7.20) and the concentrations of X and Y vs t (Figure 7.21). You could, if you wish, plot all concentrations on a single graph, but it makes the screen a bit crowded.

c. Initialize your model with the values $k_c = k_i = 0.1$, $k_{-c} = 0.0044$, and $k_{-i} = 0.067$. Set [A] to 100 and all other concentrations to 0 and examine your plots for each of the following values of k_x and k_y:

Figure 7.19

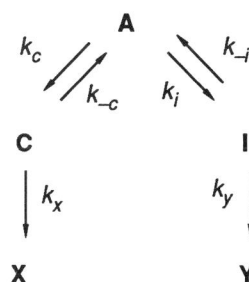

Although the conversions of A to C and A to I are reversible, equilibrium is never attained in the presence of the unidirectional reactions C → X and I → Y. Unidirectionality may result from the physical removal of products (e.g., a solvent stream) or from reactions with large equilibrium constants.

Figure 7.20

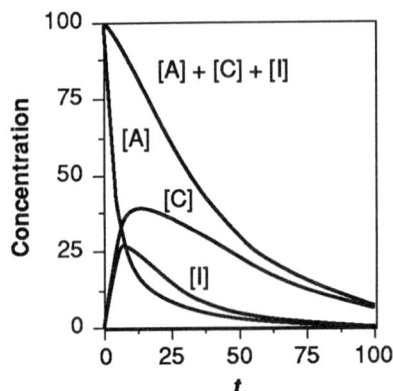

XY-PLOT

Data
Horizontal (x) axis: cells A9–A109 (time)
Vertical (y) axis: cells B9–B109, C9–C109, D9–D109, and E9–E109 (concentrations of A, C, I, and A + C + I)

k_x	k_y
0.00	0.00
0.03	0.03
0.10	0.10
0.50	0.50
1.00	1.00

e. Model 7.9 is an extension of Model 7.8. Verify that the two models behave identically when $k_x = k_y = 0$.

f. When $k_x = k_y = 0$, does [A] + [C] + [I] change with time? Why or why not?

   What happens to [A] + [C] + [I] vs t when $k_x = k_y \neq 0$? Why?

g. Examine the ratio of [X] to [Y] early in the reaction as k_x and k_y increase in value (keeping $k_x = k_y$). Explain why [X] is approximately equal to [Y] under these conditions.

h. Examine the ratio of [X] to [Y] late in the reaction at different values of k_x and k_y (keeping $k_x = k_y$). Explain why [X] and [Y] diverge as these reaction rates increase.

Figure 7.21

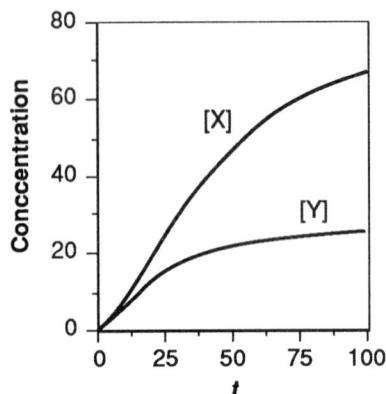

XY-PLOT

Data
Horizontal (x) axis: cells A9–A109 (time)
Vertical (y) axis: F9–F109, G9–G109, D9–D109, and E9–E109 (concentrations of X and Y)

MODELS

FORMULAS

	A	B	C	D	E	F	G	H
1	k(c) =	0.1	[A]i =	100				
2	k(-c) =	0.0044	[C]i =	0				
3	k(i) =	0.1	[I]i =	0				
4	k(-i) =	0.067						
5					|--------------	------Reaction Rates------		------------|
6	t	[A]	[C]	[I]	A --> C	A --> I	C --> A	I --> A
7	0	=D1	=D2	=D3	=B7*B1	=B7*B3	=C7*B2	=D7*B4
8	1	Formula	Formula	Formula	=B8*B1	=B8*B3	=C8*B2	=D8*B4
9	2	I	II	III	=B9*B1	=B9*B3	=C9*B2	=D9*B4
10	3				=B10*B1	=B10*B3	=C10*B2	=D10*B4

	A	E	F	G	H
106	99	=B106*B1	=B106*B3	=C106*B2	=D106*B4
107	100	=B107*B1	=B107*B3	=C107*B2	=D107*B4

VALUES

	A	B	C	D	E	F	G	H
1	k(c) =	0.1	[A]i =	100				
2	k(-c) =	0.0044	[C]i =	0				
3	k(i) =	0.1	[I]i =	0				
4	k(-i) =	0.067						
5					|----------	-----Reaction Rates-		---------|
6	t	[A]	[C]	[I]	A --> C	A --> I	C --> A	I --> A
7	0	100	0	0	10	10	0	0
8	1	80	10	10	8	8	0.044	0.67
9	2	64.714	17.956	17.33	6.4714	6.4714	0.079	1.1611
10	3	53.011	24.348	22.64	5.3011	5.3011	0.1071	1.5169

	A	B	C	D	E	F	G	H
106	99	4.6245	87.63	7.7453	0.4624	0.4624	0.3856	0.5189
107	100	4.6041	87.707	7.6888	0.4604	0.4604	0.3859	0.5151

Formula Set I Prototype Cell is B8
=B7-(**B1***B7)+(**B2***C7)-(**B3***B7)+(**B4***D7)

Formula Set II Prototype Cell is C8	Formula Set III Prototype Cell is D8
=C7+(**B1***B7)-(**B2***C7)	=D7+(**B3***B7)-(**B4***D7)

Model 7.8

BUILD and verify Model 7.8.

ENTER data using cells B1–B4 (the rate constants) and D1–D3 (the initial concentrations of reactants and products).

READ output from the model from columns A (time t), B–D (reactant and product concentrations), and E–H (forward and reverse reaction rates).

NOTICE the equations in cells B7–B107, C7–C107, and D7–D107. Each equation works on the same general principle: The present concentration (e.g., $[C]_t$) is calculated by taking the concentration at the previous time interval (e.g., $[C]_{t-1}$) subtracting losses due to reactions that deplete compound (e.g., $k_c[C]_{t-1}$) and adding gains due to reactions that increase compound (e.g., $k_c[A]_{t-1}$). In the case of [C], the concentration at each time interval can be computed using the formula $[C]_t = [C]_{t-1} + k_c[A]_{t-1} - k_c[C]_{t-1}$.

The model is restricted to 100 separate t values because some spreadsheet programs are unable to plot more than approximately 100 data points. You will nevertheless be required to extend your model in order to identify the true equilibrium concentrations of reactants and products.

Model 7.9

BUILD and verify Model 7.9.

ENTER data using cells B1–B6 (the rate constants) and D1–D5 (the initial concentrations of reactants and products).

READ output from the model from columns A (time) and B–G (reactant and product concentrations). Column E calculates the total concentrations of A + C + I.

NOTICE that the only difference between Models 7.8 and 7.9 is the addition of two reactions. The products C and I are constantly depleted by the unidirectional reactions $C \rightarrow X$ and $I \rightarrow Y$ (e.g., X and Y may be swept away in a stream of solvent).

FORMULAS

	A	B	C	D	E	F	G
1	k(c) =	0.1	[A]i =	100			
2	k(-c) =	0.0044	[C]i =	0			
3	k(i) =	0.1	[I]i =	0			
4	k(-i) =	0.067	[X]i =	0			
5	k(x) =	0.03	[Y]i =	0			
6	k(y) =	0.02					
7							
8	t	[A]	[C]	[I]	[A]+[C]+[I]	[X]	[Y]
9	0	=D1	=D2	=D3	=B9+C9+D9	=D4	=D5
10	1				=B10+C10+D10	=F9+(B5*C9)	=G9+(B6*D9)
11	2	Formula Set I	Formula Set II	Formula Set III	=B11+C11+D11	=F10+(B5*C10)	=G10+(B6*D10)
12	3				=B12+C12+D12	=F11+(B5*C11)	=G11+(B6*D11)

	A	B	C	D	E	F	G
108	99				=B108+C108+D108	=F107+(B5*C107)	=G107+(B6*D107)
109	100				=B109+C109+D109	=F108+(B5*C108)	=G108+(B6*D108)

VALUES

	A	B	C	D	E	F	G
1	k(c) =	0.1	[A]i =	100			
2	k(-c) =	0.0044	[C]i =	0			
3	k(i) =	0.1	[I]i =	0			
4	k(-i) =	0.067	[X]i =	0			
5	k(x) =	0.03	[Y]i =	0			
6	k(y) =	0.02					
7							
8	t	[A]	[C]	[I]	[A]+[C]+[I]	[X]	[Y]
9	0	100	0	0	100	0	0
10	1	80	10	10	100	0	0
11	2	64.714	17.656	17.13	99.5	0.3	0.2
12	3	52.997	23.52	22.111	98.62772	0.8297	0.5426

	A	B	C	D	E	F	G
108	99	0.6638	7.4626	1.1941	9.3205132	70.622	20.057
109	100	0.6439	7.2723	1.1566	9.0727532	70.846	20.081

Formula Set I
Prototype Cell is B10

=B9-(**B1***B9)+(**B2***C9)-(**B3***B9)+(**B4***D9)

Formula Set II
Prototype Cell is C10

=C9+(**B1***B9)-(**B2***C9)-(**B5***C9)

Formula Set III
Prototype Cell is D10

=D9+(**B3***B9)-(**B4***D9)-(**B6***D9)

REACTION RATES: FURTHER EXAMPLES

The kinetic behavior of chemical reactions is often surprising even when the rate laws are relatively simple.

In this exercise, you have the opportunity to use numeric methods to explore a number of different kinds of chemical reactions and to build insights into their kinetic behavior.

Recall the caveat noted in previous exercises. Numeric methods involve approximations, and, under some circumstances, the errors in these approximations can rapidly proliferate. You can always correct the problem by choosing a smaller value for Δt.

LESSON

Consider the reaction

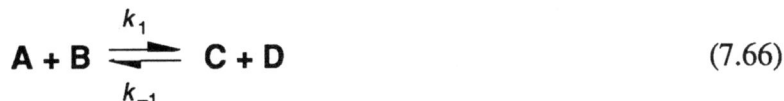

$$A + B \underset{k_{-1}}{\overset{k_1}{\rightleftharpoons}} C + D \tag{7.66}$$

The reaction is second-order overall in both the forward and reverse directions. Verify for yourself that Equations 7.67–7.70 display the contributions from both the forward and reverse reactions to the net reaction rates of each of the reactants and products:

TO GET THE MOST OUT OF THIS EXERCISE, YOU SHOULD ALREADY KNOW

The definition of the terms:
equilibrium
equilibrium constant
reaction rate
rate constant

How to find the equilibrium concentrations of reversible reactions.

How to use rate laws to calculate the concentrations of reactants and products as they change with time.

$$\frac{d[A]}{dt} = -k_1[A][B] + k_{-1}[C][D] \tag{7.67}$$

$$\frac{d[B]}{dt} = -k_1[A][B] + k_{-1}[C][D] \tag{7.68}$$

$$\frac{d[C]}{dt} = -k_{-1}[C][D] + k_1[A][B] \tag{7.69}$$

$$\frac{d[D]}{dt} = -k_{-1}[C][D] + k_1[A][B] \tag{7.70}$$

Build and verify Model 7.10. Model 7.10 repeatedly uses Equations 7.67–7.70 to simulate a reversible chemical reaction that is second-order overall in both directions. Explore the model and make sure that you understand how Equations 7.67–7.70 are implemented.

Build an on-screen graphic that plots [A], [B], [C], and [D] as a function of time (Figure 7.22).

a. Initialize your model such that $k_1 = 0.01$, $k_{-1} = 0.02$, and $\Delta t = 0.5$. Explore the behavior of [A], [B], [C], and [D] vs t for each of the starting conditions below:

[A]	[B]	[C]	[D]
2.0	2.0	0.0	0.0
2.0	3.0	0.0	0.0
2.0	4.0	0.0	0.0
2.0	2.0	0.0	0.0
3.0	2.0	0.0	0.0
4.0	2.0	0.0	0.0
2.0	2.0	0.0	0.0
2.0	2.0	1.0	0.0
2.0	2.0	2.0	0.0

Under what conditions do plots of [A] vs t and [B] vs t coincide?

Under what conditions do plots of [C] vs t and [D] vs t coincide?

b. Under each set of conditions in Lesson a, determine the ratio of [C] to [A] at equilibrium. For a first-order reaction, $[C]/[A] = k_1/k_{-1}$. Is the same true for a second-order reaction? Why or why not?

c. For a first-order reaction, the rate of approach to equilibrium is independent of initial concentrations (verify this fact using Model 7.7). Explain why, for a second-order reaction, equilibrium conditions are approached more rapidly as the initial concentrations of

Figure 7.22

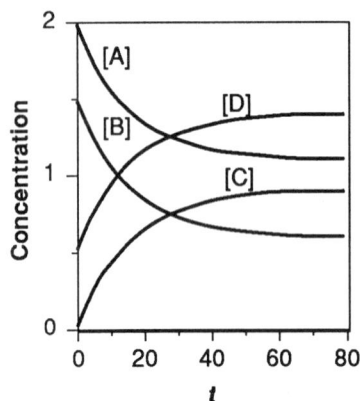

XY-PLOT

Data
Horizontal (x) axis: cells A8–A108 (time)
Vertical (y) axis: cells B8–B108 ([A]), C8–C108 ([B]), D8–D108 ([C]), and E8–E108 ([D])

[A] and [B] are increased. (Notice that this is true even though more molecules must be converted to reach equilibrium as [A] and [B] increase.)

[NOTE: Because the rate constants are multiplied by the product of [A] and [B}, this exercise may blow up (i.e., give negative values or impossibly large numbers) when the initial concentrations are increased without limit. Recall that such behavior occurs when the amount of reaction per time interval is too large. If you encounter such problems, try decreasing the concentrations, decreasing the rate constants proportionately, or decreasing Δt.]

PROBLEMS

1. In many cases in nature, the product of a reaction undergoes further reaction. This is especially true in living cells where each reaction is normally only one step in a sequence of steps. The kinetics of sequential reactions can become extremely complex. In even the simplest case — two first-order reactions in which the product of one reaction is the reactant of the next — there are many interesting possibilities.

   Consider a sequential reaction in which A is converted to B reversibly and B is converted to C irreversibly (Figure 7.23).

   Verify that Equations 7.71–7.73 correctly describe the rates of change of each of the compounds involved:

   $$\frac{d[A]}{dt} = -k_1[A] + k_{-1}[B] \tag{7.71}$$

   $$\frac{d[B]}{dt} = -k_{-1}[B] - k_2[B] + k_1[A] \tag{7.72}$$

   $$\frac{d[C]}{dt} = -k_2[B] \tag{7.73}$$

   Design, build, verify, and explore a model to simulate this sequential reaction. You may compare your design with our design, if you wish (Model 7.11). Initialize your model with $[A]_0 = 100$, $[B]_0 = 0$, $[C]_0 = 0$, $k_1 = 0.1$, $k_{-1} = 0.1$, and $k_2 = 0.03$. Plot [A], [B], and [C] vs t (Figure 7.24).

   The best way to explore a model such as this is to simply jump into it — change rate constants and initial concentrations of reactants and products in an orderly sequence. Try to picture the forces governing this reaction and vary conditions to verify your predictions.

Figure 7.23

In this sequential reaction, the first step A ↔ B is first-order and reversible, while the second step is first-order and goes to completion. Perhaps the second step involves a very large equilibrium constant, or perhaps the product C is physically swept away.

Figure 7.24

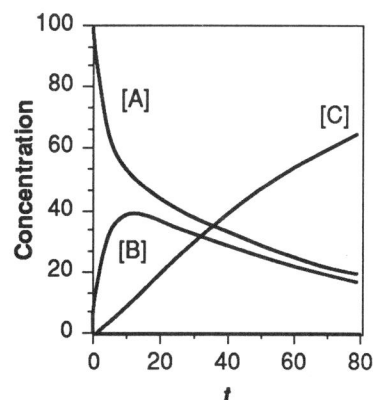

Sequential reactions lead to complex profiles of [A], [B], and [C] vs t. Use your model to determine the influence of rate constants and reactant concentrations on reaction rates.

Here are just a few fruitful paths:

a. The initial conditions noted describe a situation in which only
    A is initially present and in which A and B are interconverted at
    a somewhat faster rate than B is converted to C. Notice that
    Figure 7.24 can be divided into roughly two regions — an early
    time period in which A and B approach equilibrium with each
    other and a later time period in which [A] and [B] decrease at
    roughly the same rate. Would you expect changes in the rate
    constants k_1 and k_{-1} to have a greater influence on the early or
    later time intervals of the reaction? Verify your prediction by
    changing k_1 and k_{-1} from 0.1 to 0.5.

    Explore other combinations of k_1 and k_{-1} values.

b. Did you notice that changes in k_1 and k_{-1} appear to have little
    influence on the rate of decline of [A] and [B] at later time
    periods? Does it make sense that k_2 might play a role here?
    Increase k_2 from .03 to .1 to test your prediction.

    Explore other combinations of k_1, k_{-1}, and k_2 and verify the
    following generalizations: When $k_1 >> k_2$, [B] rises rapidly and
    then falls along a nearly first-order trajectory; if $k_2 << k_{-1}$, [A] and
    [B] come to a quasi equilibrium and then fall together as C is
    formed.[1]

c. What combination of relative values of k_1, k_{-1}, and k_2 would tend
    to keep [B] low throughout the reaction? Test your prediction.

    Under these conditions [A] vs t is very similar to a first-order
    reaction (you can prove this by plotting ln[A] vs t if you wish).
    Why?

d. Are the reactions rates of A and B influenced by [C]? Why or
    why not? Test your prediction.

e. Is it possible for [B] to exceed [A] during any part of the
    reaction? What relative values of the rate constants might cause
    such a situation? Test your prediction.

These are just a few of the paths that you can explore with this
model. You might also like to explore a model in which only B is
present initially or to build a model in which the conversion of B to
C is also reversible.

2. There are many ways in which two reactions may be related to each other. Consider the coupled reaction described in Figure 7.25.

Verify that Equations 7.74–7.76 correctly describe the rates of change of each of the compounds involved:

$$\frac{d[A]}{dt} = -k_1[A] - k_2[A][B] + k_{-1}[B] \tag{7.74}$$

$$\frac{d[B]}{dt} = -k_{-1}[B] - k_2[A][B] + k_1[A] \tag{7.75}$$

$$\frac{d[C]}{dt} = k_2[A][B] \tag{7.76}$$

Build, verify, and explore a model to simulate this coupled reaction. Compare your design with our design, if you wish (Model 7.12). Initialize your model with $[A]_0 = 100$, $[B]_0 = 0$, $[C]_0 = 0$, $k_1 = 0.05$, $k_{-1} = 0.05$, and $k_2 = 0.001$. Plot [A], [B], and [C] vs t. This model is also worthy of considerable independent exploration. Some features worth examining include

a. When this reaction is initialized with the conditions noted above, it appears to be quite similar to the reaction explored in Problem 1 — an initial period during which A and B reach a quasi equilibrium and a later phase during which A and B decline as C accumulates. But all is not as it seems.

Vary the value of k_{-1} from 0.1 to 0.001 and notice that as k_{-1} decreases the model moves rapidly to an apparent endpoint in which a large amount of [B] remains. Why?

b. Initialize your model with $[A]_0 = 100$, $[B]_0 = 200$, $[C]_0 = 0$, $k_1 = 0.05$, and $k_{-1} = 0.005$, and vary k_2 from 0.00005 to 0.005.

You will see that for some values of k_2, [B] first rises and then falls, whereas for other values, [B] falls from the beginning. Why?

c. Explore your model under conditions in which [B] > [A] initially. There are some conditions under which the reaction follows approximately second-order kinetics until the initial stock of A is consumed, after which the reaction proceeds at a slower pace following approximately first-order kinetics. Notice that the rate of second phase of the reaction is governed by k_{-1}.

Figure 7.25

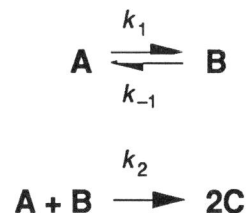

In this sequence of reactions, A first reacts to reversibly form B. The reaction observes first-order kinetics in both directions. Once measurable quantities of both A and B are simultaneously present, A and B react (second-order kinetics, overall) to form two molecules of C.

You can vary initial concentrations and the relative values of the rate constants and obtain a wide variety of patterns even with this simple model. Because of the interaction between the parameters, however, the model is quite susceptible to blowing up if the rate constants are too high. It is the relative values of the rate constants that matter in your explorations and so the need to hold them to low values will not hamper your exploration of the properties of this system.

3. The gas-phase reaction studied in Chapter 5 is recalled in Equation 7.77:

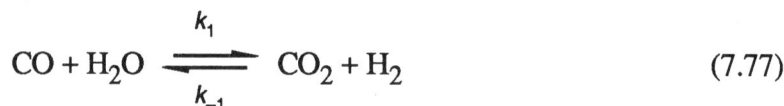

$$CO + H_2O \underset{k_{-1}}{\overset{k_1}{\rightleftharpoons}} CO_2 + H_2 \qquad (7.77)$$

where $k_1/k_{-1} = 5$.

The rate laws for the forward and reverse reactions can be written

$$\frac{d[CO]}{dt} = \frac{d[H_2O]}{dt} = -k_1[CO][H_2O] \qquad (7.78)$$

$$\frac{d[CO_2]}{dt} = \frac{d[H_2]}{dt} = -k_{-1}[CO_2][H_2] \qquad (7.79)$$

Numeric approximations to the rate laws for CO and CO_2 that take into account both the forward and reverse reactions are written

$$\begin{aligned}\Delta[CO] &= \Delta[H_2O] \\ &= -\Delta t \, k_1[CO][H_2O] + \Delta t \, k_{-1}[CO_2][H_2] \qquad (7.80)\end{aligned}$$

$$\begin{aligned}\Delta[CO_2] &= \Delta[H_2] \\ &= -\Delta t \, k_{-1}[CO_2][H_2] + \Delta t \, k_1[CO][H_2O] \qquad (7.81)\end{aligned}$$

Design, build, and verify a model and appropriate graphics to explore this reaction. (Model 7.13 is our answer.)

a. Suppose that the initial total pressure of gas in a reaction vessel is maintained at 2 atm but that the initial partial pressures of reactants and products are varied according to a table such as that below.[2] What is the equilibrium mole fraction of CO and CO_2 corresponding to each set of starting conditions?

[CO]	[H$_2$O]	[CO$_2$]	[H$_2$]
1.0	1.0	0.0	0.0
0.9	0.9	0.1	0.1
0.7	0.7	0.3	0.3
0.5	0.5	0.5	0.5
0.3	0.3	0.7	0.7
0.1	0.1	0.9	0.9
0.0	0.0	1.0	1.0

Explore other starting conditions in which the total pressure of the reaction vessel is 1 atm. Keep in mind, however, that equilibrium values of reactants and products will reproduce those above only if [CO] = [H$_2$O] and [CO$_2$] = [H$_2$].

b. Now increase the total pressure in the reaction vessel to 10 atm and re-explore the effect of starting conditions. What happens to the mole fraction of CO and CO$_2$ under these conditions?

c. Plot the mole fraction of CO and CO$_2$ at equilibrium as a function of total starting pressure. Work under conditions in which [CO] = [H$_2$O] and [CO$_2$] = [H$_2$].

Unlike the dimerization of NO$_2$ (Problem 2, Exercise 7.4), the reaction of CO with H$_2$O appears to be insensitive to reaction-vessel pressure. Why?

4. Your textbook or any physical chemistry textbook will have numerous examples of complex chemical reactions that can be explored using numeric techniques. Take the time to find such problems and to build models that simulate their behavior.

MODELS

Model 7.10

BUILD and verify Model 7.10.

ENTER data using cells B1–B4 (initial concentrations of reactants and products), B5 (Δt), and D1 and D2 (forward and reverse rate constants).

READ output from the model from columns A (time) and B–E (concentrations of reactants and products with time).

FORMULAS

	A	B	C	D	E
1	[A]i =	1.5	k1 =	0.02	
2	[B]i =	2	k-1 =	0.01	
3	[C]i =	0			
4	[D]i =	0.5			
5	Δt =	0.8			
6					
7	Time	[A]	[B]	[C]	[D]
8	0	=B1	=B2	=B3	=B4
9	=A8+B5	Formula	Formula	Formula	Formula
10	=A9+B5	I	Set II	III	IV
11	=A10+B5				

VALUES

	A	B	C	D	E
1	[A]i =	1.5	k1 =	0.02	
2	[B]i =	2	k-1 =	0.01	
3	[C]i =	0			
4	[D]i =	0.5			
5	Δt =	0.8			
6					
7	Time	[A]	[B]	[C]	[D]
8	0	1.5	2	0	0.5
9	0.8	1.452	1.952	0.048	0.548
10	1.6	1.406862	1.906862	0.093138	0.593138
11	2.4	1.36438	1.86438	0.13562	0.63562

Formula Set I Prototype Cell is B9
=B8-($B5 \cdot D1 \cdot B8*C8$)+($B5 \cdot D2 \cdot D8*E8$)

Formula Set II Prototype Cell is C9
=C8-($B5 \cdot D1 \cdot B8*C8$)+($B5 \cdot D2 \cdot D8*E8$)

Formula Set III Prototype Cell is D9
=D8-($B5 \cdot D2 \cdot D8*E8$)+($B5 \cdot D1 \cdot B8*C8$)

Formula Set IV Prototype Cell is E9
=E8-($B5 \cdot D2 \cdot D8*E8$)+($B5 \cdot D1 \cdot B8*C8$)

FORMULAS

	A	B	C	D
1	[A]i =	100	k1 =	0.1
2	[B]i =	200	k-1 =	0.1
3	[C]i =	0	k2 =	0.02
4	Δt =	0.8		
5				
6	Time	[A]	[B]	[C]
7	0	=B1	=B2	=B3
8	=A7+B4	Formula Set I	Formula Set II	Formula Set III
9	=A8+B4			
10	=A9+B4			

VALUES

	A	B	C	D
1	[A]i =	100	k1 =	0.1
2	[B]i =	200	k-1 =	0.1
3	[C]i =	0	k2 =	0.02
4	Δt =	0.8		
5				
6	Time	[A]	[B]	[C]
7	0	100	200	0
8	0.8	108	188.8	3.2
9	1.6	114.464	179.3152	6.2208
10	2.4	119.6521	171.2581	9.089843

Formula Set I
Prototype Cell is B8

=B7-(**B4*****D1***B7)+(**B4*****D2***C7)

Formula Set III
Prototype Cell is D8

=D7+(**B4*****D3***C7)

Formula Set II
Prototype Cell is C8

=C7-(**B4*****D2***C7)-
(**B4*****D3***C7)+(**B4*****D1***B7)

Model 7.11

BUILD and verify Model 7.11.

ENTER data using cells B1–B3 (initial concentrations of reactants and products), B4 (Δt), and D1–D3 (rate constants).

READ output from the model from columns A (time) and B–D (concentrations of reactants and products with time).

Model 7.12

BUILD and verify Model 7.12.

ENTER data using cells B1–B3 (initial concentrations of reactants and products), B4 (Δt), and D1–D3 (rate constants).

READ output from the model from columns A (time) and B–D (concentrations of reactants and products with time).

FORMULAS

	A	B	C	D
1	[A]i =	100	k1 =	0.1
2	[B]i =	10	k-1 =	0.12
3	[C]i =	5	k2 =	0.0002
4	Δt =	1		
5				
6	Time	[A]	[B]	[C]
7	0	=B1	=B2	=B3
8	=A7+B4	Formula	Formula	Formula
9	=A8+B4	Set	Set	Set
10	=A9+B4	I	II	III
11	=A10+B4			

VALUES

	A	B	C	D
1	[A]i =	100	k1 =	0.1
2	[B]i =	10	k-1 =	0.12
3	[C]i =	5	k2 =	0.0002
4	Δt =	1		
5				
6	Time	[A]	[B]	[C]
7	0	100	10	5
8	1	91	18.6	5.4
9	2	83.79348	25.12948	6.07704
10	3	78.00853	30.07215	6.919315
11	4	73.34716	33.79517	7.857668

Formula Set I
Prototype Cell is B8

=B7-(**B4*****D1***B7)-(**B4*****D3***B7*C7)+(**B4*****D2***C7)

Formula Set II
Prototype Cell is C9

=C7-(**B4*****D2***C7)-(**B4*****D3***B7*C7)+(**B4*****D1***B7)

Formula Set III
Prototype Cell is D8

=D7+2*(**B4*****D3***B7*C7)

FORMULAS

	A	B	C	D	E	F	G	H	I
1	Initial P[CO] =	1							
2	Initial P[H2O] =	1							
3	Initial P[CO2] =	0							
4	Initial P[H2] =	0							
5									
6	k (atm/sec) =	5							
7	k-1 (atm/sec) =	1							
8									
9	Δt (sec) =	0.02							
10									
11	t	P[CO]	P[H2O]	P[CO2]	P[H2]	MF[CO]	MF[H2O]	MF[CO2]	MF[H2]
12	0.0	=B1	=B2	=B3	=B4	0.5			
13	=A12+B9	Formula Set I	Formula Set II	Formula Set III	Formula Set IV	Formula Set V	Formula Set VI	Formula Set VII	Formula Set VIII
14	=A13+B9								
15	=A14+B9								

VALUES

	A	B	C	D	E	F	G	H	I
1	Initial P[CO] =	1							
2	Initial P[H2O] =	1							
3	Initial P[CO2] =	0							
4	Initial P[H2] =	0							
5									
6	k (atm/sec) =	5							
7	k-1 (atm/sec) =	1							
8									
9	Δt (sec) =	0.02							
10									
11	t	P[CO]	P[H2O]	P[CO2]	P[H2]	MF[CO]	MF[H2O]	MF[CO2]	MF[H2]
12	0.0	1	1	0	0	0.5	0.5	0	0
13	0.0	0.9	0.9	0.1	0.1	0.45	0.45	0.05	0.05
14	0.0	0.8192	0.8192	0.1808	0.1808	0.4096	0.4096	0.0904	0.0904
15	0.1	0.752745	0.752745	0.247255	0.247255	0.376372	0.376372	0.123628	0.123628

Model 7.13

BUILD and verify Model 7.13.

ENTER data using cells B1–B3 (initial concentrations of reactants and products), B6–B7 (rate constants), and B9 (Δt).

READ output from the model from columns A (time), B–E (concentrations of reactants and products with time), and F–I (mole fractions of reactants and products with time).

Formula Set I
Prototype Cell is B13

=B12+(**B9** ***B7** *D12*E12)-(**B9** ***B6** *B12*C12)

Formula Set V
Prototype Cell is F12

=B12/SUM(B12:E12)

Formula Set II
Prototype Cell is C13

=C12+(**B9** ***B7** *D12*E12)-(**B9** ***B6** *B12*C12)

Formula Set VI
Prototype Cell is G12

=C12/SUM(B12:E12)

Formula Set III
Prototype Cell is D13

=D12+(**B9** ***B6** *B12*C12)-(**B9** ***B7** *D12*E12)

Formula Set VII
Prototype Cell is H12

=D12/SUM(B12:E12)

Formula Set IV
Prototype Cell is E13

=E12+(**B9** ***B6** *B12*C12)-(**B9** ***B7** *D12*E12)

Formula Set VIII
Prototype Cell is I12

=E12/SUM(B12:E12)

EXERCISE **7.7**

OSCILLATING REACTIONS

TO GET THE MOST OUT OF THIS
EXERCISE, YOU SHOULD
ALREADY KNOW

The definition of the terms:
reaction rate
rate constant

How to use rate laws to calculate the
concentrations of reactants and
products as they change with time.

We end this chapter with an exercise that really tests your abilities at independent exploration. Until now you have examined chemical reactions that start and end with clear, definable concentrations of reactants and products. In recent years, many reactions have been discovered that contain intermediates that rise and fall in a cyclic fashion over extended (even indefinite) periods of time.

Such reactions are of enormous theoretical and practical interest and may even form the basis of many of the "clocks" that make up living systems. Like all clocks, oscillating reactions must be driven by a renewable source of external energy or they will run down as surely as any pendulum clock. The "spring" that supplies this energy for chemical oscillations is often a substrate that maintains its concentration via its continual renewal from some outside source.

The earliest theoretical basis for oscillating reactions came from the work of Alfred Lotka in 1910. He showed that a simple reaction with a single autocatalytic step (i.e., a substrate that promotes its own formation) could exhibit damped oscillations in the concentrations of its intermediates over a period of time (Equations 7.82a–c)

254

Step 1: $A \rightarrow X$ (7.82a)

Step 2: $X + Y \rightarrow 2Y$ (7.82b)

Step 3: $Y \rightarrow Q$ (7.82c)

Notice that Step 2 consists of the autocatalytic formation of Y. Although the oscillations exist even when A is depleted with time and replaced with Q, the oscillations are easier to observe if A is continually replenished and therefore [A] remains constant. Under these circumstances, you should be able to confirm the reaction rates for each component of the reaction (Equations 7.83a–c).

$$\frac{d[X]}{dt} = k_1[A] - k_2[X][Y] \qquad (7.83a)$$

$$\frac{d[Y]}{dt} = k_2[X][Y] - k_3[Y] \qquad (7.83b)$$

$$\frac{d[Q]}{dt} = k_3[Y] \qquad (7.83c)$$

where k_1, k_2, and k_3 are the rate constants for Steps 1, 2, and 3, respectively.

Ten years later, Lotka reported that undamped (continuous) oscillations could be produced if the mechanism were changed slightly to give two autocatalytic steps:

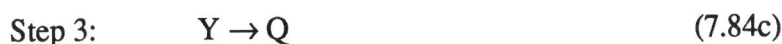

Step 1: $A + X \rightarrow 2X$ (7.84a)

Step 2: $X + Y \rightarrow 2Y$ (7.84b)

Step 3: $Y \rightarrow Q$ (7.84c)

The rate laws for this second mechanism are

$$\frac{d[X]}{dt} = k_1[A][X] - k_2[X][Y] \qquad (7.85a)$$

$$\frac{d[Y]}{dt} = k_2[X][Y] - k_3[Y] \qquad (7.85b)$$

$$\frac{d[Q]}{dt} = k_3[Y] \qquad (7.85c)$$

Figure 7.26

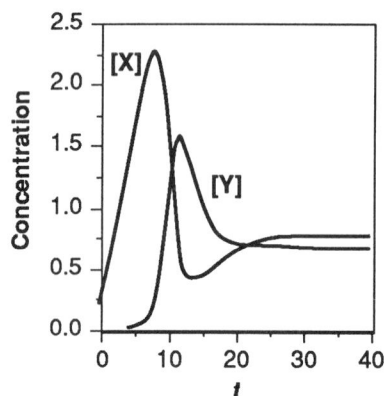

XY-PLOT

Data
Horizontal (x) axis: cells A10–A810
(time)
Vertical (y) axis: cells C10–C810 ([X]),
D10–D810 ([Y])

Figure 7.27

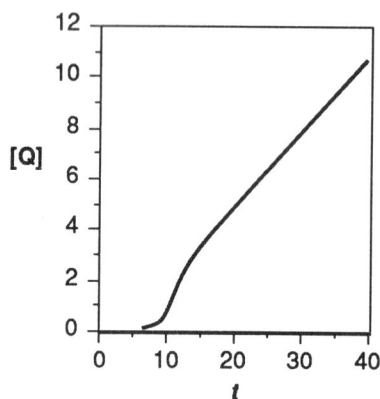

XY-PLOT

Data
Horizontal (x) axis: cells A10–A810
(time)
Vertical (y) axis: cells E10–E810 ([Q])

LESSONS

1. Build and verify Model 7.14. Model 7.14 simulates the reaction described by Equations 7.82a–c) and uses techniques that, by now, should be completely familiar. However, the model requires a large number of rows and a relatively small Δt to work satisfactorily. Until now, small deviations from reality in your reaction-rate simulations did not "feed back on themselves" to cause problems. Oscillatory motions, however, are highly susceptible to the errors in numeric calculation, and Δt must be kept quite small if the system is to approximate reality.

   a. Enter your simulated laboratory and explore Lotka's first reaction. Build on-screen graphics (Figures 7.26 and 7.27) to help in your exploration.

   What values of k_1, k_2, and k_3 lead to what kinds of behavior? How do the initial concentrations of reactants and products affect the reaction? Explain the behavior of [Q] during the course of the reaction.

   Write a report on your observations as Lotka might have done.[1] Remark particularly on those conditions that lead to extended oscillatory behavior and how conditions affect the frequency and amplitude of the oscillation.

   b. Some features of your model do not reflect reality but rather the errors induced by the numeric method. These features will decrease as Δt decreases (and the number of rows increase). Write a report on the artifacts inherent in your tools of exploration.

2. Build and verify Model 7.15. Model 7.15 simulates the reaction described by Equations 7.84a–c). Build on-screen graphics to help in your exploration. Explore this reaction as fully as you can and write a report on your observations. Contrast this reaction to that observed in Lesson 1.

MODELS

FORMULAS

	A	B	C	D	E
1	[A]0 =	1	k1 =	0.3	
2	[X]0 =	0.15	k2 =	0.6	
3	[Y]0 =	0.01	k3 =	0.4	
4	[Q]0 =	0			
5					
6	t0 =	0			
7	Δt =	0.05			
8					
9	t	[A]	[X]	[Y]	[Q]
10	=B6		=B2	=B3	=B4
11					
12					
13	Formula Set I	Formula Set II	Formula Set III	Formula Set IV	Formula Set V
14					
808					
809					
810					

Model 7.14

BUILD and verify Model 7.14.

ENTER data using cells B1–B4 (initial concentrations of reactants and products) and D1–D3 (rate constants).

READ output from the model from columns A–E (time, concentrations of reactants and products).

NOTICE that there are far more rows in this column than other models. Oscillatory reactions diverge quickly from reality unless Δt is kept very small. Depending on your spreadsheet, you may or may not be able to graph this many points on your computer screen.

VALUES

	A	B	C	D	E
1	[A]0 =	1	k1 =	0.3	
2	[X]0 =	0.15	k2 =	0.6	
3	[Y]0 =	0.01	k3 =	0.4	
4	[Q]0 =	0			
5					
6	t0 =	0			
7	Δt =	0.05			
8					
9	t	[A]	[X]	[Y]	[Q]
10	0	1	0.15	0.01	0
11	0.05	1	0.164955	0.009845	0.0002
12	0.1	1	0.17990628	0.00969682	0.0003969
13	0.15	1	0.19485394	0.00955522	0.00059084
14	0.2	1	0.20979809	0.00941997	0.00078194

	A	B	C	D	E
808	39.9	1	0.66722981	0.74899847	10.7137717
809	39.95	1	0.66723719	0.74901113	10.7287517
810	40	1	0.66724415	0.74902395	10.7437319

Formula Set I **Prototype Cell is A11**
=A10+**B7**

Formula Set III **Prototype Cell is C11**
=C10+(**D1** *B10-**D2** *C10*D10)* **B7**

Formula Set II **Prototype Cell is B10**
=**B1**

Formula Set IV **Prototype Cell is D11**
=D10+(**D2** *C10*D10-**D3** *D10)***B7**

Formula Set V **Prototype Cell is E11**
=E10+**D3** *D10***B7**

Model 7.15

BUILD and verify Model 7.15.

ENTER data using cells B1–B4 (initial concentrations of reactants and products) and D1–D3 (rate constants).

READ output from the model from columns A–E (time, concentrations of reactants and products).

NOTICE that this model is identical to Model 7.14 except for the contents of Formula Set III and the number of rows that are calculated.

FORMULAS

	A	B	C	D	E
1	[A]0 =	1	k1 =	0.3	
2	[X]0 =	0.15	k2 =	0.6	
3	[Y]0 =	0.01	k3 =	0.4	
4	[Q]0 =	0			
5					
6	t0 =	0			
7	Δt =	0.05			
8					
9	t	[A]	[X]	[Y]	[Q]
10	=B6		=B2	=B3	=B4
11					
12					
13	Formula Set I	Formula Set II	Formula Set III	Formula Set IV	Formula Set V
14					
1683					
1684					
1685					

VALUES

	A	B	C	D	E
1	[A]0 =	1	k1 =	0.3	
2	[X]0 =	0.15	k2 =	0.6	
3	[Y]0 =	0.01	k3 =	0.4	
4	[Q]0 =	0			
5					
6	t0 =	0			
7	Δt =	0.05			
8					
9	t	[A]	[X]	[Y]	[Q]
10	0	1	0.15	0.01	0
11	0.05	1	0.152205	0.009845	0.0002
12	0.1	1	0.15444312	0.00969305	0.0003969
13	0.15	1	0.15671486	0.0095441	0.00059076
14	0.2	1	0.15902071	0.00939809	0.00078164

	A	B	C	D	E
1683	83.65	1	0.00271233	0.43971556	18.8382932
1684	83.7	1	0.00271724	0.43095703	18.8470875
1685	83.75	1	0.00272286	0.42237302	18.8557067

Formula Set I Prototype Cell is A11
=A10+ **B7**

Formula Set III Prototype Cell is C11
=C10+(**D1** *B10*C10 -**D2** *C10*D10)* **B7**

Formula Set II Prototype Cell is B10
=B1

Formula Set IV Prototype Cell is D11
=D10+(**D2** *C10*D10 -**D3** *D10)* **B7**

Formula Set V Prototype Cell is E11
=E10+ **D3** *D10* **B7**

CHAPTER **8**

THERMODYNAMICS

Spontaneous processes occur only in specific directions:

- Water runs downhill.

- Gases expand to fully occupy their containers.

- Hot water cools until it reaches the same temperature as its surrounding environment.

Think about what it would mean if these processes moved spontaneously in the other direction. Have you ever seen

- Water run uphill?

- Gases contract into one corner of their containers?

- Hot water become hotter by extracting heat from its cooler surroundings?

You explored such issues in Chapter 5 when you studied chemical equilibrium. You learned that concentrations of the reactants and products in the reversible reaction

$$A \rightleftharpoons B \tag{8.1}$$

will spontaneously change until or unless Equation 8.2 is satisfied:

$$K = \frac{[B]_{eq}}{[A]_{eq}} \tag{8.2}$$

where K is the equilibrium constant of the reaction and $[A]_{eq}$ and $[B]_{eq}$ are the concentrations of the reactant and product, respectively.

That is, a system in equilibrium will remain in equilibrium, and a system displaced from equilibrium will spontaneously change until it is in equilibrium. You acquired some understanding of the physical basis of such equilibria when you explored the concept of dynamic equilibrium (e.g., the reaction described by Equation 8.1 is in equilibrium when the rate at which A converts to B equals the rate at which B converts to A).

In this chapter, you explore a quite different approach to these matters — thermodynamics. Thermodynamics stands as a theoretical foundation for reversible and irreversible processes regardless of their nature. It is as equally at home in the study of hydraulics as it is in the study of chemistry.[1]

In the exercises that follow, you will slowly build up a framework of insights into the concepts that underlie thermodynamics — energy, enthalpy, entropy, etc. — but you might find it helpful to see where you're going before you get there. By the end of Exercise 8.2, you will have a qualitative understanding of the meaning of Gibbs free energy G. G is a measure of the ability of a system to spontaneously perform work.[2] If a change in a system leads to a reduction in G (i.e., if $\Delta G < 0$), that change can occur spontaneously. If a change in a system leads to an increase in G (i.e., if $\Delta G > 0$), then the change will not occur unless an external source of energy is available to do the work.

Consider, for example, the reaction in Equation 8.1. Suppose that the initial concentrations of A and B in the reaction flask are $[A]_0$ and $[B]_0$. It can be shown that the change in free energy (per mole of reactant converted) ΔG that accompanies the transition $A \rightarrow B$ (under conditions where their concentrations do not change) is given by

$$\Delta G = RT \ln\left(\frac{Q}{K}\right) \tag{8.3}$$

where R is the gas constant, T is the temperature (in K), K is the equilibrium constant, and

$$Q = \frac{[B]_0}{[A]_0} \tag{8.4}$$

Figure 8.1 shows a plot of ΔG vs [B] (also the mole fraction of B X_B, in this case) under conditions in which $K = 1$ (and therefore $X_B = 0.5$ at equilibrium). Notice that when X_B is less than 0.5, ΔG is negative: The reaction $A \rightarrow B$ causes [A] and [B] to approach their equilibrium concentrations, and the reaction can occur spontaneously. However, when X_B is greater than 0.5, ΔG is positive: The reaction $A \rightarrow B$ causes [A] and [B] to move farther from their equilibrium concentrations, and the reaction cannot occur spontaneously. What happens when $X_B = 0.5$?

You can verify Figure 8.1 by building Model 8.1. Furthermore, you can explore other plots of ΔG vs X_B for other values of K.

Another way to look at this equilibrium is to plot G vs [B] (actually, of course, relative G) (Figure 8.2). Now it becomes clear that the system is in equilibrium when G is at its minimum value. *Systems approach equilibrium by minimizing the free energy of the system.*

You can verify Figure 8.2 using Model 8.1. How does G vs X_B change as K changes?

Figure 8.1

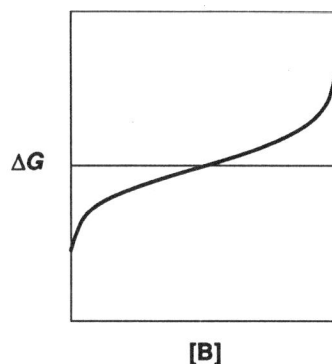

The change in free energy ΔG vs [B] tells you whether, at a given value of [B], a reaction can proceed spontaneously. Generate this plot by plotting column B vs column D in Model 8.1.

Figure 8.2

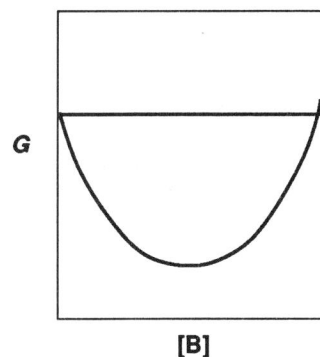

G vs [B] can also tell you whether a reaction is in equilibrium. The reaction is at equilibrium when any change in the system causes an increase in G (i.e., G is at its minimum value). Generate this plot by plotting column B vs column E in Model 8.1.

Gibbs free energy is a very useful concept, and tables of, for example, the free energy of formation of compounds constitute an important part of any chemist's stock in trade. Exercises 8.1 and 8.2 should help you understand the origins of ΔG in thermodynamic theory. Exercise 8.3 exploits ΔG in various chemical settings.

MODEL

Model 8.1

BUILD and verify Model 8.1.

ENTER data using cell B4 (equilibrium constant).

READ output from the model at cells B7–B105, D7–D105, and E7–E105 (concentration of B, change in free energy, and relative free energy).

NOTICE that ΔG is the change in G per mole of reactant or product. G, however, is the actual change in G of the system, and so Formula Set IV includes a correction for actual changes in concentration that the system undergoes.

FORMULAS

	A	B	C	D	E
1	R =	0.00198			
2	T =	300			
3					
4	K =	1			
5					
6	[A]	[B]	Q	ΔG	G
7	0.99				0
8	0.98				
9	0.97	Formula Set I	Formula Set II	Formula Set III	Formula Set IV
10	0.96				

	A
103	0.03
104	0.02
105	0.01

VALUES

	A	B	C	D	E
1	R =	0.00198			
2	T =	300			
3					
4	K =	1			
5					
6	[A]	[B]	Q	ΔG	G
7	0.99	0.01	0.01010101	-2.7295012	0
8	0.98	0.02	0.02040816	-2.3117413	-0.0231174
9	0.97	0.03	0.03092784	-2.0648026	-0.0437654
10	0.96	0.04	0.04166667	-1.887764	-0.0626431

	A	B	C	D	E
103	0.03	0.97	32.3333333	2.06480262	-0.0231174
104	0.02	0.98	49	2.31174126	6.9389E-18
105	0.01	0.99	99	2.72950119	0.02729501

Formula Set I Prototype Cell is B7	Formula Set III Prototype Cell is D7
=1-A7	=B1*B2*LN(C7/B4)

Formula Set II Prototype Cell is C7	Formula Set IV Prototype Cell is E8
=B7/A7	=E7+(D8*(A7-A8))

THE FIRST LAW: CONSERVATION OF ENERGY

Thermodynamics arose out of an apparent failure of the law of conservation of energy in classical physics. Consider an object of mass m attached to the end of a rope that passes through a single pulley (Figure 8.3). The total mechanical energy E of the mass is partitioned between potential energy U and kinetic energy K_e. Potential energy is "energy of position." A mass suspended above a flat surface has the potential to do work in proportion to its height above the surface (Equation 8.5):

$$U = mgh \qquad (8.5)$$

where g is the acceleration of gravity (9.8 m s^{-2}) and h is the height of the object above the surface.

Kinetic energy is "energy of motion" (Equation 8.6):

$$K_e = \frac{mv^2}{2} \qquad (8.6)$$

where v is the velocity of the object.

A mass that has been pulled up from its underlying surface by the pulley acquires potential energy. If the mass is suddenly released, it falls back to the surface with increasing velocity and therefore increasing kinetic energy. It is an observed fact that, if frictional forces are reduced to

TO GET THE MOST OUT OF THIS EXERCISE, YOU SHOULD ALREADY KNOW

The definition of the terms:
energy
work
pressure
volume
temperature
ideal gas

How to predict the behavior of an ideal gas using the ideal gas law.

Figure 8.3

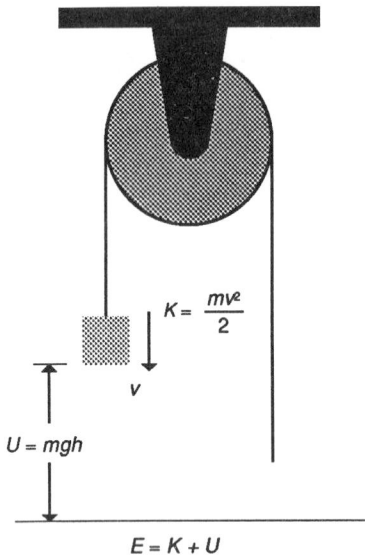

$$K = \frac{mv^2}{2}$$

v

$U = mgh$

$$E = K + U$$

If a pulley is used to raise an object of mass m above a surface, the object acquires potential energy U. If the object is permitted to fall freely back to the surface, U is coverted to kinetic energy K. If the system is frictionless, the total mechanical energy ($E = K + U$) of the system during free fall remains constant.

Figure 8.4

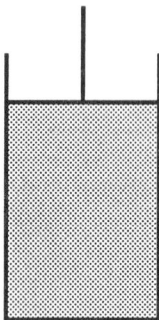

A cylinder of gas enclosed at one end with a movable pistion provides an excellent arena for studying thermodynamics.

below measurable amounts, the increasing K_e of the object is exactly offset by a decreasing U and that the total mechanical energy E of the system remains constant (Equation 8.7):

$$E = K_e + U \tag{8.7}[1]$$

Equation 8.7 constitutes the law of conservation of mechanical energy, and it can be found in any textbook of elementary physics.

Equation 8.7 fails in systems that are subject to measurable frictional forces. If the pulley, for example, is not well-oiled and resists the motion of the falling object, E decreases as the object falls because K_e does not increase as rapidly as U decreases. Thermodynamics arose out of the discovery that in cases where E is not conserved, something (e.g., the pulley) warms up. Today we understand that the "missing" kinetic energy appears as the kinetic energy of random motion in the molecules of the pulley and that total energy is conserved if thermal energy is taken into account (Equation 8.8):

$$E = K_e + U + H \tag{8.8}$$

where H is the thermal energy of the system. Equation 8.8 is called the *first law of thermodynamics*.

One of the best ways to explore thermodynamics is to return to the behavior of an ideal gas (Chapter 3) and to examine this behavior from the viewpoint of work and energy. That is what you'll do in the Lessons that follow. The exercise concludes with remarks best made after the lessons are complete.

LESSONS

1. Recall from your studies of classical physics that when a constant force F moves an object across a distance x the work W done on the object is

$$W = Fx \tag{8.9}$$

Now consider a cylinder of gas closed at one end by a movable, massless, frictionless piston (Figure 8.4). If the pressure that the gas exerts on the piston is infinitesimally greater than some constant pressure P_{ext}, then the work done by the gas on the piston as it pushes outward is

$$W = P_{ext}\Delta V \tag{8.10}$$

where ΔV is the change in gas volume caused by the expansion of the gas.[2]

Note that the work done by the gas in pushing against the piston depends on the pressure exerted by the piston on the gas. *If the gas were to expand against a piston that exerted no external pressure on the gas, no work would be done.*

a. Build and verify Model 8.2. Model 8.2 is really two models. Columns A–C calculate the work done by lifting a box a distance x against the constant force of gravity (mass times the acceleration of gravity). Columns E–G calculate the work done by an expanding gas as it moves a piston that resists the gas with a constant external pressure. Explore this model and verify that you understand how it works.

b. If the unit of force is in Newtons (N) and the unit of distance is meters, the unit of work must be Newton-meters (or Joule). What are the units of pressure and volume? Verify that the unit of work calculated as PV is also the Joule.

c. How much work must be performed on a 10-kg box to lift it from 1 m to 3 m? What about a 1-kg box? A massless box?

d. How much work must be performed by a gas to expand from 1 m³ to 3 m³ against a piston that resists the expansion of the gas with a pressure of 1.2 N m⁻²? What if the external pressure is reduced to 0.6 N m⁻²? What if the piston offers no resistance at all?

2. Build and verify Model 8.3 (or modify Model 8.2 with extensive editing). Model 8.3 performs the same calculations as Model 8.2 but can also calculate the internal pressure of the gas (column E) by using the ideal gas law discussed in Chapter 3. Notice that only a portion of column B is filled in (cells B6–B11) — the model is initialized to calculate the work performed as the gas expands from 1 L to 2 L. (You'll find a use for the blank cells in this column in a moment.)

Explore your model until its structure is clear to you.

a. If 1 mol of gas expands from a volume of 1 L to 2 L at a temperature of 300 K, how much work can the gas perform on a piston that exerts an external force of 2 atm? 4 atm? 8 atm?

12.31 atm? (Notice that all values in cells B6–B11 must be changed.)

Compare the pressure exerted by the moving piston on the gas (column B) with the pressure of the gas on the other stationary walls of the cylinder (column E) and verify that some of the gas's ability to do work is wasted.

b. From column E notice that, in a volume of 2 L, the gas exerts a pressure of approximately 12.31 atm on the walls of its container.

Suppose that the piston initially exerts an external pressure of 24.62 atm on the gas and that this pressure is suddenly reduced to 15.39 atm. Do you see that the gas will expand only until the external and internal pressures are equal? What is the new gas volume?

In Lesson 2a, you discovered that when a gas expands against a constant pressure some of its potential ability to do work is wasted. This is equally true in Lesson 2b. Let's see if we can find the maximum amount of work that the gas can perform.

c. Imagine that your piston exerts just enough pressure on 1 L of gas to keep the gas from either expanding or compressing (i.e., the gas is in equilibrium at a pressure of 24.62 atm).

Now instantaneously reduce the external pressure to 12.31 atm and allow the gas to expand to its new equilibrium volume of 2 L. (You would do this by placing the value 12.31 into cells B7–B11. You can also place 24.62 atm in cell B6 if you wish, but it will not affect your calculations.)

How much work is performed by the gas on the piston during its expansion?

d. Now imagine that the piston (initially exerting a pressure of 24.62 atm) experiences first a pressure reduction to 17.58 atm and then, after the gas achieves equilibrium under these conditions, a further pressure reduction to 12.31 atm. How much work is performed by the gas on the piston during its expansion?

(17.58 atm was chosen deliberately, of course, so that the first equilibrium volume is 1.4 L. The exercise is accomplished therefore by entering 17.58 into cells B7 and B8 and 12.31 into cells B9–B11. Pause and be sure you understand what we're doing here!)

e. If the piston were to undergo five stepwise external pressure reductions such that the gas achieved equilibrium at 1.2, 1.4, 1.6, 1.8, and 2.0 L, how much work would the gas perform on the piston?

(One way to achieve this goal would be to place the formula =E6 in cell B6 and copy it down the column.)

f. How much work does the gas perform if the piston undergoes ten stepwise external pressure reductions (such that volumes of 1.1, 1.2, ..., 2.0 L) are achieved? (You'll need to change ΔV in cell B2 to 0.1 and enter the formula =E6 into cells B7–B16.)

g. How much work does the gas perform if the piston undergoes twenty stepwise external pressure reductions (such that volumes of 1.05, 1.10, ..., 2.00 L) are achieved? (You'll need to change ΔV in cell B2 to 0.05 and enter the formula =E6 into cells B7–B26.)

h. Summarize your data in a table and confirm the fact that as the number of steps increases the amount of work performed by the gas also increases, but that the value of this work appears to approach a limit.

If the steps were to become infinitely small and if the gas were in equilibrium at every step during its expansion, the gas would perform the maximum possible amount of work on its confining piston. (Of course, it would take an infinite amount of time to perform this work!)

Processes that are at equilibrium throughout their duration are called reversible processes. In the case just explored, an infinitely small decrease in the external pressure of the piston causes an infinitely small expansion of the volume. However, an infinitely small increase in the external pressure would have the opposite effect. This reversal would not occur if the system were not in equilibrium.

If you have studied calculus, you will realize that you have been numerically approximating the work performed by a variable force (Equation 8.11):

Figure 8.5

A cylinder of gas closed at one end by a movable piston can be used to study the properties of work, heat, and energy. If a gas expands and thereby performs work against the piston, the work must be "paid for" with energy.

(a) If the gas cylinder is insulated from its surrounding environment, only the thermal energy of the gas is available to work against the piston. In this case, the work of expansion causes the temperature and internal energy of the gas to drop. This is called *adiabatic expansion*.

(b) If the gas cylinder is in thermal contact with its environment, the work of expansion is "paid back" by the transfer of thermal energy from the environment to the gas. The temperature of the gas remains constant. This is called *isothermal expansion*.

(a)   (b)

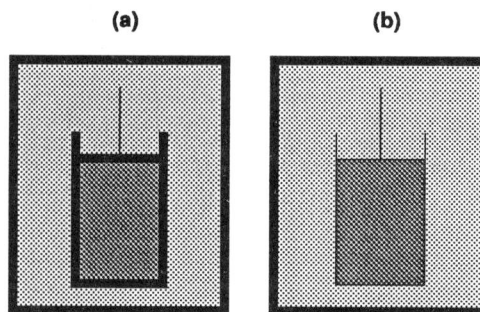

$$W = \int_{V_1}^{V_2} P(V)dV \;=\; \int_{V_1}^{V_2} nRT \, \frac{dV}{V} \;=\; nRT \int_{V_1}^{V_2} \frac{dV}{V} \qquad (8.11)$$

where $P(V) = nRT/V$ is pressure as a function of volume as calculated by the ideal gas law.

3. You now have explored the concepts of work and of a reversible process. Both are central to the framework of ideas that make up thermodynamics. Heat and energy constitute the next set of concepts, after which you'll be ready to build and explore simulations of real thermodynamic processes.

    Energy is the capacity to do work. When a gas expands against a resisting piston and performs work, the capacity to do this work comes from the thermal energy of the gas. As the work is performed, a portion of the thermal (internal) energy of the gas is depleted. Ultimately, either energy must enter the system from an outside source or the gas must cool. The accounting is performed by Equation 8.12):

$$\Delta E = q + w \qquad (8.12)$$

    where ΔE is the change in internal energy, q is the heat *absorbed by* the system, and w is the work *done on* the system. [There are several possible sign conventions for Equation 8.12 and your primary textbook may or may not use ours. Notice that our convention declares that the work done on the gas during an expansion is negative (because the gas does work on the piston).] There are two classic thought experiments in which Equation 8.12 plays a crucial role — adiabatic expansion and isothermal expansion.

- *Adiabatic expansion* occurs when the container of gas is thermally isolated from its environment (Figure 8.5a). The work of expansion depletes reserves of internal energy and the gas cools (Equation 8.13):

$$\Delta E = q + w = 0 + w = w \qquad (8.13)$$

Using our sign convention, both ΔE and w are negative numbers when a gas expands (and positive when a gas is compressed via the agency of external work), of course.

- *Isothermal expansion* occurs when the container of gas is in thermal contact with an environment maintained at constant temperature (Figure 8.5b). The work of expansion starts to deplete thermal energy, but heat is transferred from the constant temperature environment to the gas container to maintain the temperature. The energy expended in expanding the gas against its confining piston is replaced by the heat transfer from the environment, and no change in the internal energy of the gas occurs (Equation 8.14):

$$\Delta E = q + w = 0 \qquad (8.14)$$

a. Build and verify Model 8.4. Model 8.4 performs the same calculations as Models 8.2 and 8.3 but also includes calculations for ΔE and q for adiabatic (Equation 8.13) and isothermal (Equation 8.14) processes. Cells A6–E17 constitute the region of the model that simulates the isothermal expansion of a gas against a piston exerting a constant pressure (columns D and E make use of Equation 8.14). Cells G6–K17 constitute the region of the model that simulates the adiabatic expansion of a gas against a piston exerting a constant pressure (columns J and K use Equation 8.13).

Explore this model until you understand its structure. Make sure you understand the rationale supporting columns D, E, J, and K.

Also build on-screen graphics (Figures 8.6 and 8.7) to support your study of this model.

b. Use your model and its supporting graphics to monitor the work performed by a gas expanding isothermally against a pressure exerting a constant pressure of 0.1 atm.[3]

Figure 8.6

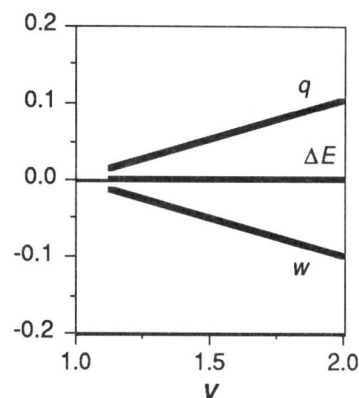

XY-PLOT

Data
Horizontal (*x*) axis: cells A9–A17 (volume)
Vertical (*y*) axis: cells C9–C17, D9–D17, and E9–E17 (*w*, *q*, ΔE)

Figure 8.7

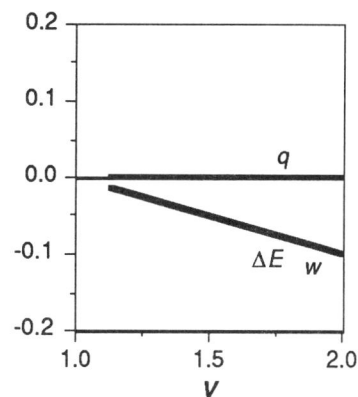

XY-PLOT

Data
Horizontal (*x*) axis: cells G9–G17 (volume)
Vertical (*y*) axis: cells I9–I17, J9–J17, and K9–K17 (*w*, *q*, ΔE)

As the gas expands, what happens to the temperature and internal energy of the gas? What is the source of the energy that supports the work performed on the moving piston?

c. How much heat enters the gas from the surrounding environment as the gas expands from 1 L to 2 L against an external pressure of 0.1 atm? 0.05 atm? 0 atm?

Write a brief paragraph explaining why heat is not transferred into the gas when the gas expands against an unresisting piston.

d. Use your model and its supporting graphics to monitor the work performed by a gas expanding adiabatically against a pressure exerting a constant pressure of 0.1 atm.

As the gas expands, what happens to the temperature and internal energy of the gas? What is the source of the energy that supports the work performed on the moving piston?

e. How much heat enters the gas from the surrounding environment as the gas expands from 1 L to 2 L against any external pressure? Why?

What is the relationship between the external pressure and the heat that enters the gas? Explain.

f. Explore on your own the behavior of an ideal gas subject to isothermal and adiabatic compression.

4. You now have explored the concepts of work and reversible processes (Lesson 2) and isothermal and adiabatic processes (Lesson 3). Now you are ready to combine these explorations and enter the heart of thermodynamics — isothermal and adiabatic *reversible* processes.

a. Build and verify Model 8.5. Model 8.5 draws on the conclusions of Lessons 2 and 3 to simulate the isothermal (columns A–F) and adiabatic (columns H–L) expansion of an ideal gas. Column I, however, introduces new mathematics. The adiabatic expansion of a gas consumes a portion of the thermal energy of the gas, and so, for a given volume increase, the pressure and temperature of the gas is lower than it would be if the gas expanded isothermally. We state without proof that for an ideal monatomic gas

$$P_2 = P_1 \left(\frac{V_1}{V_2}\right)^{5/3} \qquad\qquad (8.15)$$

where $V_1, V_2, P_1,$ and P_2 are the volumes and pressures of the gas before and after the adiabatic expansion, respectively.

Key elements of the model include the ideal gas law (column B), the definition of the work of a constant force (Equation 8.10, columns C and J), and the first law of thermodynamics as applied to adiabatic and isothermal processes (Equations 8.13 and 8.14, columns D, E, K, and L). Equation 8.15 is used in column I. Notice that the temperature of the gas (columns F and L) is calculated using the ideal gas law.

Explore Model 8.5 until you understand its structure.

b. Consider a uninsulated container designed with a movable piston at one end. Because the container is uninsulated, any gas in the container can freely exchange heat with a surrounding environment that is maintained at a constant temperature of 300 K. Imagine that the container contains 1 mol of gas in a volume of 10 L. What is the external pressure that the piston exerts on the gas?

Now imagine that the external pressure of the piston is reduced just enough to permit the gas to expand to 10.2 L. How much work does the gas perform on the piston? (Watch your signs! Remember our convention that work performed *on* the gas is positive.)

Reduce the external pressure in a stepwise fashion so that, at each step, the volume of the gas increases exactly 0.2 L. How much total work does the gas perform on the piston in expanding from 10 L to 12 L?

How much total work does the gas perform on the piston if it expands from 10 L to 12 L in increments of 0.1 L?

Notice that although your model does not simulate a true reversible process, it approximates the behavior of such a process as the ΔV of each step decreases.

c. Repeat the exercises described in Lesson 4b under adiabatic conditions. To one decimal place, how much work does the gas perform on the piston in expanding reversibly from 10 L to 12

Figure 8.8

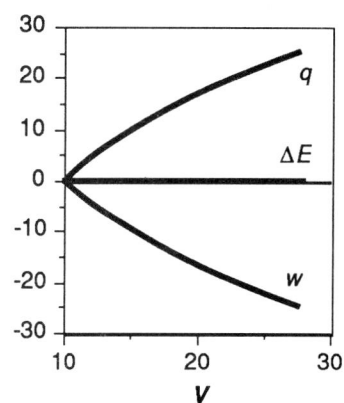

XY-PLOT

Data
Horizontal (x) axis: cells A11–A100 (volume)
Vertical (y) axis: cells C11–C100 (w), D11–D100 (q), and E11–E100 (ΔE)

Figure 8.9

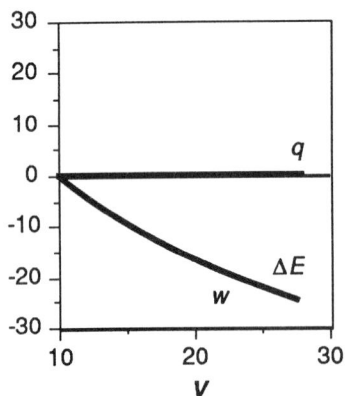

XY-PLOT

Data
Horizontal (x) axis: cells A11–A100
(volume)
Vertical (y) axis: cells J11–J100 (w),
K11–K100 (q), and L11–L100 (∆E)

Figure 8.10

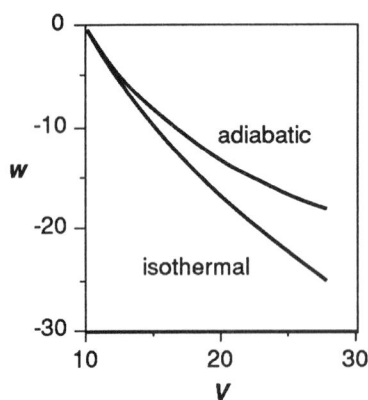

XY-PLOT

Data
Horizontal (x) axis: cells A11–A100
(volume)
Vertical (y) axis: cells C11–C100
(isothermal work) and J11–J100
(adiabatic work)

L? Is this more work, less work, or the same amount of work performed by the gas when it expands isothermally?

Explain the difference, if any, in the work performed isothermally and adiabatically by the gas.

d. Assume that your model satisfactory approximates a reversible process when the ΔV of each step equals 0.2. Use your model (and an on-screen graphic designed around Figure 8.8) to monitor the behavior of a gas expanding reversibly and isothermally against a movable piston. As the gas expands, what happens to the temperature and internal energy of the gas? What is the source of the energy that supports the work performed on the moving piston?

e. Use your model (and an on-screen graphic designed around Figure 8.9) to monitor the behavior of a gas expanding reversibly and adiabatically against a movable piston. As the gas expands, what happens to the temperature and internal energy of the gas? What is the source of the energy that supports the work performed on the moving piston?

f. Use your model (and an on-screen graphic designed around Figure 8.10) to compare the work performed by a gas on its environment when it expands isothermally and adiabatically.

Notice that, in both cases, the work performed by the gas is not a linear function of gas volume. Explain.

Why does an adiabatically expanding gas perform less work than an isothermally expanding gas?

g. Use your model (and an on-screen graphic designed around Figure 8.11) to compare the temperature of a gas as it expands isothermally and adiabatically. Explain this behavior.

h. Use your model (and an on-screen graphic designed around Figure 8.12) to compare the plot of P vs V of an isothermally expanding gas with a similar plot for an adiabatically expanding gas. Explain this difference.

i. Suppose that work is performed on the gas by the external piston and that the gas is compressed either isothermally or adiabatically. Write a brief essay comparing and contrasting the behavior of the gas in these two experiments. Discuss internal energy, temperature, heat transfer, pressure, and volume.

CONCLUSION

The internal energy of substances is a matter of great interest to chemists. Chemical reactions, changes of state, and other phenomena are usually accompanied by changes in the internal energy of the substances involved, and it is important to know what those changes are. Furthermore, we are slowly moving toward an understanding, in thermodynamic terms, of the direction of a spontaneous chemical reaction. Recall from the introduction to this chapter that the direction of spontaneous change can be determined from the change in the Gibbs free energy associated with that change. We are, in part, interested in the internal energy of a compound because it is an important component of Gibbs free energy.

It would be nice if changes in internal energy were a function solely of the absorption or release of heat, but a moment's reflection will remind you that this is not the case. Consider an uninsulated container of gas that is moved from a 300 K environment to a 350 K environment. If the container has a fixed volume, the pressure of the gas increases as the temperature rises, and the heat increase q equals the change in internal energy ΔE. This is because the volume does not change and no work is performed (Equation 8.16):

$$q = \Delta E - w = \Delta E - 0 = \Delta E \qquad (8.16)$$

If the increase in gas temperature is accompanied by a change in gas volume, however, some of the thermal energy that enters the gas is used in the work of expansion against an external pressure and does not appear as heat (Equation 8.17):

$$q = \Delta E - w = \Delta E + P\Delta V \qquad (8.17)$$

If chemists could design experiments in which the volumes of reactants and products were kept unchanged during a reaction, changes in internal energy could be measured with a thermometer, balance, and pipette. But chemists usually work under conditions of constant pressure (i.e., 1 atm) in which the volume of substances may or may not change significantly.

As a result, chemists often refer to changes in the enthalpy ΔH of a reaction rather than changes in internal energy ΔE. ΔH is simply the amount of heat absorbed or released during a chemical change under

Figure 8.11

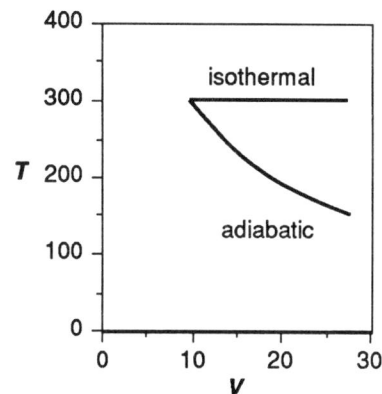

XY-PLOT

Data
Horizontal (x) axis: cells A11–A100 (volume)
Vertical (y) axis: cells F11–F100 (isothermal gas temperature) and M11–M100 (adiabatic gas temperature)

Figure 8.12

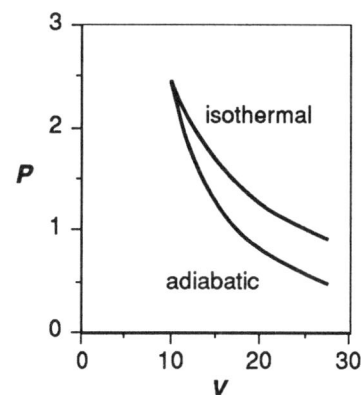

XY-PLOT

Data
Horizontal (x) axis: cells A11–A100 (volume)
Vertical (y) axis: cells B11–B100 (isothermal gas pressure), and B11–B100 (adiabatic gas pressure)

conditions of constant pressure when no work except PV work is performed. In these circumstances, Equation 8.17 becomes

$$q = \Delta E + P\Delta V = \Delta H \qquad (8.18)$$

Consider the burning of hydrogen gas to form water vapor (Equation (8.19):

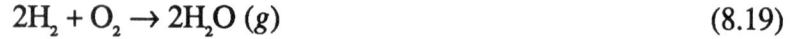

$$2H_2 + O_2 \rightarrow 2H_2O \ (g) \qquad (8.19)$$

This reaction is accompanied by the liberation of a great deal of heat and a considerable reduction in the internal energy of the system. Rather than refer to the ΔE of the reaction, however, chemical handbooks identify ΔH (–115,560 cal mol^{-1}).

It is ΔH rather than ΔE that contributes to the definition of ΔG (Gibbs free energy change), as you will see in the next exercise.

MODELS

Model 8.2

BUILD and verify Model 8.2.

ENTER data using cells B1–B3 (box mass, acceleration of gravity), F1 (external pressure), B5 and B6 (starting and incremental position), and E5 and E6 (starting and incremental volume).

READ output from the model from columns A and C (box position and work performed on the box) and E and G (gas volume, and work performed by the gas on the piston).

NOTICE that the dimensions of work are the same for $w = Fx$ as for $w = PV$.

FORMULAS

	A	B	C	D	E	F	G
1	m =	10	kg		P =	1.2	N/m^2
2	a =	9.8	m/sec^2				
3	F =	=B1*B2	kg m/sec^2 = Newtons				
4							
5	x0 =	1	m		V0 =	1	m^3
6	Δx =	0.1	m		ΔV =	0.1	m^3
7							
8	Lifting a box				Pushing a piston		
9	x	F	w		V = xA	P = F/A	w
10	(m)	(N)	(N m = J)		(m^3)	(N/m^2)	(N m = J)
11	=B5				=E5		
12							
13							
14	Formula Set I	Formula Set II	Formula Set III		Formula Set IV	Formula Set V	Formula Set VI
29				D			
30							
31							

VALUES

	A	B	C	D	E	F	G
1	m =	10	kg		P =	1.2	N/m^2
2	a =	9.8	m/sec^2				
3	F =	98	kg m/sec^2 = Newtons				
4							
5	x0 =	1	m		V0 =	1	m^3
6	Δx =	0.1	m		ΔV =	0.1	m^3
7							
8	Lifting a box				Pushing a piston		
9	x	F	w		V = xA	P = F/A	w
10	(m)	(N)	(N m = J)		(m^3)	(N/m^2)	(N m = J)
11	1	98			1	1.2	
12	1.1	98	9.8		1.1	1.2	0.12
13	1.2	98	19.6		1.2	1.2	0.24
14	1.3	98	29.4		1.3	1.2	0.36

	A	B	C	D	E	F	G
29	2.8	98	176.4		2.8	1.2	2.16
30	2.9	98	186.2		2.9	1.2	2.28
31	3	98	196		3	1.2	2.4

Formula Set I Prototype Cell is A12	Formula Set III Prototype Cell is C12	Formula Set V Prototype Cell is F11
=A11+**B6**	=C11+B12*(A12-A11)	=F1

Formula Set II Prototype Cell is B11	Formula Set IV Prototype Cell is E12	Formula Set VI Prototype Cell is G12
=B3	=E11+**F6**	=G11+F12*(E12-E11)

FORMULAS

	A	B	C	D	E	F
1	V0 =	1		n =	1	mo
2	ΔV =	0.2		R =	0.0821	l atm/mol K
3				T =	300	K
4						
5	V	P(ext)	w		P(gas)	
6	=B1	2	0			
7		2				
8		2				
9		2				
10		2				
11	Formula	2	Formula		Formula	
12	Set		Set		Set	
13	I		II		III	
14						
15						
16						
17						
18						
19						
20						
21						
22						
23						
24						
25						
26						

Formula Set I
Prototype Cell is A7

=A6+ **B2**

Formula Set II
Prototype Cell is C7

=C6+B7*(A7-A6)

Formula Set III
Prototype Cell is E6

=E1 *E2 *E3 /A6

Model 8.3

BUILD and verify Model 8.3.

ENTER data using cells B1 and B2 (starting and incremental volume) and all or part of the cell range B6–B26 (external pressure).

READ output from the model from columns A, C, and E (gas volume, work performed by the gas, and internal pressure of the gas).

VALUES

	A	B	C	D	E	F
1	V0 =	1		n =	1	mo
2	ΔV =	0.2		R =	0.0821	l atm/mol K
3				T =	300	K
4						
5	V	P(ext)	w		P(gas)	
6	1	2	0		24.617	
7	1.2	2	0.4		20.514	
8	1.4	2	0.8		17.584	
9	1.6	2	1.2		15.386	
10	1.8	2	1.6		13.676	
11	2	2	2		12.309	
12	2.2		2		11.19	
13	2.4		2		10.257	
14	2.6		2		9.4682	
15	2.8		2		8.7919	
16	3		2		8.2058	
17	3.2		2		7.6929	
18	3.4		2		7.2404	
19	3.6		2		6.8381	
20	3.8		2		6.4782	
21	4		2		6.1543	
22	4.2		2		5.8613	
23	4.4		2		5.5948	
24	4.6		2		5.3516	
25	4.8		2		5.1286	
26	5		2		4.9234	

Model 8.4

BUILD and verify Model 8.4.

ENTER data using cells B1 and B2
(starting and incremental volume) and
B4 (external pressure).

READ output from the model from
columns A–E (isothermal volume,
pressure, work, heat transfer, and
internal energy change), G–K
(adiabatic volume, pressure, work,
heat transfer, and internal energy
change).

FORMULAS

	A	B	C	D	E	F	G	H	I	J	K
1	V0 =	1	l								
2	ΔV =	0.125	l								
3											
4	P =	0.1	atm								
5											
6	Constant Pext (Isothermal, ΔE = 0)						Constant Pext (Adiabatic, ΔE = w)				
7	V	P	w	q	ΔE		V	P	w	q	ΔE
8	(l)	(atm)	(l atm)	(l atm)	(l atm)		(l)	(atm)	(l atm)	(l atm)	(l atm)
9	=B1						=B1				
10										0	
11										0	
12	Formula	Formula	Formula	Formula	Formula		Formula	Formula	Formula	0	Formula
13	Set	Set	Set	Set	Set		Set	Set	Set	0	Set
14	I	II	III	IV	V		VI	VII	VIII	0	IX
15										0	
16										0	
17										0	

VALUES

	A	B	C	D	E	F	G	H	I	J	K
1	V0 =	1	l								
2	ΔV =	0.125	l								
3											
4	P =	0.1	atm								
5											
6	Constant Pext (Isothermal, ΔE = 0)						Constant Pext (Adiabatic, ΔE = w)				
7	V	P	w	q	ΔE		V	P	w	q	ΔE
8	(l)	(atm)	(l atm)	(l atm)	(l atm)		(l)	(atm)	(l atm)	(l atm)	(l atm)
9	1	0.1					1	0.1			
10	1.125	0.1	-0.013	0.0125	0		1.125	0.1	-0.013	0	-0.013
11	1.25	0.1	-0.025	0.025	0		1.25	0.1	-0.025	0	-0.025
12	1.375	0.1	-0.038	0.0375	0		1.375	0.1	-0.038	0	-0.038
13	1.5	0.1	-0.05	0.05	0		1.5	0.1	-0.05	0	-0.05
14	1.625	0.1	-0.063	0.0625	0		1.625	0.1	-0.063	0	-0.063
15	1.75	0.1	-0.075	0.075	0		1.75	0.1	-0.075	0	-0.075
16	1.875	0.1	-0.088	0.0875	0		1.875	0.1	-0.088	0	-0.088
17	2	0.1	-0.1	0.1	0		2	0.1	-0.1	0	-0.1

Formula Set I Prototype Cell is A10	Formula Set IV Prototype Cell is D10	Formula Set VII Prototype Cell is H9
=A9+**B2**	=-C10	=**B4**

Formula Set II Prototype Cell is B9	Formula Set V Prototype Cell is E10	Formula Set VIII Prototype Cell is I10
=**B4**	=C10+D10	=I9-H10*(G10-G9)

Formula Set III Prototype Cell is C10	Formula Set VI Prototype Cell is G10	Formula Set IX Prototype Cell is K10
=C9-B10*(A10-A9)	=G9+**B2**	=I10+J10

FORMULAS

	A	B	C	D	E	F	G	H	I	J	K	L	M
1	n =	1	mo										
2	R =	0.0821	l atm/mol K	(Value in B2 is 0.0820575)									
3	T =	300	K										
4	gam =	1.6667											
5													
6	V0 =	10											
7	ΔV =	0.2											
8													
9	ISOTHERMAL (ΔE = 0)							ADIABATIC (ΔE = w)					
10	V	P	w	q	ΔE	T		V	P	w	q	ΔE	T
11	=B6							=B6					
12	Form Set I	Form Set II	Form Set III	Form Set IV	Form SetV	Form Set VI		Form Set VII	Form Set VIII	Form Set IX	0	Form Set X	Form Set XI
13											0		
14											0		

=B1*B2*B3/H11

VALUES

	A	B	C	D	E	F	G	H	I	J	K	L	M
1	n =	1	mo										
2	R =	0.0821	l atm/mol K	(Value in B2 is 0.0820575)									
3	T =	300	K										
4	gam =	1.6667											
5													
6	V0 =	10											
7	ΔV =	0.2											
8													
9	ISOTHERMAL (ΔE = 0)							ADIABATIC (ΔE = w)					
10	V	P	w	q	ΔE	T		V	P	w	q	ΔE	T
11	10	2.4617				300		10	2.4617				300
12	10.2	2.4135	-0.483	0.4827	0	300		10.2	2.3818	-0.476	0	-0.476	296.07
13	10.4	2.367	-0.956	0.9561	0	300		10.4	2.306	-0.938	0	-0.938	292.26
14	10.6	2.3224	-1.421	1.4206	0	300		10.6	2.2339	-1.384	0	-1.384	288.57

Formula Set I Prototype Cell is A12	Formula Set IV Prototype Cell is D12	Formula Set VII Prototype Cell is H12
=A11+**B7**	=C12	=H11+**B7**

Formula Set II Prototype Cell is B11	Formula Set V Prototype Cell is E12	Formula Set VIII Prototype Cell is I12
=**B1*B2*B3**/A11	=C12+D12	=I11*(H11/H12)^**B4**

Formula Set III Prototype Cell is C12	Formula Set VI Prototype Cell is F11	Formula Set IX Prototype Cell is J12
=C11-B12*(A12-A11)	=(A11*B11)/(**B1*B2**)	=J11-I12*(H12-H11)

Formula Set X Prototype Cell is L12	Formula Set XI Prototype Cell is M11
=J12	=(H11*I11)/(**B1*B2**)

Model 8.5

BUILD and verify Model 8.5.

ENTER data using cells B6 and B7 (starting and incremental volume).

READ output from the model from columns A–F (reversible, isothermal volume, pressure, work, heat transfer, internal energy change, and temperature) and H–M (reversible, adiabatic volume, pressure, work, heat transfer, internal energy change, and temperature).

THE SECOND LAW: DIRECTION

The first law of thermodynamics describes observed constraints on chemical reactions and other physical processes. It states that the total energy of the system plus its surroundings cannot change. The first law, however, does not place a constraint on the *direction* of a reaction. Nevertheless, we know from experience that there is a natural direction to many physical processes:

- A gas always expands to fill its container and never confines itself to only a portion of the available space.

- At room temperature ice melts, never freezes.

There is nothing in the first law of thermodynamics that prevents either of these processes. Energy, for example, would still be conserved if room-temperature air extracted thermal energy from water and caused it to freeze.

In developing the first law, however, we sidestepped an important observation. Although we noted that it is possible for mechanical energy to be converted to thermal energy and vice versa, we avoided any statement about the extent to which these two processes can occur. *While it is possible to totally convert mechanical energy to thermal energy, there is a limit to the degree to which thermal energy can be converted to mechanical energy. Furthermore, to fully convert the theoretical maximum amount of thermal energy to mechanical energy requires a reversible process (and an infinite amount of time). In the real world, the efficiency of this conversion is always less than 100%.*

Total conversion of mechanical energy to thermal energy occurs every time you stop your automobile by increasing the temperature of your break linings. Partial conversion of thermal energy to mechanical energy occurs every time you ignite a mixture of gasoline and air in an internal combustion engine. Perhaps you know, however, that automotive engineers are always trying to design engines that can run hotter. This is because there is a theoretical limit to the efficiency of your automobile's engine; the hotter the engine runs, the more heat can be converted to mechanical energy to move your car.

In the lessons that follow, you explore two issues. First, you verify that, even in a reversible process, not all thermal energy of a substance can be converted to mechanical energy. Second, you explore the nature of irreversibility and the properties of a state of matter called entropy.

Most textbooks, including ours, illustrate these points with an idealized engine first described in 1824 by the French engineer Sadi Carnot. This engine moves through four cycles during which it converts heat to mechanical energy. The cycle of this engine is called the Carnot cycle, and it illustrates the limits placed by nature on the conversion of thermal energy to mechanical energy. In summary, the Carnot cycle formalizes a process in which the flow of heat between two reservoirs of thermal energy (a *hot* reservoir and a *cold* reservoir) is used to produce mechanical energy.

Figures 8.13 and 8.14 illustrate the four steps of the Carnot cycle. Each individual step is already familiar from the lessons in Exercise 8.1.

- *Step 1: Isothermal Expansion.* The cycle begins with the isothermal expansion of gas in thermal contact with a hot reservoir. As the gas expands, it performs work w_1 on a movable piston and receives thermal energy q_2 from the hot reservoir. From the standpoint of the gas, w_1 is negative (the gas performs work on the piston), and q_2 is positive (the gas receives heat from the hot reservoir). Because $q_2 = -w_1$, the gas itself displays no change in its internal energy ($\Delta E = 0$).

- *Step 2: Adiabatic Expansion.* The gas expands adiabatically in a container insulated from its surroundings and performs work w_2 via the expenditure of some of its own internal energy ($\Delta E =$

Figure 8.13

The Carnot cycle is a mathematical description of an idealized heat engine consisting of four steps. Thermal energy is converted to mechanical energy by moving a piston that seals one end of a container of ideal gas.

- In Step 1, the gas expands isothermally and pushes the piston outward. The work of the expansion is "paid for" by thermal energy from a hot reservoir maintained at a constant temperature.

- In Step 2, the gas continues to expand, but now the expansion is adiabatic. The work of the expansion comes from the thermal energy of the gas, and the gas cools.

- In Step 3, the gas is placed in thermal contact with a reservoir cooled to the same, now cooler, temperature of the gas. Some of the work used to push the piston outward is used to compress the gas isothermally, and the work of the piston is transferred to the reservoir as thermal energy.

- In Step 4, the gas is returned to its original state. An additional increment of the work originally used to push the piston outward is used to compress the gas adiabatically. The work of the piston raises the temperature of the gas back to that of the hot reservoir.

At the completion of the cycle, the heat originally extracted from the hot reservoir is partitioned between two recipients. Some heat is converted to mechanical energy via the moving pistion. The remaining heat is transferred to the cool reservoir.

It is impossible to perform useful work with thermal energy unless a temperature gradient exists. Otherwise you could power ships with the thermal energy of the ocean.

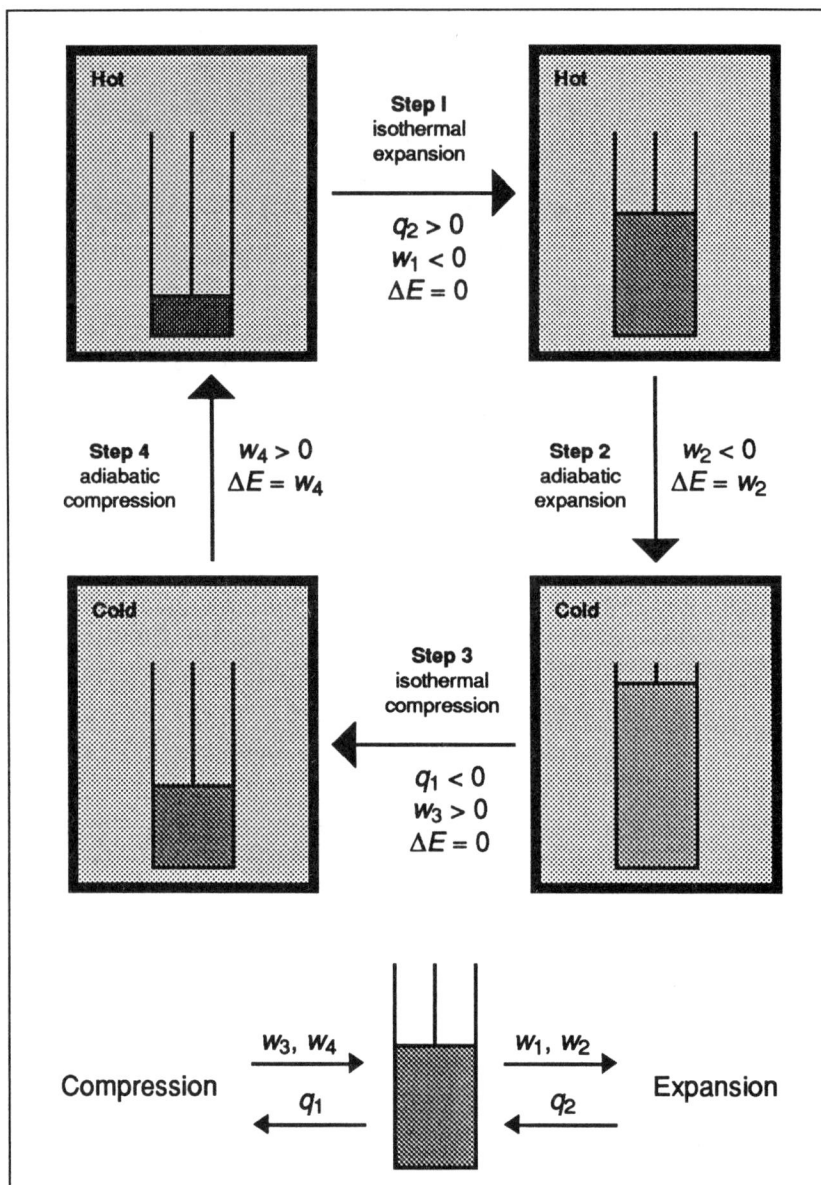

w_2). No heat exchange occurs between the gas and its environment during this stage. The expansion is stopped when the gas has come to the temperature of the cold reservoir.

- *Step 3: Isothermal Compression.* Some (but not all) thermal energy used to move the piston during Step 1 now departs the gas and enters the cold reservoir. The piston performs work w_3 on the gas as the gas is compressed, and the gas transmits thermal energy q_1 to the cold reservoir. From the standpoint of the gas, w_3 is positive (the gas is worked on by the piston), and q_1 is negative (the gas transmits heat to the cold reservoir).

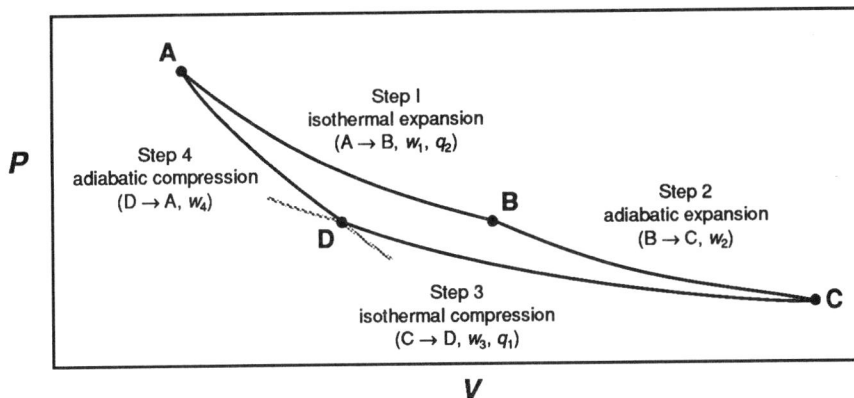

Figure 8.14

The Carnot cycle (Figure 8.13) takes an ideal gas through four separate changes of pressure and volume.:

- Step 1 increases the volume and reduces the pressure via isothermal expansion. The work of expansion against an external pressure ultimately comes from the thermal energy of "hot" environment.

- Step 2 continues along the same path as Step 1, but now the work of expansion is "paid for" out of the thermal energy of the gas itself. As the gas cools, the temperature drops, and so the P vs V curve is somewhat steeper.

- Step 3 decreases the volume and increases the pressure via isothermal compression. The work of compression comes from the piston, and the energy entering the gas is transferred to a "cold" reservoir.

- Step 4 returns the gas to its original state via adiabatic compression. The work of compression again comes from the piston, but the energy entering the gas remains in the gas and its temperature rises. The P vs V curve is therefore steeper than in Step 3.

If the four states of the gas are identified as A, B, C, and D, it is easy to overlook the fact that each of these points is *not* independently assignable. When three points are chosen, the fourth point is determined by the system. Consider the line segment AD. In this fully reversible process, this path can be used to describe adiabatic compression from line D and adiabatic expansion from line A. *If the gas is to return to its original state, its path along Step 3 must terminate at D so that Step 4 can begin.*

Again, as in all isothermal processes, the gas itself displays no change in its internal energy ($\Delta E = 0$).

- *Step 4: Adiabatic Compression.* Finally, the piston compresses the gas adiabatically and performs work w_4, using some of the mechanical energy transmitted to it during Steps 1 and 2. (You can imagine, for example, that the piston is attached to a shaft and flywheel that can store energy as kinetic energy.) The work w_4 is used to heat the gas back to the temperature of the hot reservoir so that the cycle can begin again.

As you will shortly see, the net transfers of heat and work between the elements of the Carnot cycle yield a net conversion of thermal energy to mechanical energy. You will also see that it is impossible in this arrangement to convert all thermal energy of the hot reservoir to mechanical energy (i.e., some energy always returns to the cold reservoir).

In the lessons coming up, you will discover two important properties about the Carnot cycle:

- For a reversible Carnot cycle, the heat transferred during an isothermal expansion or compression (q_2 or q_1) is a linear function of the temperature at which the isothermal expansion or compression (T_2 or T_1) occurs (Equation 8.20):

$$\frac{q_2}{T_2} + \frac{q_1}{T_1} = 0 \qquad (8.20)$$

where q_2 and T_2 are the heat transferred and temperature during isothermal expansion (Step 1) and q_1 and T_1 are the heat transferred and temperature during isothermal compression (Step 3).

- For a Carnot cycle made up of discrete, irreversible steps

$$\frac{q_2}{T_2} + \frac{q_1}{T_1} < 0 \qquad (8.21)$$

The quantity q/T plays an important role in the second law of thermodynamics. For a reversible process, it defines a property called entropy S that depends only on the actual state of matter and not on the path matter takes to reach that state:

$$\Delta S = \frac{q}{T} \qquad (8.22)$$

where ΔS is the change in entropy that occurs when heat q is transferred at absolute temperature T.

Note carefully that we said that S depends on the *state* of matter — it has the same qualities as, for example, the internal energy or volume of a gas (i.e., it is independent of the path by which the matter reached this state). q, on the other hand, is not path-independent and assumes lower values for irreversible processes. Therefore, for irreversible processes,

$$\Delta S > \frac{q}{T} \qquad (8.23)$$

You will verify this fact in the lessons that follow.

There is nothing magical about entropy, but interesting discussions about the nature of order and disorder arise when it is mentioned. We will postpone these discussions until you've had some hands-on experience with the behavior of q/T.

One last point. While working with the Carnot cycle, you will study the *efficiency* of an engine. Efficiency is simply the work produced divided by the energy input. The Carnot cycle takes a certain amount of thermal energy q_2 from a hot reservoir and converts a portion of it into mechanical work. The total work is $q_1 + q_2$ (remember that q_1 and q_2 have opposite signs). The efficiency e of the cycle therefore is

$$e = \frac{q_2 + q_1}{q_2} \qquad (8.24)$$

For a completely reversible process, Equation 8.20 tells you that Equation 8.24 could also be written

$$e = \frac{T_2 - T_1}{T_2} \qquad (8.25)$$

Equation 8.21 tells you, however, that if irreversible processes are included the equation should read

$$e = \frac{q_2 + q_1}{q_2} \leq \frac{T_2 - T_1}{T_2} \tag{8.26}$$

The efficiency calculated from temperature differences is the maximum efficiency of the cycle. The actual efficiency must be calculated from the heat transferred.

LESSONS

1. Build and verify Model 8.6. Model 8.6 uses the procedures introduced in Model 8.5 to represent all four steps of a Carnot cycle. Examine the structure of this model by the following guidance:

    a. Model 8.6 is driven by values entered into cells B1–B3 (moles of gas, gas constant, and γ). γ always has the value of 5/3 for an ideal monatomic gas.

    Cells B5 and B6 define the temperature of the hot reservoir and the volume of the gas at the beginning of Step 1 of the Carnot cycle. The temperature of the cold reservoir is not specified explicitly in the model because it is determined by the degree to which adiabatic expansion of the gas is allowed to proceed.

    The steps by which the gas volume is incremented during each stage of the cycle are specified in cells B8, I8, P8, and W8.

    b. The Carnot cycle is fully specified by the previous values, but analysis of the work and heat transferred during the cycle requires visual inspection of the model followed by the manual entry of values into two additional cells. (These cells, I4 and I5, will be explained shortly.)

    c. Step 1 (isothermal expansion) of the Carnot cycle is accomplished in columns A–F, beginning at row 10. The calculations are identical to those already demonstrated in Model 8.5. Note that the conditions at the beginning of Step 1 are the conditions defined by the values in cells B1–B3, B5, and B6 of the model.

    Step 2 (adiabatic expansion) is accomplished in columns H–M. The calculations are identical to those already demonstrated in Model 8.5. The initial conditions for Step 2 are identical to the ending conditions of Step 1. Consequently, the gas volume with which Step 2 begins (cell H12) takes its value from the gas volume at the end of Step 1 (cell A102).

Step 3 (isothermal compression) is accomplished in columns O–T. The calculations are identical to those in Step 1 except that ΔV is subtracted from the previous volume instead of added. The initial conditions for Step 3 are identical to the ending conditions of Step 2. Gas volume (cell O12) therefore is again taken from the end of the previous cycle (cell H102). Step 3 is extended beyond row 102 for these conditions for reasons that will soon become apparent.

Step 4 (adiabatic compression) of the Carnot cycle is accomplished in columns V–AA and is calculated in the same fashion as Step 2. A moment of reflection, however, is nevertheless required. Step 4 does not necessarily begin at the end of Step 3. Figure 8.14 displays the P vs V plots of all four steps of the cycle. Step 4 of the model can be represented by either an adiabatic compression that begins at some point D along Step 3, or as an adiabatic expansion that begins at point A of Step 1. Because we know the location of point A and because we don't yet know the location of point D, it is convenient to calculate Step 4 as an adiabatic expansion (from point A) rather than an adiabatic compression (from point D).

d. Every step along the Carnot cycle has therefore been calculated, but visual inspection of the P vs V curve is required to identify the location where Step 3 ends and Step 4 begins. You must explicitly identify this location because it tells you the values of w_3 and w_4. Specifically, by looking at column T of your model, you know the temperature of the cold reservoir; by looking down column AA, you can find the one value of P and V that corresponds to this temperature.

The initial conditions of Model 8.6 indicate that the cold reservoir has a temperature of 231.7 K and that at Step 4 of the Carnot cycle $V = 12.35$ l and $P = 1.732$ atm at this temperature. This is the point at which Step 3 ends and Step 4 begins. Therefore $w_3 = 12.28$, and $w_4 = 8.31$ (cells Q145 and X59, respectively). These values should be entered into I4 (w_3) and I5 (w_4) for your model to correctly summarize the values of work and heat transfer.

Do you see now why Step 3 of this model required extension for this set of conditions?

e. Once the point of intersection of Steps 3 and 4 has been identified (and cells I4 and I5 appropriately updated), cells H1–M6 provide a wealth of information about the outcome of one turn of your Carnot cycle. The work performed by the gas on the piston ($w_1 + w_2 + w_3 + w_4$) and the heat exchanged by the gas and the hot and cold reservoirs ($q_1 + q_2$) are instantly available. In addition, the values of q_2/T_2, q_1/T_1, actual $e(q)$, and maximum $e(T)$ efficiency (Equation 8.26) can be read directly.

2. Build an on-screen graphic in accord with the specifications of Figure 8.15. Examine the figure while recalling your knowledge about isothermal and adiabatic processes.

   a. Match your on-screen graphic with the annotations of Figure 8.14 and the description of the Carnot cycle at the introduction to this exercise. Does the plot of P vs V for Step 1 correspond to your understanding of the nature of an isothermal expansion? Why does pressure drop more rapidly during Step 2 (adiabatic expansion)? Isothermal compression generates the same kind of plot as Step 1 (we are, after all, approximating a reversible process), and adiabatic compression returns us to our starting conditions.

   b. While examining your on-screen graphic, recite to yourself the work done and heat transferred during each step of the cycle.

   c. If you haven't already done so, verify the values of w_3 and w_4 that have been inserted in cells I4 and I5 of the model by using the procedure described in Lesson 1d.

   d. Examine the summary data in cells H1–M6. The origins of this data should be familiar from Exercise 8.1 and the introduction to this exercise.

   Cells I2–I5 calculate the work performed on the gas during each step of the Carnot cycle. Equation 8.10 is used to perform these calculations for each step ΔV of the cycle, and the running totals are placed in cells I2–I5. Notice that w_1 and w_2 are negative numbers because, during gas expansion, work is performed by the gas on the piston. The net work on the gas is summed in cell I6.

   Cells K2 and K4 calculate the heat transfer q_2 and q_1, taking the negative value of the work performed during these steps in the

Figure 8.15

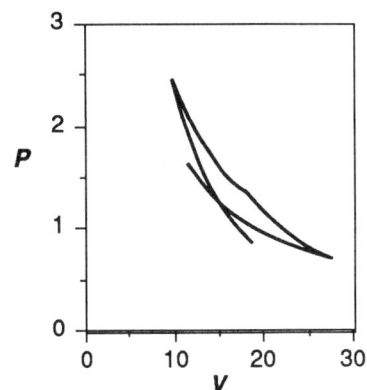

XY-PLOT

Data
Horizontal (x) axis: cells A12–A102 (volume of Step 1)
Vertical (y) axis: cells B12–B102 (pressure of step 1)

Horizontal (x) axis: cells H12–H102 (volume of Step 2)
Vertical (y) axis: cells I12–I102 (pressure of step 2)

Horizontal (x) axis: cells O12–O102 (volume of Step 3)
Vertical (y) axis: cells P12–P102 (pressure of step 3)

Horizontal (x) axis: cells V12–V102 (volume of Step 4)
Vertical (y) axis: cells W12–W102 (pressure of step 4)

NOTE: Some spreadsheets don't support multiple x-axis ranges. You may or may not be able to create this plot directly. You can always, however, create it by writing formulas into two columns of data (one each for V and P) that refer to the appropriate locations in the working part of the spreadsheet.

cycle. q_2 is the negative of w_1 (Step 1), and q_1 is the negative of w_3 (Step 3). The net heat transfer is summed in cell K6.

Cells M2 and M3 calculate q/T for Steps 1 and 3 of the cycle. If your computer could accurately simulate a perfectly reversible process, these two values would be equal (Equation 8.20). Because the computer actually simulates a stepwise process the inequality demonstrates Equation 8.21.

Cell M5 represents Equation 8.24. Cell M6 represents Equation 8.25. Cell M5 $e(q)$ is the actual efficiency of the cycle. Cell M6 is the maximum efficiency possible $e(T)$ for a fully reversible system.

e. Complete the table below:

$n =$	1	$w_1 =$	_____	$q_2 =$	_____	$q_2/T_2 =$	_____
$T =$	300	$w_2 =$	_____				
$V_0 =$	10	$w_3 =$	_____	$q_1 =$	_____	$q_1/T_1 =$	_____
$\Delta V =$	0.1	$w_4 =$	_____				
		Total work =	___	$\Delta q =$	_____	$e(q) =$	_____
						$e(T) =$	_____

f. Examine your completed table and note that the first law of thermodynamics, when applied to the ideal gas law and simulated on a computer, verifies all points made in the introduction to this chapter. Reread that introduction now and refer to your table at the appropriate times. Note especially that, after a complete cycle, the system has performed net positive work on the piston (cell I6, total work), that the two values of q/T are very close (cells M2 and M3), and that the efficiency of the thermal energy conversion is less than 1 (cell M5).

g. Explore your model by doubling the amount of gas in the system (insert a value of 2 into cell B1) or by changing the temperature at which isothermal expansion occurs (cell B5). Don't forget to manually adjust for w_3 and w_4. In both cases, the efficiencies of the system should be identical with the values in Lesson 2e if you make the adjustments correctly.

Complete a table similar to that in Lesson 2e, study your results, and compare them with the predictions at the introduction to this exercise.

h. Changing the initial volume of the gas (cell B6) does change the efficiency of the system. Verify that the increased value of $(T_1 - T_2)/T_1$ leads to an increased efficiency of the system.

i. Explore your model and attempt to find conditions in which 100% of the energy taken in during Step 1 of the Carnot cycle can be converted to mechanical energy.[1]

3. We mentioned in the introduction to this exercise, and you discovered in Lesson 2, that an irreversible process does not obey Equation 8.20 but rather follows the inequality in Equation 8.21. We noted that, for a reversible process, q/T measures the change in the entropy (Equation 8.22) but that the value of q/T for irreversible processes always underestimates the change in entropy. We further stated that this is because ΔS depends only on the initial and final states (and is therefore path-independent) but q is not a function of the state (and is therefore path-dependent).

This fact is very easy to verify using the first law of thermodynamics and the ideal gas law.

a. Build and verify Model 8.7. Also build the on-screen graphics described by Figures 8.16 and 8.17. Model 8.7 represents two alternative pathways for the isothermal expansion of an ideal gas.

Columns A and B (below row 6) simply calculate the volume and pressure of an ideal gas for the conditions specified.

Columns D–H calculate the changes in heat and entropy that occur when an ideal gas expands reversibly. Notice that the external pressure (column D) is simply taken from column B, the heat changes (columns E and F) are calculated on the basis of $P_{ext}\Delta V$ work done by the gas, and the change in entropy of the gas and the environment (columns H and I) are calculated by dividing heat change (columns E and F) by absolute temperature (cell B3). The total entropy change of the system sums columns H and I.

Columns K–P calculate the changes in heat and entropy that occur when an ideal gas expands irreversibly against an unresisting piston (i.e., the external force is zero). The external pressure (column K) is zero, the heat changes (columns L and M) are calculated on the basis of $P_{ext}\Delta V$ work done by the gas but must also be zero because $P_{ext} = 0$. *The change in the entropy of the gas is identical to the change in entropy for a reversible process, and so column N takes its values from column G.* The change in entropy of the environment must be zero because the gas did no work on the piston and no change in the environment

Figure 8.16

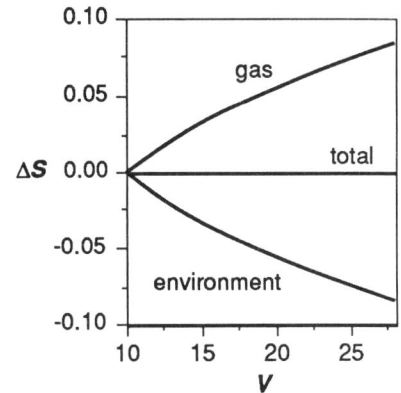

XY-PLOT

Data
Horizontal (x) axis: cells A10–A100 (volume)
Vertical (y) axis: cells G10–G100 (ΔS of gas), H10–H100 (ΔS of environment), and I10–I100 (total ΔS)

Figure 8.17

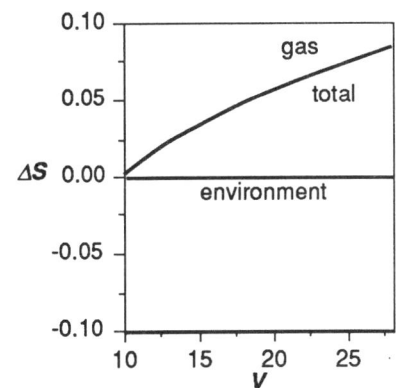

XY-PLOT

Data
Horizontal (x) axis: cells A10–A100 (volume)
Vertical (y) axis: cells N10–N100 (ΔS of gas), O10–O100 (ΔS of environment), and P10–P100 (total ΔS)

ensued. The total entropy change of the system sums columns N and O.

b. Examination of your on-screen graphics should tell you that something remarkable is happening here! The entropy of the gas increases as it expands irreversibly, but the entropy of the environment doesn't change at all. Entropy depends on the *state* of matter, and the change of the entropy of the expanding gas is unchanged by a change in path. ΔS for the irreversible expansion is the same as for the reversible expansion. The environment, however, experienced no change in state because the piston did not resist the expansion of the gas, no work was done, and no heat transfer occurred. Thus, the complete system (the gas and its surroundings) experienced a net increase in entropy.

Although entropy is a function of the state of a gas and its surroundings, it is not conserved. For an irreversible process, it is inevitable that when the entire universe is taken into account, entropy will increase!

CONCLUSION

You are now in a position to state the second law of thermodynamics in one of its clearest forms:

When all parts of a process are taken into account, entropy must either remain the same (for a reversible process) or increase (for an irreversible process).

Reversible processes are always in equilibrium (by definition); irreversible processes always have a natural direction. The second law of thermodynamics states that the direction of a spontaneous, irreversible process will always be in the direction that yields an increase in entropy.

We hope that you enjoyed discovering the second law of thermodynamics by using your computer, the first law of thermodynamics, and the ideal gas law, but we suspect that you would also like to know what's happening with the molecules. Specifically, if entropy depends only on the state of the matter in the system, to what aspect of matter does entropy refer?[2]

Entropy can be understood simply and intuitively in terms of the capacity of a material to contain thermal energy. Although your textbook probably explains entropy differently, you will see shortly that the approach in this book is entirely consistent with other, more traditional, approaches. Consider a gas made up of a single species of atom that are not covalently linked. You know that the temperature of the gas corresponds to the average speed of these atoms and that the thermal energy of the gas corresponds to the total kinetic energy of these atoms. If a gas is cooled to absolute zero, its atoms have neither speed nor kinetic energy. If the temperature of the gas is raised by a tiny amount, the atoms acquire both speed and kinetic energy. As the average speed of the atoms of gas increases further, so does kinetic energy. The thermal energy E of a system is represented in thermodynamics by the expression

$$E = ST \tag{8.27}$$

where T is the temperature of the gas and S is a constant that unites them. S, in fact, is entropy. S *is a measure of the capacity of matter to hold thermal energy* E.[3]

If 2 mol of a monatomic gas are in a 1-L container at a temperature of 300 K, they have a fixed amount of thermal energy. If 2 mol of this monatomic gas combine to form 1 mol of diatomic gas (which is then stored at 300 K in a 1-L container), it will have less thermal energy than the previous example. This is because the covalent bonds between each pair of atoms restricts the motion (and therefore the capacity to hold thermal energy) of the atoms. Referring to Equation 8.27, as the structure of matter becomes more complex, S decreases. As S decreases, E/T decreases.

> *As the state of organization of matter increases, entropy decreases.*

This is a loose and very qualitative description of entropy, but it holds up remarkably well in a variety of situations. It certainly gives you the ability to see entropy as dependent only on the *state* of matter. It is a state that identifies the capacity of matter to hold thermal energy that in turn is related to the degree of order of the matter.

You must be very careful, however, about how you define order. It is fashionable in some circles to say that the entropy of an organized library of books is less than the entropy of a library in disarray. If this is so, what about the entropy of a library of books alphabetized according to the twenty-third word on the twelfth page of each book. Is the entropy of this library lower than that of a random arrangement? Does your ability to calculate the entropy of a system depend on your knowledge of some secret ordering scheme?

We finish this exercises by repeating ourselves: Entropy is a driving principle behind spontaneous change. In an isolated system,

- The change in entropy of all parts of a reversible process sums to zero. Reversible processes are always in equilibrium and exhibit no tendency to spontaneous change.

- The change in entropy of all parts of a spontaneous, irreversible process sums to a positive number.

Finally, because most laboratory experiments proceed under conditions of constant pressure, it is useful to express q in terms of enthalpy change ΔH as well as thermal energy change ΔE. Recall from the conclusion of Exercise 8.1 that q depends on the conditions of measurement. From the first law of thermodynamics,

$$\Delta E = q + w \tag{8.28}$$

where ΔE is the change in internal (thermal energy) of a system, q is the heat transferred *into* the system, and w is the work done *on* the system.

For a system at constant volume, Equation 8.28 becomes

$$\Delta E = q \tag{8.29}$$

because, if the volume of matter does not change, then no PV work is done.

For a system at constant pressure, Equation 8.28 becomes,

$$\Delta E = q - P\Delta V \tag{8.30}$$

or

$$q = \Delta E + P\Delta V = \Delta H \tag{8.31}$$

where ΔH is the change in the enthalpy of the system.

Both ΔE and ΔH are measures of the capacity of matter to hold thermal energy, and entropy can be calculated in terms of either of these properties. Under conditions of constant volume,

$$\Delta E \leq T\Delta S \tag{8.32}$$

Under conditions of constant pressure,

$$\Delta H \le T\Delta S \qquad (8.33)$$

You should take this discussion and look back at the lessons of the exercise. Look especially at those reversible processes that exhibit increases or decreases in q/T. In each case, see if you can picture changes in q/T in terms of changes in order. Keep in mind that the concept of order in kinetic theory may not always correspond to the concept of order as applied to the conduct of human affairs.

MODELS

FORMULAS

	A	B	C	D	E	F	G	H	I	J	K	L	M	N
1	n =	1	mol						------------Data			---Conclusions---		
2	R =	0.082	l atm/mol K					w1 =	=C102	q2 =	=I2	q1/T1=	=K4/T12	
3	gam=	=5/3						w2 =	=J102			q2/T2=	=K2/F12	
4				Enter intersect value ->				w3 =	12.28	q1 =	=I4			
5	T =	300	K	Enter intersect value ->				w4 =	8.306			e(q)=	=(K2+K4)/K2	
6	V0 =	10	l					Tot =		Δq =		e(T)=	=(F12-T12)/F12	
7				=SUM(I2:I5)										
8	ΔV =	0.1	l					ΔV =	=B8			=K2+K4		
9														
10	Step 1: Isothermal Expansion							Step 2: Adiabatic Expansion						
11	V	P	w1	q2	ΔE	T		V	P	w2	q	ΔE	T	
12	=B6							=A102	=B1*B2*B5/H12					
13											0			
14											0			
15	Form Set I	Form Set II	Form Set III	Form Set IV	Form Set V	Form Set VI		Form Set VII	Form Set VIII	Form Set IX	0	Form Set X	Form Set XI	N
							G				K			
100											0			
101											0			
102											0			

FORMULAS (continued)

	O	P	Q	R	S	T	U	V	W	X	Y	Z	AA
1													
2													
3													
4													
5													
6													
7													
8	ΔV =	0.1	l					ΔV =	0.1	l			
9													
10	Step 3: Isothermal Compression							Step 4: Adiabatic Compression					
11	V	P	w3	q1	ΔE	T		V	P	w4	q	ΔE	T
12	=H102							=A12	=B1*B2*B5/V12				
13											0		
14											0		
15	Form Set XII	Form Set XIII	Form Set XIV	Form Set XV	Form Set XVI	Form Set XVII	U	Form Set XVIII	Form Set XIX	Form Set XX	0	Form Set XXI	Form Set XXII
											Y		
100											0		
101											0		
102											0		
173													
174													
175													

Model 8.6

BUILD and verify Model 8.6.

ENTER data using cells B1 (moles of gas), B5 and B6 (temperature and starting volume), and B7 (incremental volume). Also enter the values of w_3 and w_4 into cells I4 and I5 after visually inspecting the model to find the intersection point on the P vs V curve for Steps 3 and 4 of the Carnot cycle. The procedure for doing this is explained in the text.

READ output from the model below row 9 at columns A–F, H–M, O–T, and V–AA. Each set of columns contains the same set of information for Steps 1, 2, 3, and 4 of the Carnot cycle, respectively (volume, pressure, work performed on the gas, heat transferred to the gas, gas internal energy, and gas temperature). Also read summary data at cells I2–I5 (w_1, w_2, w_3, w_4, and total work on the gas); K2, K4, and K6 (q_2, q_1, and net heat transfer on the gas); M2 and M3 (q_1/T_1 and q_2/T_2); and M5 and M6 (actual and maximum efficiencies of the heat engine).

Model 8.6 (continued)

VALUES

	A	B	C	D	E	F	G	H	I	J	K	L	M	N
1	n =	1	mol						------------Data------------				-Conclusions--		
2	R =	0.082	l atm/mol K					w1 =	-15.7	q2 =	15.74	q1/T1=	-0.053	
3	gam =	1.667						w2 =	-8.38			q2/T2=	0.052	
4				Enter intersect value ->				w3 =	12.28	q1 =	-12.3			
5	T =	300	K	Enter intersect value ->				w4 =	8.306			e(q)=	0.22	
6	V0 =	10	l					Tot =	-3.54	Δq =	3.463	e(T) =	0.228	
7														
8	ΔV =	0.1	l					ΔV =	0.1	l				
9														
10	Step 1: Isothermal Expansion							Step 2: Adiabatic Expansion						
11	V	P	w1	q2	ΔE	T		V	P	w2	q	ΔE	T	
12	10	2.462				300		19	1.296				300	
13	10.1	2.437	-0.24	0.244	0	300		19.1	1.284	-0.13	0	-0.13	299	
14	10.2	2.413	-0.49	0.485	0	300		19.2	1.273	-0.26	0	-0.26	297.9	
15	10.3	2.39	-0.72	0.724	0	300		19.3	1.262	-0.38	0	-0.38	296.9	

	A	B	C	D	E	F	G	H	I	J	K	L	M	N
100	18.8	1.309	-15.5	15.48	0	300		27.8	0.687	-8.24	0	-8.24	232.8	
101	18.9	1.302	-15.6	15.61	0	300		27.9	0.683	-8.31	0	-8.31	232.2	
102	19	1.296	-15.7	15.74	0	300		28	0.679	-8.38	0	-8.38	231.7	

VALUES (continued)

	O	P	Q	R	S	T	U	V	W	X	Y	Z	AA
1													
2													
3													
4													
5													
6													
7													
8	ΔV =	0.1	l					ΔV =	0.1	l			
9													
10	Step 3: Isothermal Compression							Step 4: Adiabatic Compression					
11	V	P	w3	q1	ΔE	T		V	P	w4	q	ΔE	T
12	28	0.679				231.7		10	2.462				300
13	27.9	0.681	0.068	-0.068	0	231.7		10.1	2.421	0.242	0	0.242	298
14	27.8	0.684	0.137	-0.137	0	231.7		10.2	2.382	0.48	0	0.48	296.1
15	27.7	0.686	0.205	-0.205	0	231.7		10.3	2.343	0.715	0	0.715	294.1

	O	P	Q	R	S	T	U	V	W	X	Y	Z	AA
100	19.2	0.99	7.188	-7.188	0	231.7		18.8	0.86	12.6	0	12.6	196.9
101	19.1	0.995	7.287	-7.287	0	231.7		18.9	0.852	12.69	0	12.69	196.3
102	19	1	7.387	-7.387	0	231.7		19	0.845	12.77	0	12.77	195.6

	O	P	Q	R	S	T
173	11.9	1.597	16.31	-16.31	0	231.7
174	11.8	1.611	16.47	-16.47	0	231.7
175	11.7	1.625	16.64	-16.64	0	231.7

Formula Set I
Prototype Cell is A13

=A12+**B8**

Formula Set XII
Prototype Cell is O13

=O12-**P8**

Formula Set II
Prototype Cell is B12

=**B1*****B2*****B5**/A12

Formula Set XIII
Prototype Cell is P12

=**B1*****B2*****M102** /O12

Formula Set III
Prototype Cell is C13

=C12-B13*(A13-A12)

Formula Set XIV
Prototype Cell is Q13

=Q12-P13*(O13-O12)

Formula Set IV
Prototype Cell is D13

=-C13

Formula Set XV
Prototype Cell is R13

=-Q13

Formula Set V
Prototype Cell is E13

=C13+D13

Formula Set XVI
Prototype Cell is S13

=Q13+R13

Formula Set VI
Prototype Cell is F12

=(A12*B12)/(**B1*****B2**)

Formula Set XVII
Prototype Cell is T12

=(O12*P12)/(**B1*****B2**)

Formula Set VII
Prototype Cell is H13

=H12+**I8**

Formula Set XVIII
Prototype Cell is V13

=V12+**W8**

Formula Set VIII
Prototype Cell is I13

=I12*(H12/H13)^**B3**

Formula Set XIX
Prototype Cell is W13

=W12*(V12/V13)^**B3**

Formula Set IX
Prototype Cell is J13

=J12-I13*(H13-H12)

Formula Set XX
Prototype Cell is X13

=X12+W13*(V13-V12)

Formula Set X
Prototype Cell is L13

=J13

Formula Set XXI
Prototype Cell is Z13

=X13

Formula Set XI
Prototype Cell is M12

=(H12*I12)/(**B1*****B2**)

Formula Set XXII
Prototype Cell is AA12

=(V12*W12)/(**B1*****B2**)

Model 8.6 (continued)

Model 8.7

BUILD and verify Model 8.7.

ENTER data using cells B3 (temperature of isothermal expansion) and B5 and B6 (starting and incremental volume).

READ output from the model below row 7 at columns A and B (volume and pressure of an ideal gas). Columns D–I and K–P calculate the external pressure, heat transferred into the gas, heat transferred into the environment, change in gas entropy, change in environment entropy, and change in total entropy for an isothermal reversible expansion and an isothermal irreversible expansion, respectively. The irreversible expansion takes place such that the gas expands against an unresisting piston.

FORMULAS

	A	B	C	D	E	F	G	H	I	J	K	L	M	N	O	P
1	n =	1	mol													
2	R =	0.082	I atm/mol K	(Value in B2 is 0.0820575)												
3	T =	300	K													
4																
5	V0 =	10	I													
6	ΔV =	0.2	I													
7				Reversible								Irreversible expansion against 0 pressure				
8	Ideal Gas				Gas	Env.	Gas	Env.	Total			Gas	Env.	Gas	Env.	Total
9	V	P		Pext	q	q	ΔS	ΔS	ΔS		Pext	q	q	ΔS	ΔS	ΔS
10	=B5										0					
11	Form Set I	Form Set II		Form Set III	Form Set IV	Form Set V	Form Set VI	Form Set VII	Form Set VIII		0	Form Set IX	Form Set X	Form Set XI	0	Form Set XII
12											0				0	
13											0				0	
14											0				0	

VALUES

	A	B	C	D	E	F	G	H	I	J	K	L	M	N	O	P
1	n =	1	mol													
2	R =	0.082	I atm/mol K	(Value in B2 is 0.0820575)												
3	T =	300	K													
4																
5	V0 =	10	I													
6	ΔV =	0.2	I													
7				Reversible								Irreversible expansion against 0 pressure				
8	Ideal Gas				Gas	Env.	Gas	Env.	Total			Gas	Env.	Gas	Env.	Total
9	V	P		Pext	q	q	ΔS	ΔS	ΔS		Pext	q	q	ΔS	ΔS	ΔS
10	10	2.462		2.462							0					
11	10.2	2.413		2.413	0.483	-0.48	0.002	-0.002	0		0	0	0	0.002	0	0.002
12	10.4	2.367		2.367	0.956	-0.96	0.003	-0.003	0		0	0	0	0.003	0	0.003
13	10.6	2.322		2.322	1.421	-1.42	0.005	-0.005	0		0	0	0	0.005	0	0.005
14	10.8	2.279		2.279	1.876	-1.88	0.006	-0.006	0		0	0	0	0.006	0	0.006

Formula Set I Prototype Cell is A11	Formula Set V Prototype Cell is F11	Formula Set IX Prototype Cell is L11
=A10+ **B6**	=-E11	=L10+K11*(A11-A10)

Formula Set II Prototype Cell is B10	Formula Set VI Prototype Cell is G11	Formula Set X Prototype Cell is M11
=**B1·B2·B3**/A10	=E11/**B3**	=-L11

Formula Set III Prototype Cell is D10	Formula Set VII Prototype Cell is H11	Formula Set XI Prototype Cell is N11
=B10	=F11/**B3**	=G11

Formula Set IV Prototype Cell is E11	Formula Set VIII Prototype Cell is I11	Formula Set XII Prototype Cell is P11
=E10+B11*(A11-A10)	=G11+H11	=N11+O11

GIBBS FREE ENERGY: THE DIRECTION AND EXTENT OF A REACTION

The first and second laws of thermodynamics apply to all physical processes and are not confined to predicting the behavior of an ideal gas. The demonstration of this fact lies beyond the domain of this book but is central to introductory courses in physical chemistry. We begin this final exercise therefore by simply stating that the concluding equations of the previous exercise are general laws of nature and apply to all physical processes:

$$\Delta E \leq T\Delta S \quad \text{at constant } V \qquad (8.34)$$

$$\Delta H \leq T\Delta S \quad \text{at constant } P \qquad (8.35)$$

where ΔE (changes in internal energy) and ΔH (changes in enthalpy) identify the heat transferred under conditions of constant volume and constant pressure, respectively, T is absolute temperature, and ΔS is the change in the entropy of the system.

If $T\Delta S$ is subtracted from both sides of Equations 8.34 and 8.35, one obtains

$$\Delta E - T\Delta S \leq 0 \qquad (8.36)$$

TO GET THE MOST OUT OF THIS EXERCISE, YOU SHOULD ALREADY KNOW

The definition of the terms:
thermal energy
enthalpy
entropy
reversible reaction
spontaneous reaction

How to use rate laws to calculate the concentrations of reactants and products as they change with time (Chapter 7).

How to calculate the equilibrium concentrations of reactants and products from a knowledge of initial conditions and the equilibrium constant (Chapter 5).

$$\Delta H - T\Delta S \leq 0 \qquad\qquad (8.37)$$

$\Delta E - T\Delta S$ and $\Delta H - T\Delta S$ define changes in two very important functions — the Helmholtz function $A = E - TS$ and the Gibbs function $G = H - TS$ (Equations 8.38 and 8.39):

$$\Delta A = \Delta E - T\Delta S \leq 0 \qquad\qquad (8.38)$$

$$\Delta G = \Delta H - T\Delta S \leq 0 \qquad\qquad (8.39)$$

These equations can be interpreted in words to mean that:

- *Under conditions of constant volume, spontaneous change is always accompanied by a decrease in the value of the Helmholtz function.*

- *Under conditions of constant pressure, spontaneous change is always accompanied by a decrease in the value of the Gibbs function.*

Chemists almost invariably work under conditions of constant pressure, and so it is the Gibbs free-energy function that attracts the greatest attention.

Consider a chemical reaction of the form

$$A \rightleftharpoons B \qquad\qquad (8.40)$$

It can be shown that, for the reaction specified in Equation 8.40, the change in Gibbs free energy (per mole of reactant converted) ΔG is

$$\Delta G = RT \ln\!\left(\frac{Q}{K}\right) \qquad\qquad (8.41)$$

where K is the equilibrium constant of the reaction (Equation 8.42)

$$K = \frac{[B]_{eq}}{[A]_{eq}} \qquad\qquad (8.42)$$

and where

$$Q = \frac{[B]}{[A]} \qquad\qquad (8.43)$$

$[A]_{eq}$ and $[B]_{eq}$ are the concentrations of A and B at equilibrium, and $[A]$ and $[B]$ are the actual concentrations at some stage in the reaction.

Although we will save a detailed understanding of ΔG for your classes in physical chemistry, we thought it would be useful to conclude your studies in this book by getting a feel for G and ΔG by observing their values in several chemical settings.

LESSONS

1. Build and verify Model 8.8. Also build the two on-screen graphics as described in Figures 8.18 and 8.19.

   Explore the structure of Model 8.8 using the following guidance:

   a. Columns A and B (below row 6) generate a table of values for [A] and [B].

   b. Column D calculates ΔG by using Equation 8.41. The fraction Q/K in Equation 8.41 is calculated separately in column C.

   c. Column E calculates the relative values of G as a function of [A] and [B] by arbitrarily assigning a value of zero to G at the lowest value of [B] and then incrementing G from a knowledge of ΔG. Note, however, that in this case ΔG is the change in free energy *per mole* of reactant transformed. ΔG must be multiplied by Incr[B] (cell B4) before adding it to the previous value of G.

   d. Column F normalizes G by setting its minimum value to zero. The minimum value of G_{rel} (column E) is found using the MIN function in cell E4, and this value is subtracted from each value in column E to find the normalized values in column F.

      Column F will be used to identify the value of G in the exercises that follow. (Note, however, that G has no fixed value because all values are relative to some specified condition.)

2. Examine your plots of G and ΔG when your model contains its initial values.

   a. ΔG is calculated for the reaction in the direction A → B. What values of [B] allow the reaction to proceed spontaneously?

   b. Identify the value of G when A ⇌ B is in equilibrium. What values does G assume when A ⇌ B departs from equilibrium in either direction?

Figure 8.18

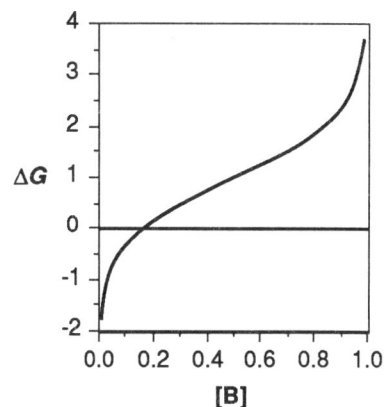

XY-PLOT

Data
Horizontal (*x*) axis: cells A7–A105
([B])
Vertical (*y*) axis: cells D7–D105 (ΔG)

Figure 8.19

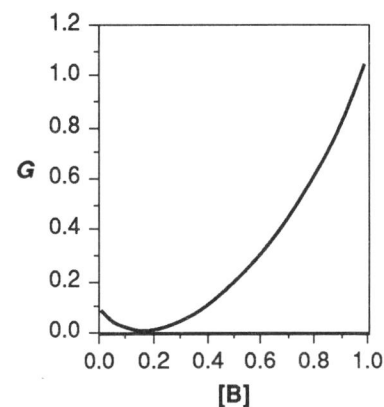

XY-PLOT

Data
Horizontal (*x*) axis: cells A7–A105
([B])
Vertical (*y*) axis: cells F7–F105
(normalized G)

3. Vary the equilibrium constant K across a range from 0.1 to 10 (smaller or larger values are impractical for the scale of your graph).

   a. As K increases, what happens to the value of [B] corresponding to a minimum value for G (Figure 8.19)?

   b. As K increases, what happens to the value of [B] corresponding to $\Delta G = 0$?

   c. Use the on-screen graphic corresponding to Figure 8.19 and pick a value of [B] (say, 0.5) that is to the right of minimum G when $K = 0.2$. Observe the slope of G vs [B] (the angle of a line tangent to G vs [B] at [B] = 0.4).

   Do you see that the initially positive (increasing to the right) slope approaches and passes zero as [B] decreases?

   Compare the slope of G vs [B] with values of ΔG at the same [B] and confirm that ΔG is negative when the slope of G vs [B] is negative, $\Delta G = 0$ when the slope of G vs [B] = 0, and ΔG is positive when the slope of G vs [B] is positive.

   Explain this relationship between the slope of G vs [B] and ΔG.

4. It is often convenient to define standard states of elements and compounds so that they can be tabulated and compared. A rearrangement of Equation 8.41 to Equation 8.44 suggests an appropriate definition for such standard states:

$$\Delta G = -RT \ln(K) + RT \ln(Q) \qquad (8.44)$$

Notice that if the concentrations of all reactants and products of a reaction are $1M$ then the second term in Equation 8.44 is zero.

$$\Delta G = -RT \ln(K) + RT \ln(1) = -RT \ln(K) \qquad (8.45)$$

The value of ΔG under conditions when both reactants and products are in their standard states (i.e., when $Q = 1$) is called the standard Gibbs energy change $\Delta G°$.

   a. Set the concentration of [A] + [B] to $2M$ and locate the point on your on-screen graphic corresponding to Figure 8.18 where [B] = [A] = $1M$. What is the value of ΔG (i.e., $\Delta G°$) at this point?

   b. What is the value of $\Delta G°$ when $K = 1$? $K = 3$? $K = 7$?

c. Does the value of $\Delta G°$ change with changes in the total concentration [A] + [B] of the solution when the ratio of [A] to [B] remains constant?

PROBLEMS

Many of the models in Chapter 7 show the concentrations of reactants and products as they approach equilibrium. Add a column to several of these models that calculate ΔG and watch ΔG as the reaction approaches equilibrium.

There should be no surprises.

MODEL

FORMULAS

	A	B	C	D	E	F
1	K =	0.2	R =	0.00198	kcal/mol	
2	[B] + [A] =	1	T =	300	(°K)	
3	Starting [B] =	0.01				
4	Incr [B] =	0.01		Min of Rel G =	=MIN(E7:E105)	
5						
6	[B]	[A]	Q/K	ΔG	G(rel)	G(norm)
7	=B3				0	
8						
9						
10	Formula Set I	Formula Set II	Formula Set III	Formula Set IV	Formula Set V	Formula Set VI

FORM						
103						
104						
105						

VALUES

	A	B	C	D	E	F
1	K =	0.2	R =	0.00198	kcal/mol	
2	[B] + [A] =	1	T =	300	(°K)	
3	Starting [B] =	0.01				
4	Incr [B] =	0.01		Min of Rel G =	-0.0762	
5						
6	[B]	[A]	Q/K	ΔG	G(rel)	G(norm)
7	0.01	0.99	0.05	-1.7735	0	0.07623
8	0.02	0.98	0.10	-1.3557	-0.0136	0.06267
9	0.03	0.97	0.15	-1.1088	-0.0246	0.05158
10	0.04	0.96	0.21	-0.9318	-0.034	0.04226

VALUES

	A	B	C	D	E	F
103	0.97	0.03	161.67	3.02081	0.89465	0.97087
104	0.98	0.02	245.00	3.26775	0.92733	1.00355
105	0.99	0.01	495.00	3.68551	0.96418	1.04041

Formula Set I Prototype Cell is A8	Formula Set IV Prototype Cell is D7
=A7+**B4**	=**D1** ·**D2** ·LN(C7)

Formula Set II Prototype Cell is B7	Formula Set V Prototype Cell is E8
=**B2**- A7	=E7+(**B4** ·D8)

Formula Set III Prototype Cell is C7	Formula Set VI Prototype Cell is F7
=A7/(**B1** ·B7)	=E7-**E4**

Model 8.8

BUILD and verify Model 8.8.

ENTER data using cells B1–B4 (equilibrium constant, total concentration, starting [B], and incremental [B]).

READ output from the model at cells A7–A105, D7–D105, and F7–F105 ([B], ΔG, and normalized G).

Exercise 1.1

5   **pK = 4.8**

[B–]	[HB]	pH
0.20	0.01	6.10
0.19	0.02	5.78
0.18	0.03	5.58
0.17	0.04	5.43
0.16	0.05	5.31
0.15	0.06	5.20
0.14	0.07	5.10
0.13	0.08	5.01
0.12	0.09	4.92
0.11	0.10	4.84
0.10	0.11	4.76
0.09	0.12	4.68
0.08	0.13	4.59
0.07	0.14	4.50
0.06	0.15	4.40
0.05	0.16	4.29
0.04	0.17	4.17
0.03	0.18	4.02
0.02	0.19	3.82
0.01	0.20	3.50

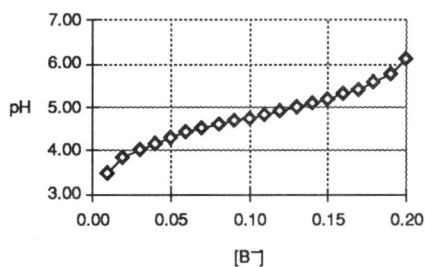

pH changes least near pH = 4.8.

Exercise 2.1

1a H_2O

1b Cell D2 contains the formula =B2*C2. Other formulas are similar.

1c Cell D21 contains the formula =SUM(D2:D19).

2a 190.32

2b 159.8

2c 72.082

2d 282.18

3a 344.38

3b 666.21

3c 598.758

3d 676.59

4a 0.00105 mol

4b 0.00243 mol

4c 1 g of $Al(CH_3)_3$ contains 0.0139 mol.
    0.5 g of UCl_3 contains 0.00145 mol.
    Therefore, 1 g of $Al(CH_3)_3$ contains more molecules.

4d 0.72 g

4e 583 g

4f 0.01 mol of $YBa_2Cu_3O_7$ contains 6.66 g.
    0.01 mol of UCl_3 contains 3.44 g.
    Therefore 0.01 of $YBa_2Cu_3O_7$ contains the larger mass.

5   $CrCl_2 \cdot 4H_2O$, $CrCl_2 \cdot 3H_2O$, $CrCl_2 \cdot 2H_2O$, $CrCl_2$

Exercise 2.2

1e Cell G10 contains the formula =B10*C10. Cell I8 contains the formula =B8*E8. Other cells contain similar formulas.

1f Cell G20 contains the formula =SUM(G4:G18). Other cells in this set contain similar formulas.

1h Cell G22 contains the formula =6.023E+23*G21. Other cells in this set contain similar formulas.

1i Cell G23 contains the formula =G21/SUM(G21:J21). Other cells in this set contain similar formulas.

1j The formula in cell G24 should be =SUM(G23:J23).

2 The mole fraction of O_2 in the mixture is 0.18. The mole fraction of N_2 in the mixture is 0.82.

3 The mole fraction of ethyl alcohol in the mixture is 0.88. The mole fraction of water in the mixture is 0.12.

Exercise 2.3

1a Cell I4 contains the formula =B4*C4. Other cells in this set contain similar formulas.

1b See Model 2.3.

1c Cell I20 contains the formula =SUM(I4:I18). Other cells in this set contain similar formulas.

1d Cell I21 contains the formula =IF(ISERROR(C1/I20),0,C1/I20) in the language of Excel. The actual calculation, of course, is simply =C1/I20 and this formula will work just fine (it simply returns an error message instead of 0 when I20 = 0). We have used the more complex calculation in anticipation of Problem 1. A different syntax for the IF() statement is used by other spreadsheets. Other cells in this set contain similar formulas.

Exercise 2.4

1a C is 40%, H is 7%, and O is 53% of the molecule.

1b Na is 8%, C is 70%, H is 12%, and O is 10% of the molecule.

2a 0.92 g of C, 0.15 g of H, 1.22 g of O

2b 0.7 g of Na, 6.8 g of C, 1.0 g of O

3 Percent composition values go down as NaI contamination increases. Because NaI contains none of the elements present in $KClO_3$, however, the ratios of O to Cl, O to K, etc., remain unchanged.

Exercise 2.5

2 $C_7H_5O_3NS$

Exercise 3.1

1b Cell A9 picks up the initial volume from cell B5. Cells A10–A18 increment the contents of the previous cell with the value found in cell B6.

2a As the volume increases from 1 L to 10 L, the pressure decreases from 24 atm to 2.4 atm. As the volume increases from 0.1 L to 1 L, the pressure decreases from 240 atm to 24 atm. As the volume increases from 0.01 L to 0.1 L, the pressure decreases from 2400 atm to 240 atm.

As the volume approaches zero, the pressure approaches infinity.

2b As the amount of gas in the container increases, the pressure of the gas (at a specific volume) also increases. The shape of the P vs V curve, however, remains unchanged. As the temperature of the gas in the container increases, the pressure of the gas (at a specific volume) also increases. The shape of the P vs V curve, however, remains unchanged.

3a The pressure increases in a linear fashion from 24.6 atm to 39.4 atm.

6b As the volume decreases, the behavior of the van der Waal's gas departs from the behavior of the ideal gas. Your general chemistry text explains why.

6c As the volume increases, the behavior of the van der Waal's gas approaches that of the ideal gas. Your general chemistry text explains why.

7c As the temperature drops, the van der Waal's gas departs from ideal behavior.

Exercise 3.2

1 You can verify your formulas by examining the Formula Set of Model 3.5.

2b 2 g of H_2 makes more of a contribution to the total pressure than 8 g of Cl_2 because 2 g of H_2 contains more molecules than 8 g of Cl_2.

2c The pressures drop proportionately, and the ratio of the partial presure of Cl_2 to H_2 do not change.

2d Same as 2c

3b Because, according to Dalton's law of partial pressures, $n_i/n_{tot} = P_i/P_{tot}$.

3c The plots do not change.

3d The plots do not change.

Exercise 3.3

3 As the mass of the gas molecules increases, the distribution of speeds shifts to the left. Temperature is a measure of the kinetic energy of the gas (sum of $mv_i^2/2$ for each gas molecule i), and as the mass of the gas molecules increases, velocity must decrease if kinetic energy is to remain constant.

4c As the volume increases, the pressure drops. The speed of the gas molecules, however, remains unchanged. The pressure drops because, as the volume increases, the number of collisions of gas molecules with the walls of the container decreases.

4d The speed of the gas molecules decreases as the molecular weight of the gas increases. The pressure, however, remains constant because the kinetic energy of the gas molecules is unchanged.

Exercise 4.1

1 You can verify the formulas supporting Figure 4.3 by examining Model 4.1.

2a As the amount of benzene increases, the the vapor pressure of benzene decreases less rapidly.

2b Although the actual vapor pressure of napthalene is not zero, this is a correct prediction of the model which assumes napthalene to be totally nonvolatile.

3a 0.078 atm

3b The vapor pressure increases, and therefore the number of molecules of nonvolatile solute must be decreasing. The molecular weight of the unknown substance is greater than that of napthalene (221).

4 Verify your formulas by examining Model 4.2.

5c The number of molecules of benzene and methylbenzene is unequal. Yes, this might form the basis of a separation technique.

5d As the mole fraction of benzene decreases (and the mole fraction of napthalene increases), the vapor pressure of the binary solution decreases towards zero. For pure napthalene, the predicted vapor pressure is zero just as in Lesson 2b (and for the same reasons).

5e When the vapor pressures of the two pure substances are identical, the compositions of all possible binary solutions and their vapors are identical and, obviously, no separation based on their vapor pressures is possible.

6a In binary solutions of acetone and chloroform, the vapor pressure of acetone is less than that predicted by Raoult's law.

6b In binary solutions of acetone and chloroform, the vapor pressure of chloroform is less than that predicted by Raoult's law.

Exercise 4.2

2a Substance B (benzene) exhibits the higher vapor pressure when equal numbers of molecules of benzene and toluene are present in solution. Substance B must therefore also contain more molecules in the vapor than substance A (ideal gas law).

Exercise 4.3

1 You can verify your formulas using Model 4.8.

2 You can verify your formulas using Model 4.8.

3 You can verify your formulas using Model 4.8.

4c Only the vapor phase of CO_2 is stable at room temperature and pressure.

4d The dry ice never melts. Rather solid CO_2 sublimes directly into its vapor phase.

Exercise 5.1

b 4.64 is the value of x that yields a physical solution. The equilibrium partial pressures of NO_2 and N_2O_4 are 0.72 atm and 4.64 atm, respectively.

c Initial pressures of NO_2 of 10, 5, 3, and 1 atm lead to equilibrium mole fractions of N_2O_4 of 0.87, 0.82, 0.77, 0.64. Storage of NO_2 at high pressures therefore increases the yield of N_2O_4.

Exercise 5.2

b

Initial Pressures		Equil. Mole Fract.	
CO	H_2O	CO	CO_2
1.0	1.0	0.15	0.35
2.0	2.0	0.15	0.35
3.0	3.0	0.15	0.35
4.0	4.0	0.15	0.35
5.0	5.0	0.15	0.35
6.0	6.0	0.15	0.35
7.0	7.0	0.15	0.35
8.0	8.0	0.15	0.35

Notice that the data plots are two straight, horizontal lines.

Exercise 5.3

1a Increasing $[A]_0$ ($[F]_0$ held constant) decreases the fraction of A that is converted to B.

1b $[G]_{eq}$ increases as $[A]_0$ increases.

1c Your table should show you that the percent conversation of A increases as $[F]_0$ increases. You can increase the percent yield of A to any arbitrary level with increases in $[F]_0$, but the solubility of F and other factors will prevent you from increasing $[F]_0$ indefinitely.

1d Indeed, you can increase the percent conversion of A to any arbitrary level by decreasing $[A]_0$. Unfortunately, as your percent yield of product increases, the *amount* of product decreases. There is a limit to which you can exploit this approach, therefore.

Exercise 5.4

2a 0.0001 g L^{-1} of aspirin is dissolved in the aqueous phase while 0.0014 g L^{-1} of aspirin is dissolved in the organic phase.

2b The equilibrium concentrations of aspirin in the aqueous and organic phases are the same as in Lesson 2a. You should be able to explain this result by recalling the principles of dynamic equilibrium.

2c Aspirin is more soluble in the organic phase.

2d About 2 L of organic solvent will be required.

4a Less than 175 mL

4b Less than 130 mL

Exercise 6.1

1b A weak acid buffers best in the range of pH values near the value of pK.

2b Glycine has a pK nearest the pH value of 9.4.

2c At pH = 2 almost all of acetic acid is in its conjugate acid form. At pH = 9 almost all of acetic acid is in its conjugate base form. Most acetic acid is converted from its conjugate acid to its conjugate base between pH = 3 and pH = 7.

3c H_3PO_4 buffers best at pH values near its three pK values (i.e., 2.12, 7.28, 12.32).

3d Maximum buffering occurs in those pH ranges where one or more of the ionic forms of H_3PO_4 are undergoing change. These pH ranges correspond to values near the three pK values of the weak acid.

3e The conversion of $H_2PO_4^-$ to HPO_4^{2-} buffers pH values near 7.2.

3f No. Very large changes in pH are accompanied by very small additions of strong acid or base in the region near pH values of 7.2. Unlike phosphoric acid, arginine has no dissociable proton with a pK near 7.2.

Exercise 6.2

1a Columns B and C represent the Henderson–Hasselbalch equation.

1b Columns D, E, and F calculate the mole fraction of each ionic form of the weak acid.

1c Column G calculates the total H^+ dissociated from or associated to the weak acid with changes in pH. Column H calculates the change in the net charge on the weak acid as H^+ arrives or departs.

2a $H_2Glycine^+$ is the predominant form below pH 1. As pH moves through a pH of 2.3, $H_2Glycine^+$ is converted to HGlycine. HGlycine is the predominant form over a wide range of values from pH = 4 to pH = 8. As pH moves through a pH of 9.8, HGlycine is converted to $Glycine^-$. Above pH 12, $Glycine^-$ is the predominant form.

2b pH 6 is very near the isoelectric point of glycine.

2c HGlycine is the predominant ionic form near its isoelectric point. Other forms are also present in very small concentrations, however.

Exercise 6.3

1c Glycine

1d You might wish to find a buffer with a pK slightly below 6.5 so that, as H+ is released, the pH of the reaction moves in the direction of maximum buffering strength.

1e You might wish to find a buffer with a pK slightly above 6.5.

3   These answers, of course, are identical to those in Lesson 3 of Exercise 6.1.

Exercise 6.4

1a Most of the acetic acid (as HAc) is in the organic phase at pH 2.

1b Most of the acetic acid (as Ac-) is in the aqueous phase at pH 7 because HAc has donated its proton to water and its negative charge makes it insoluble in the organic phase.

1c The solute does not equilibrate equally between the phases because the conjugate acid is not equally soluble in the two phases.

1d 4 mL of organic solvent (if 10 mL of H_2O is present).

Exercise 7.1

3   1/[CH_3CHO] vs t appears linear. On that basis the reaction seems to be second-order, but other possible mechanisms cannot be ruled out.

Exercise 7.4

e   The forward and reverse reactions exhibit first-order kinetics but the *net* reactions are more complex. You only time you see a linear plot of ln[A] vs t, for example, is when the reverse reaction is negligible.

f   Your model should verify this hypothesis.

g   The two plots do indeed correspond when $k_1 \gg k_{-1}$. When the reverse reaction is negligible, the net reaction is first-order.

Exercise 7.5

1a [I] achieves its maximum value at $t = 9$. [I] has a value of 32.8 (in arbitrary units of concentration) at this time. At equilibrium [I] = 5.9.

1b The forward reaction rates for the conversion of A to I and A to C are equal. The reverse reaction rates are not. Early in the reaction, prior to the accumulation of significant amounts of I and C, [I] and [C] should change at similar rates. Later in the reaction, when the reverse reactions become significant, [I] and [C] change at different rates.

2a As [A] increases, rates of the reactions A $\rightarrow$ I and A $\rightarrow$ C increase proportionally. As long as the rate constants k_i and k_c are the same these two reactions will proceed at the same rate.

2b The reaction I $\rightarrow$ A is faster than C $\rightarrow$ A.

2c When $k_{-i} < k_{-c}$, I accumulates continuously, and C accumulates temporarily to above its equilibrium concentrations (i.e., C and I exchange roles). When $k_{-i} = k_{-c}$, neither product accumulates to above equilibrium concentrations.

2d When the forward reactions A $\rightarrow$ I and A $\rightarrow$ C are equal, both accumulate product at equivalent rate. When the reverse reaction I $\rightarrow$ A is faster than C $\rightarrow$ A, the accumulation of I is only temporary. When the reverse reaction C $\rightarrow$ A is faster than I $\rightarrow$ A, the accumulation of C is only temporary.

3 See the answer to Lesson 2d as you consider your answer.

Exercise 7.6

a [A] and [B] have the same concentrations at all time points when [A] = [B] initially. [C] and [D] have the same concentrations at all time points when [C] = [D] initially.

b [C] and [A] do not uniquely determine equilibrium concentrations. [B] and [D] have something to say about it as well.

c You can see that the approach to equilibrium is influenced by the initial concentrations of [A] and [B] by making [A] = [B] and examining [A] vs t, [B] vs t, etc., for values of [A] and [B] such as 1, 2, 3, 8, and 10. Why this is so we leave for a group discussion.

Exercise 8.1

1b Pressure has units of N m^{-2} while volume has units of m^3. (N m^{-2}) x (m^3) = N m = J.

1c 196 J, 19.6 J, 0 J

1d 2.4 J, 1.2 J, 0 J

2a 2 L atm, 4 L atm, 8 L atm, 12.31 L atm

2b 1.6 L is the new gas volume.

2c 12.31 L atm

2d 14.42 L atm

2e 15.89 L atm

2f 16.46 L atm

2g 16.76 L atm

3b The temperature and internal energy of the gas remain constant. Heat is transferred into the gas from its surroundings to maintain the internal energy of the gas.

3c 0.1 L atm, 0.05 L atm, 0 L atm

No heat moves into the gas when the gas expands against an unresisting piston because no work is done by the gas and no thermal energy gradient is ever established.

3d The temperature and internal energy of the gas decreases. The work performed on the moving pistion comes from the internal energy of the gas, causing it to decrease.

3e No heat enters the gas from the surroundings because, in adiabatic expansion, the gas is thermally insulated from its surroundings. There is no relationship between the external pressure and heat transfer — no heat transfer exists.

4b The initial pressure of the gas is 2.46 atm.

0.483 L atm of work is performed on the piston (the work *on* the gas is negative) as the gas expands 0.2 L.

4.45 L atm of work is performed on the piston as the gas expands from 10 to 12 L in increments of 0.2 L.

4.47 L atm of work is performed on the piston as the gas expands from 10 to 12 L in increments of 0.1 L.

4c 4.2 L atm of work is performed on the piston as the gas expands adiabatically from 10 to 12 L in increments of 0.2 L. Less work is performed than during isothermal expansion because, as the gas cools, its pressure at a given temperature and volume is less. Therefore, the pressure exerted on the gas by the piston (during reversible work) is less.

4d The reversible isothermal expansion of a gas is accompanied by no change in temperature or internal energy. The source of the energy supporting the work is the heat in the surroundings of the gas.

4e The reversible adiabatic expansion of a gas is accompanied by a decrease in the temperature and internal energy of the gas. The source of the energy supporting the work performed is the internal energy of the gas itself.

4f As the gas volume increases, the pressure of the gas drops and the external pressure needed to support reversible work is less. An adiabatically expanding gas performs less reversible work than an isothermally expanding gas because the pressure drops more quickly in an adiabatically expanding gas.

4g A reversibly expanding gas in thermal contact with its surrounds stays at constant temperature (drawing heat from its surroundings). A reversibly expanding gas thermally isolated from its surrounds must perform work from its own stores of internal energy.

4h A gas that reversibly expands under adiabatic conditions will cool. Its pressure will drop therefore more rapidly than a gas expanding under isothermal conditions.

Exercise 8.2

2a Adiabatic expansion is accompanied by a loss of gas internal energy; therefore, pressure drops more rapidly.

2e You can read these values directly from the initial conditions of Model 8.6.

Exercise 8.3

2a [B] < 0.16 permit the forward reaction to occur spontaneously.

2b As the model is built, the reaction is at equilibrium when $G = 0$. But remember that G has been normalized. Other models might standardize the G scale at some other value. Remember that G assumes its minimum value at equilibrium and that G rises with any departure from equilibrium.

3a As K increases, the value of [B] associated with minimum G also increases.

3b As K increases, the value of [B] associated with $\Delta G = 0$ also increases.

4a 0.96

4b 0, −.65, −1.16

4c No

ENDNOTES

CHAPTER 1

Exercise 1.1

1   There are so many excellent electronic spreadsheet programs on the market that we can't cover all syntax variations here. We will confine our specific comments to Lotus 1-2-3 and Microsoft Excel.

2   One notable exception to the rule is Microsoft's Multiplan, an electronic spreadsheet program that defines both rows and columns with numbers. In Multiplan, cell B3 would be written R3C2 (for row 3, column 2). Sorry about that!

Exercise 1.2

1   Notice that this notation also provides for expressions in which a variable copies as an absolute address when moved across columns but relative when copied down columns (e.g., $B1). Can you think of a use for this feature?

2   Although you can literally type the appropriate formulas, you should try to get into the habit of preparing formulas by pointing — a feature found in most spreadsheet programs. Once you have started a formula by typing the first character (=, +, etc.), all cell addresses can be entered by simply moving your cursor to the cell you wish to reference. Only the operators need be typed from the keyboard. Conversion of cell addresses from relative to absolute is accomplished using a function or other special key sequence. After a little practice, you will find that this method of preparing formulas is not only intuitive but also almost foolproof.

Exercise 1.3

1   Scatter plots can cause other forms of confusion, however. In Excel, scatter plots that are generated without using the PASTE SPECIAL command don't know where to find x-axis data. Indeed, what you thought was your x-data often gets plotted as an additional set of y-data. Be sure to follow the instructions in Lesson 1b exactly.

CHAPTER 2

Exercise 2.2

1   Most spreadsheets have similar IF statements. However, different spreadsheets identify errors differently. Perhaps your spreadsheet has a built-in function similar to ISERROR(). ISERROR(C1/G20) is evaluated, and the function returns "True" if "C1/G20" generates an error and returns "False" if "C1/G20" fails to generate an error. Or perhaps your spreadsheet uses a statement such as: "C1/G20=@ERROR." The results are the same.

Exercise 2.3

1   You would add the water gradually because a gas, acetylene (HC_2H), is released during the reaction. Moreover, acetylene is flammable. At one time, this reaction was used as the source of acetylene in the lamps on miners' helmets.

2   This reaction is used to find the oxygen content of meteorites containing SiO_2 as the only source of oxygen. In practice it would be foolish to make BrF_3 the limiting reagent. Why?

3   Recall the syntax of the IF() statement from Chapter 1. The equation J25=MIN(I25:K25) is a predicate that is either true or false (i.e., it is either true or false that the value in J25 equals the minimum value to be found in the range of cells I25–K25).

Exercise 2.5

1   It happens that, in this instance, H_2O also represents the molecular formula since each molecule of water contains two atoms of hydrogen and one atom of oxygen.

CHAPTER 3

Exercise 3.1

1   n^2a/V^2 is a term used to correct for the attraction between molecules, and nb is a term used to correct for the volume occupied by molecules. When these terms are missing, van der waal's equation reduces to the ideal gas law.

2   You can also COPY cell A13 into the range A13 – A21. Most spreadsheets permit you to COPY a cell to itself, and the keystroke sequence is sometimes more efficient. Even more efficient, some spreadsheets support commands such as FILL DOWN or FILL RIGHT.

Exercise 3.2

1 Recall that the ideal gas law assumes no particle volume and no interparticle interactions.

Exercise 3.3

1 *Elastic collisions* are collisions in which the total kinetic energy of the system is the same before and after the collisions have occurred.

CHAPTER 4

Introduction

1 Solutions are sometimes defined as systems of different substances that are thoroughly mixed so that all parts are chemically and physically identical. Although solutions can be mixtures of gases, mixtures of liquids, or mixtures of solids, this chapter deals with only liquid solutions. Dalton's law of partial pressures has already given you insight into the nature of gaseous solutions.

Exercise 4.2

1 A liquid boils when its vapor pressure equals the external pressure.

2 Indeed, if the enthalpy of vaporization, ΔH_{vap}, of the components of the solution are identical, this ratio does not change at all.

3 The model assumes that ΔH_{vap} is the same for both components.

Exercise 4.3

1 Some solids can exist in more than one crystalline form (e.g., diamond and graphite). Although each form qualifies as a new phase, this interesting complication will not be discussed here.

2 You can use any convenient quantity of a substance in practice to construct a phase diagram.

CHAPTER 5

Introduction

1 It's approximately $55M$.

Exercise 5.1

1 By international convention, equilibrium constants are protected from dimensions by considering activities to be ratios of actual to standard concentrations and, therefore, dimensionless. It is sometimes convenient to ignore this convention.

Exercise 5.2

1   As the twentieth century draws to a close, such chaotic behavior is rapidly
becoming the topic of serious study in its own right. You might enjoy plotting
Equation 5.34 over long intervals under conditions in which it does not converge
to a "correct" solution. You may be surprised at what you see.

CHAPTER 6

Introduction

1   What is the pOH of a $10^{-3}M$ aqueous solution of HCl?

Exercise 6.1

1   Assume that your NaOH solution is sufficiently concentrated that the volume
increase in your titrated solution is negligible.

2   The Henderson-Hasselbalch equation applies strictly only to equilibrium concentrations. However, because the amounts of dissociation of proton from a conjugate acid and protonation of a conjugate base are small and tend to balance each
other, the Henderson-Hasselbalch equation can be used with initial concentrations of conjugate acid and base to give a good approximation to the final pH.

Exercise 6.2

1   An α-amino acid is an amino acid in which the amino group is attached to the
second, or α, carbon in a chain of carbons counting from the carboxyl end.

Exercise 6.3

1   Plot changes in y vs changes in x to prove this to yourself.

CHAPTER 7

Introduction

1   Any possible reverse reaction is, of course, ignored in this equation.

Exercise 7.1

1   Actually, Letort and others later showed that, under most circumstances, the
reaction approaches an order of 1.5, owing to a mechanism involving free
radicals. We won't examine that interesting but complex outcome.

Exercise 7.3

1   As Δt decreases, however, your computer will ultimately introduce rounding
errors to your calculations because of the limited size of the representation of
numbers in memory.

Exercise 7.4

1   In the real world, of course, you cannot change rate constants of a reaction. Such studies would have to be performed by examining the behaviors of a number of different reactions of the same type.

Exercise 7.5

1   The values of the rate constants of a reaction determine what constitutes a "short" vs a "long" reaction time.

2   There are no rate *constants* for an enzymic reaction because the reaction rate is dependent on the concentration of the enzyme catalyst. For purposes of this exercise therefore, the rate and equilibrium constants are defined in arbitrary time units. As you will see, each unit of time will correspond to one row of your spreadsheet.

Exercise 7.6

1   This kind of behavior is formally similar but not identical to enzyme-catalyzed reactions in living systems (i.e., Michaelis kinetics).

2   Notice that H_2O is a gas in this reaction.

Exercise 7.7

1   Actually, Lotka would have had a very difficult time writing the kind of report you will be able to prepare. He didn't have a personal computer!

CHAPTER 8

Introduction

1   It can even be argued that chemical thermodynamics flowered prior to the time when the molecular basis of chemistry was firmly established.

2   Remember the concept of work in physics? Work is the force applied to an object times the distance the object is moved by the force. A compressed spring has the ability to move a mass through a distance and do work. A resting spring does not.

Exercise 8.1

1   Actually, the law of conservation of energy is a direct consequence of Newton's laws of motion.

2   The word "infinitesimally" is used advisedly. If the gas presses on the piston with a force different than P_{ext}, then the massless piston would exhibit infinite acceleration. What actually happens is this: If the velocity of the piston is zero then P_{ext} equals the pressure of the gas. If the gas is expanding, P_{ext} is less than the pressure of the gas because the net speed of the gas molecules striking the face of the piston (relative to the face of the piston) is less than when the piston is not moving. If the gas is compressing, P_{ext} is greater than the pressure of the gas

because the net relative speed (with respect to the piston) is greater than when the piston is not moving.

3   You've probably noticed that temperature and amount of gas are unspecified by this model. As an additional exercise, determine what amounts and temperatures of gas generate internal pressures that permit the gas to expand against an external pressure of 0.1 atm.

Exercise 8.2

1   Phone us collect if you find such conditions!!!

2   The greatest strength of thermodynamics is also its greatest weakness. It is possible to weave marvelous abstractions and make accurate predictions without once gaining insight into the physical basis for the theory.

3   Thermal energy is zero when $T = 0°K$. If energy were constant with T, internal energy would be linearly proportional to temperature and the equation $E = ST$ could be differentiated to yield $dE = SdT$. That relationship is not valid over a wide range of temperature because S, in fact, changes with T. Over the temperature ranges ordinarily encountered in the laboratory, however, the change is very small, $dE = SdT$ is very nearly true, and so provides a useful guide to intuition. At constant temperature, integration of $E = ST$ yields $dE = TdS$, which is an even more useful guide to intuitive understanding. For any chemical or physical change at constant temperature $T\Delta S$ is merely the change in internal energy that is required to maintain constant temperature. For example, when ice melts energy must be put into the system to maintain the less restrained water molecules in the liquid phase at 0°C.